Survival Analysis with Long-Term Survivors

Survival Analysis with Long-Term Survivors

Ross A. Maller
Department of Mathematics,
University of Western Australia,
Nedlands, W A 6907, Australia

and

Xian Zhou
Department of Mathematics,
University of Western Australia,
Nedlands, W A 6907, Australia.

JOHN WILEY & SONS
Chichester • New York • Brisbane • Toronto • Singapore

Baffins Lane, Chichester,
West Sussex PO19 IUD, England

National (01243) 779777
International (+44) 1243 779777

e-mail (for orders and customer service enquiries): cs-books© wiley.co.uk
Visit our Home Page on http://www.wiley.co.uk
or http://www.wiley.com

Other Wiley Editorial Offices

John Wiley & Sons, Inc., 605 Third Avenue,
New York, NY 10158-0012, USA

Jacaranda Wiley Ltd, 33 Park Road, Milton,
Queensland 4064, Australia

John Wiley & Sons (Canada) Ltd, 22 Worcester Road,
Rexdale, Ontario M9W 1L1, Canada

John Wiley & Sons (Asia) Pte Ltd, 2 Clementi Loop #02-01,
Jin Xing Distripark, Singapore 0512

Library of Congress Cataloging-in-Publication Data

Maller, Ross A.
Survival analysis with long term survivors / Ross A. Maller and Xian Zhou
p. cm. — (Wiley series in probability and statistics.
Applied probability and statistics.
Includes bibliographical references and index.
ISBN 0-471-96201-5 (HB : alk. paper)
1. Survival analysis (Biometry) I. Zhou, Xian. II. Title.
III. Series
[DNLM: 1. Immunity. 2. Survival Analysis. 3. Survivors. 4. Time
Factors. QW 540 M252s 1996]
R853.S7M355 1996
519.5′4—dc 20
DNLM/DLC
for Library of Congress 96-26979
CIP

British Library Cataloguing in Publication Data

A catalogue record for this book is available from the British Library

ISBN 0 471 96201 5

Typeset in 10/12pt Times from the authors' disk by Laser Words, Madras, India
Printed and bound in Great Britain by Biddles, Guildford, and King's Lynn.
This book is printed on acid-free paper responsibly manufactured from sustainable forestation, for which at least two trees are planted for each one used for paper production.

Dedicated to the memory of Edna G. Ball

Contents

Preface

This book is an introduction to the analysis of survival data with particular emphasis on the case when 'immune' or 'cured' individuals — 'long-term survivors', by comparison with other individuals — may be present in the population from which the data is taken. An 'immune' individual means one who is not subject to the event under study — death by or contraction of a disease, or a return to prison, having been released from it, for example — and so is effectively 'immortal' as far as that event is concerned. Such immortals can only be observed in censored data, where they will be manifested as relatively large censored survival times. The primary subject of this book is whether immunes exist in the population, and if they do, how to analyse the survival times (the times till the occurrence of the event) for such things as the effects of covariates. A detailed examination of these questions opens up or places emphasis on some fascinating fields of study. For example, one question is that of the amount of 'follow-up' we have in a set of survival data — when is it 'sufficient', in some sense, and how confident can we be of this?

We develop and exemplify statistical tests for such questions, and for the presence of immunes in data. We also present a technology which takes them into account in a subsequent statistical analysis if they are thought to exist. Furthermore, we present a theoretical underpinning for our methodology, which is based for the most part on the large-sample theory of the estimators and tests we propose. We also examine the properties of those tests and estimators by simulations.

Survival data analysis with immunes, and the theory underlying it, is indeed the focus of this book, but the presence of immune individuals in the data is not essential to our development. Statisticians, biostatisticians, epidemiologists, criminologists and others with an interest in the analysis of *any* survival data will, we hope, find some profit in using the methods set out here and the philosophy developed, because it also offers a different and useful perspective on the analysis of 'ordinary' survival data. Our suggested approach to survival data analysis is to test for the presence of immunes in the data. If the (statistical) decision is that none are present, or that the precision in the data is insufficient for us to be confident of their presence, we recommend proceeding with an 'ordinary' survival analysis in any of the 'usual' ways. Often a formal test for the presence of immunes will be unnecessary, when it is obvious that none exist, but in general we only stand to gain in our understanding of 'ordinary' survival analyses by first looking at the data in this way. And, if in fact a set of survival data is better modelled by assuming that immunes are present, we may obtain wrong or misleading answers from an analysis which ignores them.

The philosophy of our approach is set out in more detail in Chapter 1, where we formalise the questions we believe are of basic interest in the analysis of survival data with (or even without) immunes, and expand on their ramifications. The remainder of Chapter 1 presents examples of data of the kind in which we are interested. These have been gleaned from the literature, especially from medical and epidemiological sources, and also from our own practice, especially from our criminological work. In order to display or 'look at' censored survival data, we advocate the use of the Kaplan–Meier empirical distribution function estimator. A brief introduction to this statistical object of great interest is also given in Chapter 1.

In Chapter 2 we begin by formulating the questions of interest in terms of hypothesis tests. This requires an understanding of some aspects of survival data — such as the mechanism which produced the censoring in the data — and also of some of the properties of the distributions of the survival times, and of the censoring times, under what we will call the 'independent and identically distributed' (i.i.d.) censoring model. The result of this formulation is two hypotheses — we call them H_{01} and H_{02} — which suggest to us how to proceed.

Chapter 3 contains a study of those aspects of the Kaplan–Meier estimator (KME) which we will need to use. In particular we are interested in the value of the KME at its right extreme, or equivalently, its maximum value. This statistic is a potential estimator for the proportion of 'susceptible' individuals — those not immune — in the population, and in order to understand its properties we need some special properties of the KME itself. These are established (often relying on our predecessors' work) in some detail in Chapter 3. We are particularly concerned with the consistency of the KME *on the whole of its domain*, since we need to use its properties at the extreme limit of observation. Fortunately for us, the basic results concerning the KME that we need have already been established by previous researchers. A second major question of interest to us is when the observed KME is likely to be a 'proper' distribution function; that is, for it to take the value 1 at its right extreme. This question is thoroughly discussed in Section 3.2, and further in an appendix note to Chapter 3. Other aspects of the KME — especially its bias and large-sample distribution — are also investigated in Chapter 3.

In Chapter 4 nonparametric methods of estimation of quantities of interest (especially the proportion of immunes) and methods of testing the basic hypotheses are formulated and their properties established using the theory of Chapter 3. We show that the right extreme of the KME, $\hat{p}_n$, is a consistent estimator of the proportion of susceptibles in the population *provided follow-up in the sample is sufficient* in a way which we can quantify. We can also use $\hat{p}_n$ to test for the presence of immunes, that is, to test the hypothesis H_{01}; we present the large-sample theory needed to justify this testing, and some simulations to look at the small-sample properties of the test. (Tables of percentage points of this test statistic and various others we propose are included at the end of the book.) The other main facet of our nonparametric investigation is a test for sufficient follow-up, that is, of the hypothesis H_{02}, and this is also developed and studied in Chapter 4.

A theory of the parametric analysis of survival data with immunes is presented in Chapters 5 and 6, and exemplified on data sets from medical and criminological areas. Chapter 5 is restricted to the single sample problem, while Chapter 6 deals with covariates and the assessment of treatment or other differences. We give examples of two-factor 'analyses of deviance', including a test for interaction between the factors. Most of our parametric analysis is based on fitting the Weibull distribution or its submodel, the exponential distribution, to survival data by the likelihood method. These models have so far worked well for a wide range of the survival data we have seen, and we have restricted ourselves to this case, though the methods are amenable to generalisation.

Likewise, when we present the large-sample theory of the parametric estimation and testing we propose, the mathematical analysis is again restricted to the case of the exponential distribution. Again, this is mainly for convenience and for brevity of exposition. Thus the large-sample theory of the maximum likelihood estimators (MLEs) of the parameters in ordinary exponential and exponential mixture models is set out in Chapters 7 and 8, and their appendix notes. Included in Chapter 7 is a proof of the uniqueness with probability approaching 1 of the MLEs of the parameters in the exponential mixture model, while Chapter 8 deals with large-sample properties of mixture models with covariates, including the analysis of data with a group structure.

Finally, in Chapter 9, we discuss a number of related topics of interest. In particular, we consider the problem of competing risks, showing that some of our theory carries over to this area. Chapter 9 also gives examples of estimating the probability that an individual is immune, and of estimating the censoring distribution from a set of data.

Technical Note

Many or most practitioners in the field of survival analysis will know that a martingale theory is firmly established as the foundation of much of the subject. This book does *not* cover any martingale theory; the theoretical developments we present require a substantially lesser technical apparatus. Nevertheless, some sections of the book, particularly Chapters 7 and 8, contain some fairly detailed theoretical calculations, along the lines of those needed to establish consistency and asymptotic normality of estimators in nonlinear models, for example. Also Chapter 3 uses some fairly intensive, though elementary, probability calculations.

Given that the martingale theory is hardly mentioned further in the book, it is appropriate for us at this point to pay homage to its originators and appliers in survival analysis: without their results (especial reliance on Gill's [1980] monograph, and subsequent results depending on it, will be discerned throughout), our own theory and application would not have been possible. Yet, having been given those firm foundations, we are able simply to read off many of the results we need, such as the large-sample properties of the Kaplan–Meier estimator, as we apply them especially in Chapters 3 and 4.

Despite the theoretical sections of this book, our treatment is aimed very much at the practitioner who wishes to analyse survival data with (or even without) immunes. So we have included what may seem to some an unusual mixture of results, whereby we discuss in detail some of the statistical ideas we need (some as basic as unbiasedness, consistency, the maximum likelihood method, etc.), whilst applying without much further comment a number of probabilistic tools (chiefly, the law of large numbers, the central limit theorem and the sorts of distinctive manipulations used in studying order statistics, as well as the aforementioned martingale theory). This practice is not uncommon in many areas of modern statistics.

We hope that those who want the book for its practical methods can accept the use of the theoretical tools without worrying too much about the technicalities. For example, it is enough in much of the book to think of the law of large numbers as saying, briefly, that the average of a sample of observations on a random variable becomes close to the expected value of that random variable as the sample size becomes large, and similarly, that the average is approximately normally distributed around that expected value in large samples, by the central limit theorem. But for those who wish to follow our working, we have tried to be rigorous in those technical results presented in detail, although other results are merely noted or sketched, or cited with reference to other publications.

Our overall aim is to suggest and to give examples of a systematic methodology for analysing survival data with immunes or long-term survivors, rather similar to the way in which practitioners are currently using generalised linear models. It is also to present in reasonable rigour and detail the theoretical underpinnings of the methods, at least as far as the asymptotic theory of estimators and hypothesis tests can be used as a foundation for everyday data analysis. Although some sections of the book — Chapters 3, 7 and 8 — are for the specialist, the others — Chapters 1–2, 4–6 and 9 — are meant for practising statisticians, biostatisticians, epidemiologists, and in general, any researchers or graduate students who are interested in the analysis of time-to-event data. As our subject is still in vigorous development, some of our recommendations must currently be tentative, but sufficient theory and practice now exists, we feel, to justify the implementation of our methods as a valuable addition to the (bio)statistician's armoury of techniques for the analysis of survival data. The data analysis, we believe, is the ultimate object of the exercise.

Acknowledgements

We are grateful to Brenda Churchill, Cath O'Neill and Jenny Harris for typing sections of the manuscript so cheerfully. Thanks also to our editor, Helen Ramsey, for her help, and also to Stephen Saunders for some useful editing. Dr John Stanton of the Department of Anthropology, University of Western Australia, kindly supplied photographs from which the cover was produced.

For permission to reproduce data sets we are grateful to the following:

1. The Biometrika Trustees, for permission to use the leukaemia data in Figure 1.1 and Section 1.3.
2. Drs A. B. Cantor and J. J. Shuster of the Paediatric Oncology Group of the University of Florida for permission to use the paediatric cancer data in Figure 1.2 and Table 1.2.
3. Prof. R. Harding, Dr R. Broadhurst and the Crime Research Centre of the University of Western Australia for permission to use the prison and arrest data mentioned in many places throughout the book.
4. Drs J. H. Kersey and A. I. Goldman of the University of Minnesota Cancer Center and Department of Biostatistics for permission to use the leukaemia data in Figure 4.3.
5. Drs A. J. Leathem and S. A. Brooks of the Medical School, University College, London, for permission to use the breast cancer data in Figure 6.1.
6. Dr T. M. Therneau of the Mayo Clinic, Rochester, Minnesota, for permission to use the ovarian cancer data in Table 6.2.

The idea for the turtle example which concludes Chapter 1 came from a hint of Prof. Terry Speed's. Thanks also to the other colleagues and friends, including in particular Dr Mohamed Ghitany, but otherwise too many to name, who contributed encouragement and ideas for improvements to the book. The second author also wishes to thank the Hong Kong Institute of Education for its support during the final stages of completing this book. Finally, our heart-felt thanks to our wives, Julie and Jing Li, for their loving care and support in all ways.

CHAPTER 1

Introduction

1.1 INTRODUCTION

The reader who has opened this book probably has at hand, or knows of, a set of survival time data (or time-to-event data of some sort), consisting of the times to occurrence of some event of interest — the death of a patient, the occurrence or recurrence of a disease, the return to prison or the rearrest of a released prisoner — and wishes to explore methods of display and analysis of that data. He or she may furthermore suspect the presence of *immune* or *cured* individuals or units in the data. By *immune* individuals we simply mean some who are not subject to the event of interest — death by the disease under study, or return to prison, for example — and so are effectively *immortal* as regards death or failure from the event under study. Their lifetimes, or the time till the occurrence of the event of interest for these individuals, will continue up to the limit of observation of the experiment or study, and this means that immune individuals will always be manifested as *censored* survival times in the data. They will be indistinguishable in general from the censored survival times which often occur in survival data due to withdrawal of individuals from the population under study, or due to limitation of study follow-up time — except that there will be a tendency for immune individuals to show up as relatively large censored survival times. Rather than refer to immunes, we may sometimes use the terminology *cured* individuals, in the context of medical trials, meaning those who will not suffer a recurrence of the disease. Once again the lifetimes of such individuals, if they are present in the population, will be censored at the limit of follow-up of the study.

The reader may wish to test statistically for the presence of immunes in the data, and to take them into account in any subsequent statistical analysis if they are thought to exist. Such data analysis, and the theory underlying it, is indeed the focus of this book, but nevertheless the presence of immune individuals in the data is not essential to its development. Statisticians, biostatisticians, epidemiologists, criminologists and others with an interest in the analysis of *any* survival data may profit from the methods set out here and the philosophy offered, and they may gain a different and useful perspective on such analyses from the new methods we suggest. Crucially, if some survival data are in fact modelled better by assuming that immunes are present, we may obtain wrong or misleading answers from an analysis which ignores them. Examples of this will be given.

Our philosophy is statistical. We develop tests for the *presence* of immunes in the data, but if we decide that none are present, we can proceed with an 'ordinary' survival analysis in any of the 'usual' ways. We only stand to gain, in our understanding of the 'ordinary' analyses, by first looking at the data in this new way. Another aspect we will consider is whether *follow-up* in the data is adequate or *sufficient*, and this is a concept which applies whether or not we decide that the data contains immune individuals. Again, we can only improve our understanding even of 'ordinary' survival analyses by this extra information.

Consequently, the methods we develop (building on the work of our predecessors) can be seen as being an enrichment or enhancement of 'ordinary' survival analysis, as much as being an exposition of the more specialised subject of immunes in survival data—simply because, if immunes are not in fact present, the problems reduce to the usual considerations of survival analysis, but with the benefit of extra degrees of understanding of the data.

Example: Gehan–Freireich leukaemia recurrence time data

At this stage, it will help to look at a set of survival time data so as to fix attention on the issues we wish to consider. Figure 1.1 gives a representation of

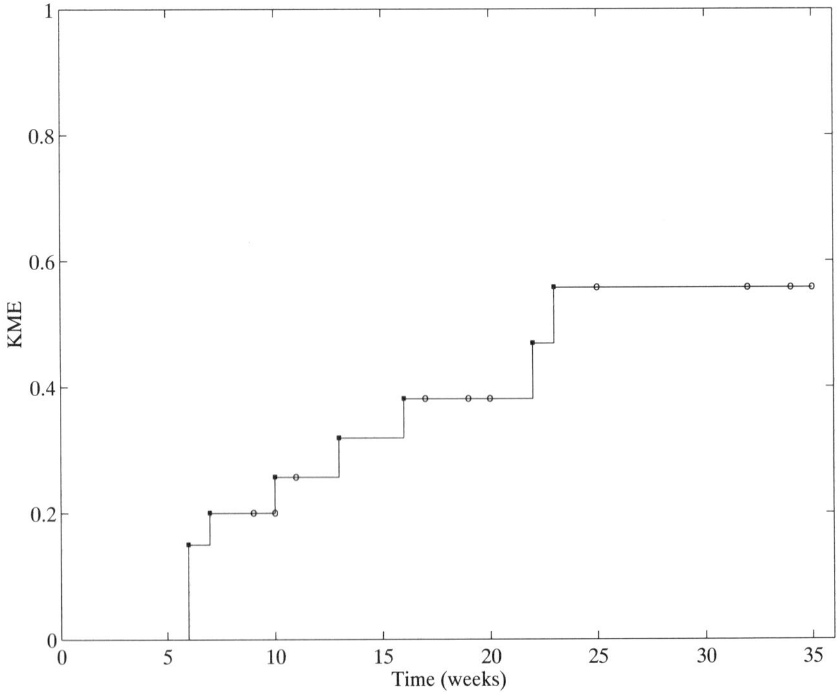

Figure 1.1 Kaplan–Meier estimate for leukaemia remission time data.

the distribution of the times to recurrence of leukaemia for 21 patients receiving a drug treatment. This data happens to be from a medical source, but as we will demonstrate later, there are many other areas of application of our methodology. In Section 1.2 we will give a detailed discussion of the particular distribution function estimator used in Figure 1.1 (the Kaplan–Meier estimator), but at this stage we need only know that it allows us to read, on the y-axis, the estimated proportion of patients (out of a total possible proportion of 1) who suffered a recurrence of the disease by a specified time, in weeks, as marked on the x-axis. In other words, we have a visually evocative display of the recurrence time distribution.

Some of the observations in this data set are *censored*; due to incomplete follow-up, a recurrence of the disease for these individuals did not occur within the time of observation. These types of observation are indicated in Figure 1.1 (and throughout the book unless otherwise specified) by an open circle, whereas the dark square indicates an *actual* or *uncensored* failure time. A feature of the Kaplan–Meier estimator is that it increases (jumps) only at actual survival times, not at censored observations. Note that we plot *all* of the censored (as well as the uncensored) observations on the graph, whereas some practitioners truncate the graph of the Kaplan–Meier estimator at the rightmost uncensored observation. We do not follow this practice because there is valuable information contained in the positioning of the censored observations towards the right-hand end of the Kaplan–Meier estimator, as we shall see.

Again, a more detailed discussion of censoring (how it arises and how to correct for the bias it may cause) will be given later. For now, we need only know that the distribution function estimator displayed in Figure 1.1 is designed to compensate for the censoring, and according to a number of criteria, it is a good estimator of the distribution of recurrence time for these individuals. (See Chapter 3 for a detailed discussion of some important properties of the Kaplan–Meier estimator.)

As a matter of fact, some of the *largest* recurrence times in this data set are censored, and this is reflected in the fact that the Kaplan–Meier estimator tends to level off at a value strictly less than 1 at its right extreme. Can we take this as indicating the presence of some individuals for whom the disease will not recur? We realise that the data set plotted in Figure 1.1 is very small, and we should not be surprised if a statistical test failed to reveal (significant) evidence of their presence — this of course is a consequence of the statistical philosophy to which we adhere. On the other hand, if significance occurs, we again admit the possibility of an erroneous decision (Type I error), as is usual in statistics. But how do we set about formulating these sorts of questions in a statistical framework?

These considerations suggest the various questions we wish to address.

The fundamental questions

Question 1 Basic to our analysis, is: are there immune or cured individuals present in the population from which my data is a sample? 'Immunes' never die, so, if present, we might expect them to be manifested in the data as large censored observations, occurring in particular at the right-hand end of the Kaplan–Meier empirical distribution function estimator, which will consequently tend to level off at a value strictly less than 1 at its right extreme, as in Figure 1.1.

Question 2 Also important is: do we have sufficient follow-up to be confident of having detected immunes, if present? The Kaplan–Meier estimator for the leukaemia recurrence times in Figure 1.1 appears to level off at a value below 1, suggesting both the presence of immunes, and perhaps enough follow-up to be fairly confident of their existence. But has the Kaplan–Meier estimator levelled off 'enough', in some sense, and how do we quantify this? For a statistical development, terms like 'fairly confident' and 'enough follow-up' must be made precise, so a major feature of our approach will be to formulate the main questions of interest as statistical hypotheses and to devise tests for them. Another major part will be to study by various means, and in some detail, the properties of those tests. But many examples of data analysis will be presented too, and our theoretical development will always be guided by the practical considerations of data analysis and interpretation.

Another question of interest relates only to data which can be divided into two or more groups, or more generally, has associated with each survival time some concomitant or 'covariate' information.

Question 3 Are there differences between subgroups in the data, and are these differences in respect to the failure rates of the individuals, as measured in some way, or in respect to immune proportions, if present, or to a combination of both?

In answering this question, it also turns out to be useful to consider whether or not we wish to *parametrise* the distribution of the data by fitting one of a variety of distributional models to the data. Calculating and plotting the Kaplan–Meier estimator (looking at the data) for each of the subgroups in the data already constitutes a simple *nonparametric* analysis of the data—and an informative *display* of a set of data should always be the first step in its analysis. But for more detailed information and comparisons, we may attempt to describe the distribution of the data more formally by one of the parametric models used in 'ordinary' survival data analysis—such as an exponential, Weibull, Gompertz or other distribution—together with some allowance for the presence of immunes in the population.

Excellent expositions of the fitting of 'ordinary' survival models are in Collett (1994) and Aitkin *et al.* (1989), for example. Readers not currently familiar with these books will profit greatly from perusing them before attempting the detailed analyses we describe later. See also the text by Harris and Albert (1991).

We will spend some time in Chapters 5 and 6 in developing the theory and application of a parametric approach to the analysis of survival data with immunes, and discuss in some detail its advantages and disadvantages. At this stage, we merely foreshadow our opinion that the best approach is a combination of nonparametric methods (at least for data display, and possibly also for addressing Questions 1 and 2) and a parametric model, if a suitable model can be found (especially for Question 3). There are many aspects to consider with either approach, and attempting to resolve them will, as we shall see, give us a great deal of insight into the analysis of any kind of survival data.

Questions 1, 2 and 3 above are not of course the only ones we are interested in. And we need not approach them in the above order; but they seem to us to be fundamental and to provide a useful way to begin to think about the analysis of survival data with or without immunes. We will proceed by considering these questions, and thereby will be led to formulate a useful philosophy and to devise an approach to the theoretical questions arising from it.

With this basic paradigm in place, we will then proceed to address the more specific questions of interest in data analysis — in particular, the problems associated with estimation and precision of estimation, from either a nonparametric or parametric model; the goodness of fit of a parametric model, if used; and a method of dealing with covariate information, if present. We will be able to handle these important issues with some confidence by the time we reach Chapter 9.

1.2 DATA DISPLAY AND THE KAPLAN–MEIER ESTIMATOR

One of the first things we should do with our (censored or uncensored) survival data is to *look* at it. Figure 1.1 amply demonstrates the power of some visual representation of the *distribution* of survival time, or whatever occurrence time variable is under study. There are a number of useful ways to display survival data; in this book we tend to favour a representation of the cumulative distribution function of the survival times using the Kaplan–Meier estimator. Throughout the book we will denote this important estimator by its initials, KME. (See also our discussion of hazard functions in Chapter 9.)

The KME is probably known to many readers, since it is one of the basic tools in the analysis of 'ordinary' survival data, by which we mean data in which immunes are not present or whose presence has not been considered. Its usefulness stems from the fact that it is an estimator of the distribution F of the survival times and an estimator which is easily displayed graphically (e.g. Figure 1.1 or the figures in Section 1.3, which deal with 'real' data). Furthermore, the KME possesses good statistical properties, as an estimator, under some reasonable assumptions on the processes which produced the data.

For readers not already familiar with the KME, we will provide a brief introduction below, and we will derive some of its more specialised properties in Chapter 3. Appendix Note 1 at the end of Chapter 3 also gives some less technical background to the KME, together with some references to the large literature on it.

For our immediate purposes, we need only know how to calculate the KME for a censored sample, and some of its basic properties. Many illustrations of how to use it, for data display and assessment, will be given in the pages which follow.

Visual assessment of data imparts some basic but important information. Firstly, the shape of the survival distribution: approximately exponential, perhaps Weibull, or lognormal, or something more exotic? We will discuss briefly the 'shapes' of some of the standard distributions in Section 5.1. Secondly, the main feature of interest to us is that the KME shows whether the distribution of the survival times appears to level off below 1 at its right extreme, thus signalling the possible presence of immunes. All this book's analyses, theoretical and practical, are motivated by these fundamental issues, and one other: whether there are differences between subgroups in the data in respect of their survival distributions, or more generally, how covariate information affects the survival time of the individuals under study.

The empirical distribution function estimator

Before introducing the KME we need briefly to review the *empirical distribution function estimator* (EDF) of a survival distribution, calculated from a sample of observations from that distribution. (By a *survival distribution* we mean the distribution of any nonnegative random variable.) Suppose then that $t_1^*, \ldots, t_n^*$ is a sample of size n from a distribution F on $[0, \infty)$, which we wish to estimate. By a *sample* from F we will understand that the t_i^* are *independent random variables with cumulative distribution function* F, and we will often fall into the common abuse of notation that the same variables $t_1^*, \ldots, t_n^*$ also denote observations, i.e. the data at hand, on independent random variables with distribution F. We are supposing at this stage that the observations are not censored. Then the estimator $\tilde{F}_n(t)$ defined for each $t \geq 0$ as the proportion of observations t_i^* not exceeding t, or equivalently, by

$$\tilde{F}_n(t) = \frac{1}{n} \sum_{i=1}^{n} I(t_i^* \leq t), \tag{1.1}$$

is an estimator of $F(t)$ which has some good properties; in particular, it is unbiased and consistent for $F(t)$. (Throughout we will use the notation $I(A)$ to denote the indicator function of an event A, which takes value 1 if A occurs and 0 otherwise.) See Sections 3.1 and 3.3 for a brief discussion of the concepts of unbiasedness and consistency. The EDF is of course a *statistic*; it can be calculated from the sample values. Considered as a function of $t \geq 0$ it is a monotonically nondecreasing step function whose only jumps are of magnitude $1/n$ at each of the data points $t_1^*, \ldots, t_n^*$. It may equivalently be defined by

$$\tilde{F}_n(t) = \frac{1}{n} \sum_{i=1}^{n} I(t_{(i)}^* \leq t) \tag{1.2}$$

where $t_{(1)}^* \leq t_{(2)}^* \leq \cdots \leq t_{(n)}^*$ denotes the *ordered sample*, i.e. the values $t_1^*, t_2^*, \ldots, t_n^*$ rearranged into ascending order. From (1.2) we see that the jump in $\tilde{F}_n(t)$ at $t_{(i)}^*$ is indeed

$$\Delta\tilde{F}_n(t^*_{(i)}) = \tilde{F}_n(t^*_{(i)}) - \tilde{F}_n(t^*_{(i)}-) = \frac{1}{n}. \tag{1.3}$$

Before we leave the EDF, note that when there are *tied* observations among the t^*_i, i.e. values which are equal, the jumps in the EDF simply accumulate. Supposing there are $d_i \geq 2$ observations tied at $t^*_{(i)}$, it is easy to see that (1.1) implies

$$\Delta\tilde{F}_n(t^*_{(i)}) = \frac{d_i}{n}. \tag{1.4}$$

(This of course is true for $d_i = 1$, also.) If there are exactly m distinct values among $t^*_1, \ldots, t^*_n$, $t^{(d)}_{(1)} \leq \cdots \leq t^{(d)}_{(m)}$, say, then we have $d_1+d_2+\cdots+d_m = n$, and

$$\sum_{i=1}^{m} \Delta\tilde{F}_n\left(t^{(d)}_{(i)}\right) = 1. \tag{1.5}$$

Equation (1.5) reflects the fact that the EDF is a *proper* distribution function estimator; it has total mass 1.

There is a vast theory concerning the behaviour of the EDF and estimators related to it; see for example the volume by Shorack and Wellner (1986) and its references. However, our main interest in it here is simply to motivate discussion of the KME; this is because what we usually observe in survival data are not the (true) survival times t^*_i, but censored versions of them, and the EDF constructed from the censored observations is an unsuitable estimator of the 'true' survival distribution $F(t)$. It is biased for $F(t)$ (although the same is true of the KME, but the bias is usually smaller for the KME, and disappears in large samples; see the discussion in Sections 3.3 and 3.4) and the EDF is also inconsistent for $F(t)$, whereas the KME is consistent, under certain reasonable conditions. We need a modification of the EDF to allow for the censoring, and this is the KME, to which we turn in the next section.

Calculating the KME

The KME, like the EDF, is a step function estimator of the survival distribution $F(t)$ which generated the unobserved 'true' survival times $t^*_1, \ldots, t^*_n$, but unlike the EDF, it takes into account the fact that the observed survival times may be censored. To define the KME, suppose a sample consists of the observed survival times $t_1, \ldots, t_n$. The effect of censoring is to limit the length of a survival time, so for censored observations we know only that $t^*_i \geq t_i$, whereas for uncensored observations we have $t^*_i = t_i$. We know *which* observations are censored: we also observe censor indicators $c_1, \ldots, c_n$, such that

$$c_i = \begin{cases} 1 & \text{if observation } t_i \text{ is uncensored} \\ 0 & \text{otherwise.} \end{cases}$$

We think of the t_i as being observations on 'true' lifetimes t^*_i which we would have observed, but for the censoring.

Sort the t_i into their *order statistics*

$$t_{(1)} \leq t_{(2)} \leq \cdots \leq t_{(n)} \tag{1.6}$$

(not necessarily distinct) and let

$$c_{(i)} = c_j \qquad \text{if } t_{(i)} = t_j.$$

There may be tied (i.e. equal) observations among the t_i, and we will adopt the following convention concerning the indexing of the ordered observations: ties between uncensored observations, or between censored observations, may be ordered arbitrarily; but an uncensored observation tied with a censored one is indexed before the censored one. In other words if

$$t_{(j)} = t_{(j+1)} = \cdots = t_{(j+k+l)}$$

are tied observations, of which $k + 1$ are uncensored and l are censored, then according to our convention, we know that they have been indexed so that

$$c_{(j)} = \cdots = c_{(j+k)} = 1 \text{ and } c_{(j+k+1)} = \cdots = c_{(j+k+l)} = 0.$$

With this convention, the KME can be defined simply as

$$\hat{F}_n(t) = 1 - \prod_{i:t_{(i)} \leq t} \left(1 - \frac{c_{(i)}}{n - i + 1}\right), \tag{1.7}$$

where the product over any empty set is taken as 1. The function $\hat{F}_n(t)$ defined by (1.7) is a right continuous nondecreasing function, as befits a distribution function estimator. It does not necessarily have total mass 1, however.

More convenient for calculations than (1.7) is the following form. Condense the ordered failure times into the subclass of *distinct* ordered failure times, denoted by

$$t_{(1)}^{(d)} < t_{(2)}^{(d)} < \cdots < t_{(m)}^{(d)}. \tag{1.8}$$

Thus there are m distinct failure epochs (at some of which, all survival times may be censored). Define for $i = 1, 2, \ldots, m$

$$d_i = \text{number of individuals failing at time } t_{(i)}^{(d)},$$

$$a_i = \text{number of individuals censored at time } t_{(i)}^{d},$$

and

$$n_i = \text{number of individuals `at risk' at times } t_{(i)}.$$

By the *number at risk* at time $t_{(i)}$ we mean the number of individuals present in the data (not having previously died or been censored) at a time just prior to $t_{(i)}$. In other words, $n_1 = 1$ and for $i = 2, 3, \ldots, m$,

$$\begin{aligned} n_i &= n - (d_1 + \cdots + d_{i-1}) - (a_1 + \cdots + a_{i-1}) \\ &= (d_i + \cdots + d_m) + (a_i + \cdots + a_m). \end{aligned} \tag{1.9}$$

Then the KME $\hat{F}_n(t)$ may equivalently be defined at times $t_{(i)}^{(d)}$, $1 \le i \le m$, by

$$\hat{F}_n\left(t_{(i)}^{(d)}\right) = 1 - \prod_{j=1}^{i}\left(1 - \frac{d_j}{n_j}\right), \tag{1.10}$$

and at times t, $t \le t_{(m)}^{(d)}$, by

$$\hat{F}_n(t) = \begin{cases} 0 & \text{if } t < t_{(1)}^{(d)} \\ \hat{F}_n(t_{(i)}^{(d)}) & \text{if } t_{(i)}^{(d)} \le t < t_{(i+1)}^{(d)}, \qquad 1 \le i \le m-1. \end{cases} \tag{1.11}$$

Thus, like the EDF, the KME is constant except perhaps at the points $t_{(i)}^{(d)}$. At these points its jump is of magnitude

$$\Delta\hat{F}_n\left(t_{(i)}^{(d)}\right) = \hat{F}_n\left(t_{(i)}^{(d)}\right) - \hat{F}_n\left(t_{(i)}^{(d)}-\right)$$
$$= \prod_{j=1}^{i-1}\left(1 - \frac{d_j}{n_j}\right) - \prod_{j=1}^{i}\left(1 - \frac{d_j}{n_j}\right) = \frac{d_i}{n_i}\prod_{j=1}^{i-1}\left(1 - \frac{d_j}{n_j}\right). \tag{1.12}$$

(This formula is correct for $1 \le i \le m$ if we interpret $\Pi_{j=1}^{0}$ as 1, as we shall do throughout). Equation (1.12) shows that $\hat{F}_n$ has positive jumps only at those points $t_{(i)}^{(d)}$ for which $d_i > 0$, i.e. at points for which at least one of the failure times is uncensored. In this respect it differs from the EDF.

Since $\Delta\hat{F}_n(t) \ge 0$ for all $t \ge 0$, as shown by (1.12), we see that $\hat{F}_n(t)$ is indeed a monotone nondecreasing step function, as foreshadowed above. When $i = m$ we have from (1.10) that

$$\hat{F}_n\left(t_{(m)}^{(d)}\right) = 1 - \prod_{j=1}^{m}\left(1 - \frac{d_j}{n_j}\right) = 1$$

if and only if $d_m/n_m = 1$, equivalently, if and only if all of the tied largest observations (or *the* largest observation, if there is only one) are *uncensored.*

If $d_m < n_m$, so that at *least one of the tied largest observations is censored,* then $\hat{F}_n(t_{(m)}^{(d)}) < 1$, the KME has total mass less than 1, and the question arises as to how to define $\hat{F}_n(t)$ for values of $t > t_{(m)}^{(d)}$. Some writers, uncomfortable with the idea of an improper distribution, even one derived from the sample, advocate (re)defining $\hat{F}_n(t) = 1$ for $t \ge t_{(m)}^{(d)}$. Readers should recognise that this is the crux of our development; a value of $\hat{F}_n(t_{(m)}^{(d)}) < 1$ signals the possible presence of immunes, as in the examples given in Section 1.1, and we do not want to define away this information! So we will certainly let $\hat{F}_n(t_{(m)}^{(d)})$ remain as whatever value is calculated from (1.10).

For most purposes we need not even define $\hat{F}_n(t)$ for values of t larger than $t_{(m)}^{(d)}$, because we will never observe it there—we can only calculate $\hat{F}_n(t)$ for values of t up to the largest survival time, $t_{(m)}^{(d)}$, observed in the sample! It is important to keep this in mind. The originators of the KME, Kaplan and Meier (1958), adopted the approach of leaving $\hat{F}_n(t)$ undefined for $t > t_{(m)}^{(d)}$. However, for purposes merely of convenience in the formulation of some of our results, it will occasionally be useful to have an expression for $\hat{F}_n(t)$ on the interval $(t_{(m)}^{(d)}, \infty)$ and we will simply define

$$\hat{F}_n(t) = \hat{F}_n\left(t_{(m)}^{(d)}\right), \qquad \text{for } t > t_{(m)}^{(d)}. \tag{1.13}$$

Wellner (1985), for example, does the same, and proves that the resulting estimator has some optimality in a certain sense.

Formula (1.10) is the most convenient for calculation of the KME of a given set of data. We simply condense the data, as suggested, into distinct classes tied at times $t_{(i)}^{(d)}$, $1 \le i \le m$, then count the number of censored and uncensored observations tied at these times to get the d_i and a_i. Then the n_i and $\hat{F}_n(t)$ are easily obtained from (1.9) and (1.10). This condensing of the data into what is effectively a smaller number of grouped observations also helps to speed up calculations (e.g. to fit parametric models, as in Chapters 5 and 6). For small data sets, it is easy to compute the KME on a hand calculator; see for example the calculations in Kalbfleisch and Prentice (1980, Table 1.3, p. 14). For large sets, one of many available survival analysis computer packages can be used.

We can also use formula (1.7) to get the same result as in (1.10), but remembering that, if (1.7) and (1.10) are to agree, we have to invoke the convention that an uncensored observation precedes a censored, tied, observation. It is obvious, too, that (1.7) and (1.10) are the same when the distribution of the possibly censored observations t_i is continuous, because in that case there can be no tied observations in the data (almost surely) and so $m = n$, the $t_{(i)}^{(d)}$ equal the $t_{(i)}$, $d_i = c_{(i)} = 1 - a_i$, and the number at risk at time $t_{(i)}$ becomes

$$n_i = n - i + 1, \qquad 1 \le i \le n.$$

When there are no censored observations, then $a_i = 0$ for all i, so the numbers at risk become

$$n_1 = n \text{ and } n_i = n - d_1 - \cdots - d_{i-1}, \qquad 2 \le i \le m.$$

Formula (1.12) then gives the telescoping product

$$\Delta\hat{F}_n\left(t_{(i)}^{(d)}\right) = \left(\frac{d_i}{n - d_1 - \cdots - d_{i-1}}\right)\left(1 - \frac{d_1}{n}\right)\left(1 - \frac{d_2}{n - d_1}\right) \cdots \left(1 - \frac{d_{i-1}}{n - d_1 - \cdots - d_{i-2}}\right)$$

$$= \frac{d_i}{n}, \qquad 1 \leq i \leq m,$$

which is the same as for the EDF; compare (1.4). Consequently the KME $\hat{F}_n(t)$ reduces exactly to the EDF $\tilde{F}_n(t)$ when there is no censoring at all, and in this case it is also a proper distribution function estimator.

The precision of the KME

The KME $\hat{F}_n(t)$ is an estimator of the cumulative distribution function $F(t)$ of the 'true' (uncensored) survival times. It is a statistic—a random variable whose value can be calculated from the sample at hand. As such, it has an associated variance which represents the precision with which it estimates $F(t)$.

It is important in our analysis of the sample data to be able to estimate this variance, too. There are various ways of doing this, of which the most straightforward is known as Greenwood's method (Greenwood 1926). This was originally derived as a 'life table' estimator; see Kalbfleisch and Prentice (1980, p. 14) for a discussion. In the notation of expressions (1.8) to (1.10) for the KME, Greenwood's variance estimate is given by

$$\hat{v}_1(t) = \text{Estimated Var } (\hat{F}_n(t)) = \left(1 - \hat{F}_n(t)\right)^2 \sum_{j=1}^{i} \frac{d_j}{n_j(n_j - d_j)}. \qquad (1.14)$$

Here $t = t_i^{(d)}$, where $t_i^{(d)}$ is a time at which at least one individual fails or is censored; we usually need only calculate the variance estimate of the KME at such times. Kaplan and Meier (1958) derived the above expression by an argument involving a representation of the KME using conditional binomial random variables.

We can use $\hat{v}_1(t)$ to find a *pointwise* 95% confidence interval for the true c.d.f. $F(t)$, at time t, of the form:

$$\hat{F}_n(t) - 1.96\sqrt{\hat{v}_1(t)} \leq F(t) \leq \hat{F}_n(t) + 1.96\sqrt{\hat{v}_1(t)}. \qquad (1.15)$$

This is based on the fact that the KME is, for each t, approximately normally distributed in large samples under certain conditions (see Section 3.4).

While the confidence interval in (1.15) is reasonable when n is large, for small n and/or for values of t near 0 or $t_{(n)}$, the largest observed censored or failure time, the interval in (1.15) may include 1 or 0, whereas of course $F(t)$ must be in the interval [0,1]. A variant form of confidence interval which is always in [0,1] can be based on the complementary log-log transform of Kalbfleisch and Prentice (1980, p. 15). It takes the form

$$1 - \left(1 - \hat{F}_n(t)\right)^{e^{-1.96\sqrt{\hat{v}_2(t)}}} \leq F(t) \leq 1 - \left(1 - \hat{F}_n(t)\right)^{e^{1.96\sqrt{\hat{v}_2(t)}}} \qquad (1.16)$$

where, for $t = t_i^{(d)}$,

$$\hat{v}_2(t) = \frac{\sum_{j=1}^{i} d_j/(n_j(n_j - d_j))}{\left[\sum_{j=1}^{i} \log(1 - d_j/n_j)\right]^2}. \tag{1.17}$$

Throughout the book, (1.16) is the formula we will use for calculating a confidence interval on $F(t)$ from the KME.

How good is this formula? In a study of the validity of (1.16) and a similar interval based on an arcsine transform, Borgan and Liestøl (1990) found in simulations that, for sample sizes as low as $n = 25$, (1.16) performed acceptably in terms of its error rate — the proportion of times that the true cumulative distribution function fell outside the confidence interval calculated from (1.16) — and better than (1.14). They used Weibull and exponential distributions to model the survival and censoring distributions of the data.

Borgan and Liestøl (1990) also simulated some 'simultaneous' confidence bounds based on the KME. In contrast to the pointwise confidence intervals, which are valid for each fixed t, the simultaneous limits bound the values of $F(t)$ for all $t \geq 0$, with a given confidence. One must pay for this increased knowledge by accepting a wider band. Borgan and Liestøl (1990) show that the simultaneous confidence interval calculated according to a method of Hall and Wellner (1980) gives good results.

1.3 SOME EXAMPLES OF SURVIVAL DATA

Remission durations of 42 leukaemia patients

Table 1.1 gives the remission durations (t_i, in weeks) of 42 leukaemia patients (Gehan 1965; Freireich *et al.* 1963). This is a venerable data set, often used for illustration in survival analysis expositions (e.g. Kalbfleisch and Prentice 1980, p. 246; Lawless 1982, p. 5; Cox and Oakes 1984, p. 8; Andersen *et al.* 1993, p. 23), but we will find something new to say about it here.

The data is from a prospective study in which 21 patients were allocated to a drug treatment group and 21 to a control group. (Actually the data was collected on matched pairs of individuals, but we have ignored this aspect here.) Patients have been numbered arbitrarily, from 1 to 42.

The time variable of interest is the time spent in remission by each patient. Some remission durations were censored due to limitations of follow-up (many patients were alive with no recurrence of the disease at the end of the study), and censored recurrence times are denoted in Table 1.1 by a value of the censor indicator c_i of 0, whereas recurrences of leukaemia are denoted by $c_i = 1$. The variable i (throughout most of this book) takes the values $i = 1, 2, \ldots, n$, where

Table 1.1 Remission durations (t_i, in weeks) of 42 leukaemia patients.

Controls				Drug treatment			
No.	t_i	c_i	x_i	No.	t_i	c_i	x_i
1	1	1	1	22	6	0	2
2	1	1	1	23	6	1	2
3	2	1	1	24	6	1	2
4	2	1	1	25	6	1	2
5	3	1	1	26	7	1	2
6	4	1	1	27	9	0	2
7	4	1	1	28	10	0	2
8	5	1	1	29	10	1	2
9	5	1	1	30	11	0	2
10	8	1	1	31	13	1	2
11	8	1	1	32	16	1	2
12	8	1	1	33	17	0	2
13	8	1	1	34	19	0	2
14	11	1	1	35	20	0	2
15	11	1	1	36	22	1	2
16	12	1	1	37	23	1	2
17	12	1	1	38	25	0	2
18	15	1	1	39	32	0	2
19	17	1	1	40	32	0	2
20	22	1	1	41	34	0	2
21	23	1	1	42	35	0	2

n is the total number in the sample; here $n = 42$. Also given in Table 1.1 is the value of a treatment group indicator x_i which takes the value 1 for each patient in the control group, and 2 for each patient in the drug treatment group.

The remission durations have been ordered from smallest to largest down the page, for each group, as a visual aid. Immediately striking is the fact that all control group patients suffered a recurrence of the disease within the period of the clinical trial, whereas 12 (more than 50%) of the 21 treated patients did not. Furthermore, several of these censored observations occur at the largest times under observation for this group.

The Kaplan–Meier estimator for Group 2 was plotted earlier, in Figure 1.1, and we mentioned in Section 1.1 that it shows signs of what we are interested in; a levelling of the KME at its right extreme, possibly indicative of the presence of immune or cured individuals. For the control group of patients, however, there is no such indication, because *all* the observations, in fact, are uncensored, not just the largest ones.

Remission durations of 78 paediatric cancer patients

The data set in Table 1.2 is given in Cantor and Shuster (1992). It consists of the remission times, in years, of 78 paediatric cancer patients. The patients were

Table 1.2 Remission durations (t_i, in years) of paediatric cancer patients: an asterisk denotes an *uncensored* observation (i.e. the time of a recurrence of the disease).

t_i for Group 1					t_i for Group 2			
0.90*	3.33*	5.32	6.88	8.47	0.10*	5.87	6.37	8.12
1.06*	3.40	5.52	7.97	8.82	2.15*	5.96	6.61	8.23
1.12*	3.48	5.60	8.03	8.85	2.31	5.96	6.71	9.32*
1.98	3.89	5.69	8.16	9.10	4.89	6.01	6.76	9.38
2.05*	4.39	6.11	8.20	9.31	5.42	6.06	6.90	9.46
2.41	4.41	6.27	8.25	9.36	5.50	6.14	7.42	9.79
2.44	4.55	6.31	8.27		5.53	6.14	7.92	9.85*
2.58	4.63	6.63	8.36		5.58	6.29	8.05	10.33
3.18	5.06	6.66	8.45		5.61	6.34	8.10*	10.65

randomised into two treatment groups, 42 in the first group and 36 in the second. Only 5 patients (denoted by an asterisk) in each group suffered a recurrence of the disease within the followup time.

In Group 1 all recurrences occurred within the first 4 years; the remaining 37 patients in this group remained free of the disease until the end of the

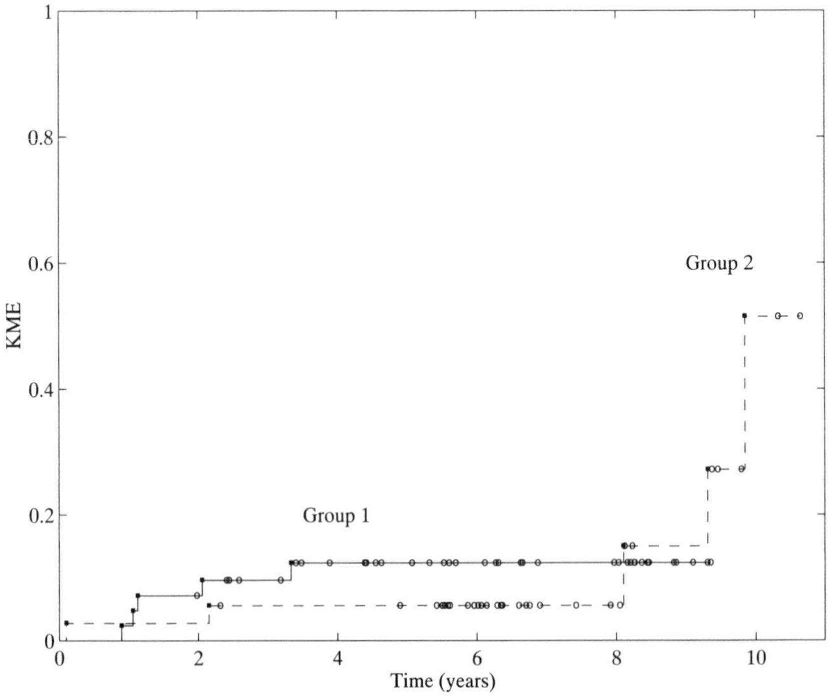

Figure 1.2 Remission durations of paediatric patients: (_____) KME for Group 1, (- - -) KME for Group 2.

follow-up period, almost 10 years from the commencement of the study. In Group 2 there were again only 5 recurrences of the disease, but they occurred at various times throughout the trial, in particular, near the end. The KMEs for each group are plotted in Figure 1.2, and we guess by inspection that there is a good indication of 'immune' individuals in Group 1, but not in Group 2.

The spread of an epidemic

The following example helps explain some of the terminology we will use throughout. In a closed population of individuals, the *susceptibles* may catch a disease, whereas the *immunes* will not. We observe the population from time 0, recording the times at which individuals are infected. Infected individuals are obviously susceptibles, but we never know which are the immunes. Eventually we cease observation at some point and plot the cumulative numbers infected against the times of infection, as in Figure 1.3, which is based on data (Table 1.3) from a smallpox epidemic analysed by Bailey and Thomas (1971); see also Becker (1983).

The curve levels off well below 1, indicating a substantial proportion of the population who are unlikely to contract the disease.

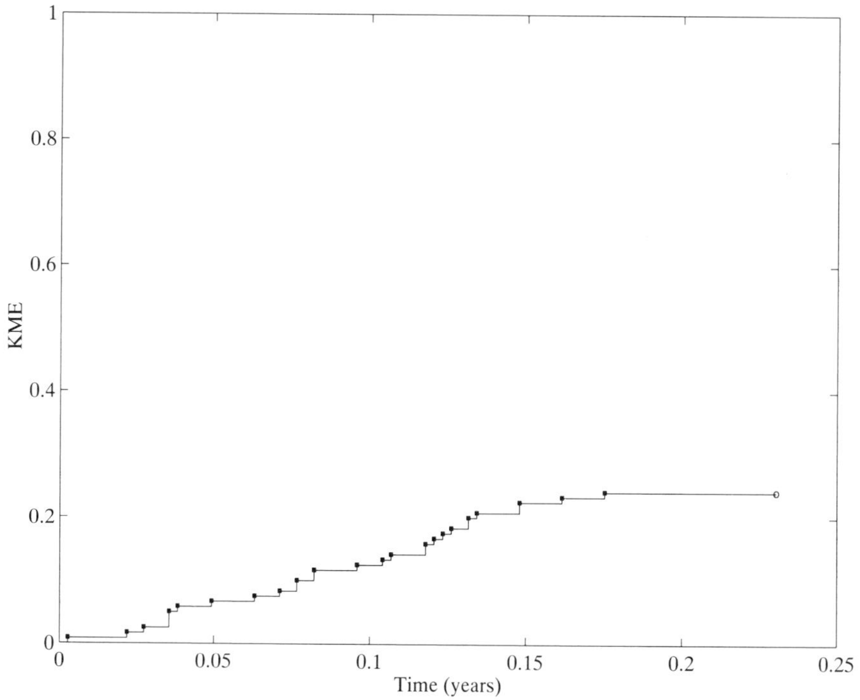

Figure 1.3 Times to infection by smallpox.

Table 1.3 Times to infection (t_i, in days) of a smallpox epidemic.

No.	t_i	c_i	No.	t_i	c_i	No.	t_i	c_i
1	1	1	2	8	1	3	10	1
4	13	1	5	13	1	6	13	1
7	14	1	8	18	1	9	23	1
10	26	1	11	28	1	12	28	1
13	30	1	14	30	1	15	35	1
16	38	1	17	39	1	18	43	1
19	43	1	20	44	1	21	45	1
22	46	1	23	48	1	24	48	1
25	49	1	26	54	1	27	54	1
28	59	1	29	64	1	30–120	84	0

If the curve has levelled *sufficiently* at its right-hand end, we can be fairly confident that most susceptibles have caught the disease and the remainder are immune. The data in Figure 1.3 suggest that around 25% of the population consists of susceptibles, and that most of them have been infected during the time of observation. Note that, in Figure 1.3, no censored observations occur in the body of the curve; all censoring occurs at the final time (the limit of observation), and all censored times have the same value for this data set.

We will use the terminology 'immunes' in a generic sense in this book, but nothing immunological is thereby implied; individuals may fail to contract a disease such as smallpox simply because they do not come into contact with it, although they may well have the physiological potential to contract it.

The recidivism of a released prisoner

Following his/her release from a prison or correctional institution, an offender may never return to that prison again, or any other; this is one aim of correctional or rehabilitative incarceration. One way to measure the success or otherwise of incarceration is to estimate the proportion of offenders who eventually return to prison, illustrated in Figure 1.4.

Figure 1.4 shows the Kaplan–Meier estimator for the recidivism times (times to return to prison) of a number of prisoners released from Western Australian prisons over a recent period. The return times are censored by the need to restrict follow-up of released prisoners to a finite time: the 'cutoff' date of the analysis. Since prisoners were released at different times, however, their recidivism times may be censored at any point from the time of their initial release till the cutoff time. As usual, these are indicated by open circles on Figure 1.4, and return times (the uncensored times) are indicated by black boxes.

There is clearly some evidence of a component who are 'immune' to whatever process results in the recidivism — from the graph we can see that a proportion of around 30% would not be expected to return.

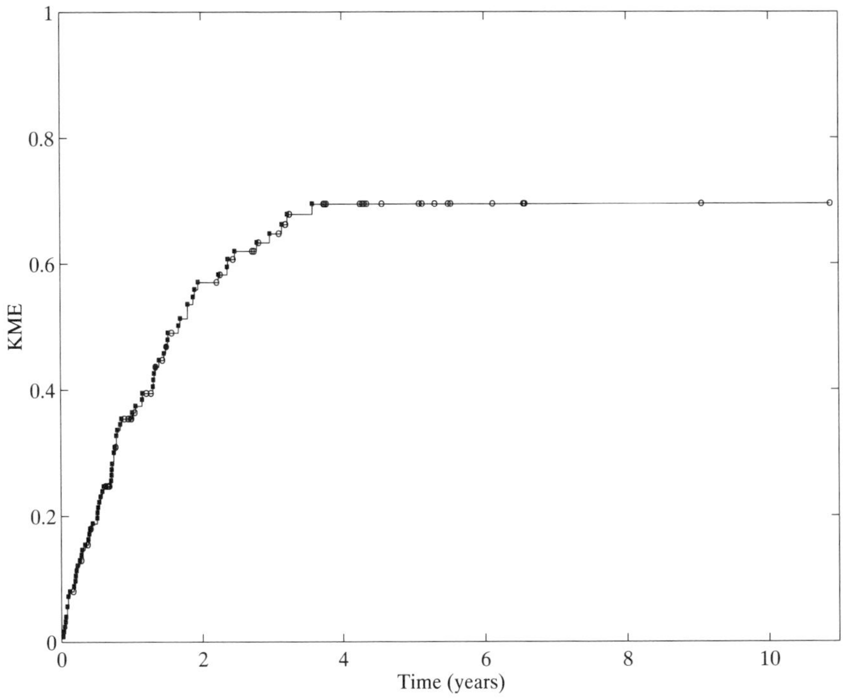

Figure 1.4 Time to return to prison for 121 Western Australian releasees.

1.4 AREAS OF APPLICATION AND PRECURSORS

The idea of allowing for 'long-term survivors', 'immortals', 'cured' individuals or 'immunes' in censored survival data is an old one. As far as we know, Boag, in the United Kingdom, and Berkson and Gage, in the United States, were the first to give explicit formulations of models for data which may have contained a cured component. They worked virtually contemporaneously on the problem of estimating the sizes of cured proportions of individuals in cancer data. Many other workers, not only in the medical area, have taken up the idea since then. In this section we list a number of contributions and their areas of application.

Medical applications: Boag, and Berkson and Gage

Boag (1949) was the first to publish, in a discussion paper of the Royal Statistical Society. He used the method of maximum likelihood to estimate the proportion of cured breast cancer patients in a population represented by a data set of 121 women from an English hospital. The follow-up time for each woman varied up to a maximum of 14 years.

Readers may have guessed, or know, that one approach to the analysis of survival data with immunes is to fit a parametric model which is a mixture of two distributions, one representing the survival or failure time distribution for the *susceptibles* — those who fail — and the other a degenerate distribution allowing for the essentially infinite survival times of the immunes. We will refer to these kinds of models as *mixture models*. Boag fitted such a model; in fact, he fitted a slightly more complicated mixture model because he allowed for a component of patients who were still alive at the end of follow-up but who suffered a recurrence of the cancer rather than a death from the cancer. Boag then estimated a parameter he called c, the proportion 'cured of' the breast cancer.

Boag's approach was to assume a lognormal distribution as the survival distribution of the susceptibles (although, curiously, he noted that an exponential distribution is in fact a better fit to the particular set of data analysed) and to treat deaths from causes other than the cancer under consideration as a censoring mechanism. As his discussants noted, Boag's contribution was revolutionary, opening the way to a whole new range of techniques in the design of clinical trials and the analysis of the resulting data. But, as they also noted, it also had its problems.

In contrast to Boag, Berkson and Gage (1952) used a model consisting of a mixture of the exponential distribution and a degenerate distribution, to allow for a cured proportion; they fitted this model to a large data set consisting of 2682 patients from the Mayo clinic who suffered from cancer of the stomach. The follow-up time on some of their patients was as much as 15 years. They noted that a correct interpretation of the existence of patients 'cured' of the disease should be that the death rates for individuals with long follow-up drop to the baseline death rate of the population. In their case this was obtained from US life tables, and they allowed for it in their model by a simple scaling of the data. Using the method of least squares, they then fitted their 'exponential mixture model' to the scaled proportions dying in intervals of one year.

Berkson and Gage obtained a very good fit of the model to their data (which was further subdivided into groups representing more or less advanced stages of the disease), a very good fit to another set of data on breast cancer, and a very good fit to Boag's data set of 121 women (again, with an adjustment for the baseline mortality of the population).

Mould and Boag (1975) expanded on Boag's (1949) study by fitting a variety of mixture models to patients suffering from cervical cancer. They obtained over 3000 case histories from five English hospitals and one hospital in Oslo, covering the period from 1925 (in the case of one hospital) to 1962. The mixture models consisted, as usual, of a parameter to model the proportion of long-term survivors, and the survival distribution of the susceptibles was selected from a suite of failure distributions, including the lognormal and the exponential. (We note, for later reference, that they did not include the Weibull distribution as a possibility.) The models were fitted separately to women whose disease had reached one of three 'stages' at the time of diagnosis: Stages I, II or III.

Mould and Boag found that, in almost all cases, the lognormal distribution was best among those they compared, at least as judged by its ability to predict survival at various specified times (e.g. after a 10 year follow-up), but that estimates of the cured proportion differed little between any of the fitted models, suggesting good robustness of this parameter to a fairly wide variety of models.

There have been numerous subsequent analyses of medical data sets using the idea of a cured or immune component, following the pioneering efforts of Boag, and Berkson and Gage. Without attempting to be exhaustive, we now list a number of studies we have found which advocate (or sometimes criticise) the use of models with immunes for the analysis of survival data from a variety of areas. They are separated into groups by area of application.

Other medical and epidemiological applications

Haybittle (1959, 1965) fitted a two-parameter Gompertz model to three data sets containing survival times of patients with localised cancer of the cervix or breast (he called the model an 'extrapolated actuarial' model). The Gompertz model as used by Haybittle (1959) is not a mixture model in the sense of the models of Boag or Berkson and Gage, which contain a separate parameter representing a proportion of immunes. Rather, for certain values of the parameters, it is an *improper* distribution (has total mass less than 1) and so a proportion of the population effectively have infinite survival times. Thus a 'cured' component is built into the model, in this sense. (See Section 5.6 for further discussion of the Gompertz distribution.) Haybittle, like Berkson and Gage, also allowed in his model for what he calls the 'normal' mortality of the patient, i.e. the expected mortality rate when the diseases are not present.

Haybittle found a good fit of the Gompertz model to each of his data sets, assessing goodness of fit by a chi-square test which compared observed and predicted numbers dying in various time intervals. He did not make a direct test for the presence of immunes in his data, and left it presumably to be inferred from his values of the cured proportions and their standard errors (estimated by the maximum likelihood method) that they were indeed present. The same is true of the work of Boag, and Berkson and Gage discussed above.

In a later paper, Brinkley and Haybittle (1975) investigated the curability of breast cancer by plotting survival distributions, again standardised for 'normal mortality'. They concluded that, after about 20 years' follow-up, there was some evidence that the survival curves had asymptoted to those of the 'normal' population.

Farewell (1977a) applied a Weibull mixture model with allowance for immunes to a *prospective* study on breast cancer. Information on various factors was collected at a certain time for approximately 5000 women, of whom 48 subsequently developed breast cancer. Farewell's approach illustrates the idea that, of a population at risk, only a proportion (relatively small in this case) will eventually develop the disease; the remainder are immune to it.

Farewell wished to estimate that proportion and to investigate how it may be influenced by risk factors, as well as to investigate how risk factors might affect the *time* to development of the cancer, if this occurred. His model — a Weibull distribution for the time taken (after age 30 years) for a woman to develop breast cancer, mixed with a degenerate distribution representing those immune to the cancer — is one we will use later in a variety of contexts. Like Farewell, we will allow the covariates associated with individuals (the risk factors in Farewell's case) — length of breast-feeding, age at menarche, family history of the disease, etc. — to relate to the probability of being immune in a logistic linear fashion, and to relate to the scale parameter of the Weibull survival distribution in a log-linear way. There is a clear analogy here with the generalised linear models proposed by Nelder and Wedderburn (1972) for the analysis of nonnormal data. (See Chapter 6 for further discussion of this analogy.)

Farewell (1977b) published a companion paper to Farewell (1977a) in which he discussed the efficiency of his method compared to a mixed logistic exponential model with complete follow-up. In Farewell (1982), he turned to the analysis of survival data as such, and applied the same Weibull mixture model with covariates to some fish toxicology data. It produced a somewhat better description of the data, he argued, than could be achieved by models without long-term survivors. A little later, Struthers and Farewell (1989) used a Weibull mixture model with covariates to model progression to the disease AIDS (acquired immunodeficiency syndrome) from persons who were HIV-positive, meaning that they showed a positive exposure to certain viral infections of the immune system at some time. Struthers and Farewell's model allowed for a possible proportion of HIV-positive individuals to be immune to the ultimate contraction of AIDS, with the probability of immunity related to the covariates via a logistic–linear transformation.

Farewell (1986) analysed some time-to-relapse or time-to-death curves for 139 patients with breast cancer, investigating again the question of whether there was evidence of cured individuals. He obtained quite different results, by comparison with standard techniques (such as the Cox proportional hazards model, see Chapter 9), when testing for the effects of the treatments in the study, an outcome which he described as intriguing and clearly worthy of serious further investigation. He went on to discuss possible problems associated with the mixture model approach, pointing out in particular that a long-tailed survival curve could mimic the effect of a nonzero probability of no relapse. Furthermore, he noted that a test for the presence or absence of immunes (which we could phrase as the question of whether or not the proportion of susceptibles is 1) is a 'nonstandard' inference problem, involving testing at the boundary of the parameter space. We will address these well-founded criticisms in some detail in later chapters, taking a statistical approach to what are clearly statistical questions: how confident are we that immunes really are present, and how is this related to the size of our data set and the amount of follow-up we have?

Bithell and Upton (1977) fitted the exponential mixture model with covariates (see Chapters 6 and 8) to data on 487 British children suffering from neuroblastoma or ganglioneuroma. They found significant differences in survival rates and in probabilities of long term survival between subgroups of the children which were defined according to the age of the child at the time of treatment.

Langlands *et al.* (1979) made a retrospective analysis of 3878 cases of breast cancer from 1954 to 1964 in Edinburgh, United Kingdom. Women diagnosed as being in Stages I to IV of the cancer were analysed separately. The method of Langlands *et al.* was to plot year-by-year mortality rates separately for women diagnosed as being at Stages I to IV, and to compare them with expected mortalities calculated from standard life tables. Comparisons were made by elementary statistical methods and no models were fitted. Langlands *et al.* came to the conclusion that, even after as much as 20 years' follow-up, there was still excess mortality (though slight) from the breast cancer, so that the survival curves for these women had not completely levelled in that time. See also Pocock *et al.*, 1982.

Goldman (1984), like Berkson and Gage, studied the exponential mixture model for survival, which she formulated as a mixture of a parameter π representing a proportion of cured individuals in the population, and an exponential distribution with rate parameter λ representing the survival distribution of susceptibles. She obtained a good fit of the model to a group of 33 patients with Stage II testicular cancer from the University of Minnesota Hospitals between 1973 and 1979, and presented the results of Monte Carlo studies which examine the precision of estimation of the cure proportion π. She applied those results to estimate the power associated with accrual schemes in clinical trials.

In a later paper (Goldman 1991), Goldman used the largest value of the Kaplan–Meier estimator of the survival distribution, calculated from the data, to estimate the immune proportion (called there the 'event free fraction'). We will study this estimator in Chapter 4. Goldman used Monte Carlo methods to examine the effect of follow-up on the bias and precision of this estimator. Data on 61 bone marrow transplant patients divided into two groups—with and without GVHD (graft versus host disease)—were analysed.

In Woods *et al.* (1990) and Kersey *et al.* (1987), Goldman applied the 'cure' model (the exponential mixture model) again to compare autologous and allogeneic transplant treatments for leukaemia patients. In the second paper this model revealed significant differences between the groups in rates of failure but not in 'cured' proportions.

Kimber and Crowder (1984) fitted a Weibull mixture model to the distribution of times to infection by whooping cough of 343 children and their siblings. The 343 children had a positive diagnosis of the disease, and the aim was to follow the progress of the infection among the siblings in order to examine the effect of vaccination on the spread of the disease. The immune component in the model allowed for the fact that some individuals were never infected. Kimber and Crowder obtained a reasonably good fit of the model, and were able to analyse for the effects of covariates such as family social status, age, illness length, etc.

Rutqvist (1985) fitted the three-parameter lognormal mixture model of Boag to data on 8170 female breast cancer patients obtained from the Swedish Cancer

Registry in 1961–1963, and compared it with similar (1971–1973) data (10 655 cases). He also distinguished three age-groups of patients (< 50 years, 50–69 years and > 70 years) and found good fits of the three-parameter model in most cases. He did not give formal tests for differences between groups or for the presence of a cured component. A similar investigation is reported in Rutqvist and Wallgren (1985).

In Rutqvist *et al.* (1984) the lognormal mixture model is compared with exponential and Weibull mixture models, with the 'extrapolated exponential' of Haybittle (the Gompertz) and with an 'exponential with shoulder' mixture distribution. The lognormal mixture model gave the best fit to a data set containing information on 14 731 cases of breast cancer reported to the Cancer Registry of Norway from 1953 to 1967. 'Normal' mortality risk was allowed for.

Greenhouse and Wolfe (1984) fitted the Weibull mixture model, and related mixture models, to the famous Stanford heart transplant data (Crowley and Hu 1977). They used the survival times of 69 patients who received heart transplants from 1968 to 1974, and obtained good fits of their models. Chen *et al.* (1985) fitted the Weibull and exponential mixture models to 29 patients with cancer of the endometrium during the years 1955–1974 (Group B, tumour partially penetrating wall of uterus), and a further 21 leukemia patients from a different study (Group C). The graphical evidence presented by Chen *et al.* suggests a good fit of the Weibull mixture model to the leukaemia data but not to the endometrium data. In a spirited discussion of the pros and cons of allowing for a cured proportion via a mixture model of some kind, one of the authors (B. M. Hill) argues that the technology is an important step in improving 'the statistical models ... being used for the analysis of survival data.'

A number of more recent papers have considered various aspects of models with immune or cured proportions applied to data from the health sciences. Halpern and Brown (1987) gave the results of simulation studies which compared the powers of log-rank and Wilcoxon statistics when used to detect a difference between two groups whose survival times come from an exponential mixture model. Gamel *et al.* (1990) fitted the lognormal mixture model of Boag (1949) to the survival times of 4335 patients with intraocular melanoma, stratifying the sample by tumour size.

One point we must consider carefully, especially when the data consists of the lifetimes of patients in a medical trial, is the meaningfulness of the concept of long-term survivors or 'cured' individuals. On the one hand, sufficient follow-up is required to detect the possible existence of an immune proportion in the data; on the other hand, only if the follow-up time is relatively short with respect to the lifetimes of the patients will we see an absolute levelling of the KME—otherwise, deaths from other causes will be observed with a nonnegligible frequency as well. The survival curves displayed in Milner and Watts (1987) are of interest in this regard. One possible resolution of this issue, if it appears to be significant in a particular set of data, lies in the use of competing risks models (see Section 9.3).

These considerations may lead us to the fitting of a mixture model in which the second component is designed to describe the survival times of patients who die of causes other than the disease under study. Gordon (1990), for example, fitted a Gompertz mixture model to the survival function for 812 Stage I and II breast cancer patients from a 1974–1979 prospective study at Case Western Reserve University. The mixture model allowed for deaths from the disease and deaths from other causes. Gordon denoted the latter as 'cured' of the cancer and discussed other concepts of 'cure' in a medical context. Unlike the study of Haybittle (1965), Gordon's versions of the Gompertz distributions were 'proper', thus they do not contain an immune proportion in the sense that we use the term. For the data discussed in Table 1.2, Cantor and Shuster, by contrast, used the Haybittle formulation of an *improper* Gompertz. We will return in later chapters to discuss various aspects of these papers and their data analyses.

Criminology

A major area of interest in criminology is the 'recidivism' of individuals who have been released from prison, or from some form of detention such as arrest, etc. Of great interest is the proportion who ultimately return to prison, having been released from it, or are rearrested after the first arrest, say, and also the rate of return or rearrest for those that do return to prison or are rearrested. Also of interest is the complement to the recidivism probability — the probability that an individual does *not* return to prison after being released from it. Estimation of this proportion and information on whether it differs between subgroups of interest (races, sexes, etc.) is of great interest and importance to criminology. If we take the 'survival' time or 'lifetime' of an individual to be the time that he/she remains out of prison, or the time elapsing before a rearrest, we can fit this kind of investigation into the area of survival analysis.

We need to recognise the existence of an immune component in most populations of interest, since it is (fortunately) the case that, for many individuals, one experience of prison is sufficient; they never reoffend. Others reoffend a number of times then finally desist. In a data set with limited follow-up, these long-term survivors will appear as relatively large censored observations — those that have been released from prison but have not returned by the cutoff date for analysis of the data — whereas other survival times will be censored by the need to restrict follow-up in any practical study.

The first to model criminological data allowing for immunes seems to have been Partanen (1969), who fitted the exponential mixture model to a data set consisting of the times to first return to Finnish prisons of 606 Finnish convicts, following their first release from prison during the period 1949–1951. Using the method of maximum likelihood, a good fit of the mixture model to the data was obtained. The fit was demonstrably much better than using the ordinary exponential model. There was estimated to be a large component of immunes in the population, who in this context could be called 'nonrecidivists'.

Next were Maltz and McCleary (1977), who also recognised the need for a model which allowed for long-term survivors, and fitted the exponential mixture model to data on prisoners released on parole from correctional institutes in Illinois. Figure 2 of Maltz and McCleary shows clearly the superior fit obtained by the mixture model, as compared with the ordinary exponential model.

Soon after this, Bloom (1979) suggested a Gompertz model (in its two-parameter or improper form) and fitted it to the same data set used by Maltz and McCleary, obtaining a slightly better fit. As Maltz (1984, Ch. 8) shows, however, both models depart somewhat from the data for early failures (see Maltz' Fig. 8.3).

Somewhat later, Schmidt and Witte (1988) fitted a variety of mixture models (they called them 'split population models') to a large set of data consisting of the times to return to North Carolina prisons of around 19 000 men and women released between 1977 and 1980. They found that the lognormal mixture distribution gave the best fit to their data, but judging from the likelihood values they give, there was little difference in fit between the lognormal model and a Weibull mixture model. Schmidt and Witte did not distinguish between the cardinality of the return — whether time to first, second, etc., return to prison. Lumping together all times to return in this way distorts the distribution of the recidivism times, since, as we next discuss, the distribution of the time to first recidivism usually differs greatly — both in respect to the rate of failing and in the immune proportion of nonrecidivists — from that of the time to second recidivism, following the first. And in general, the distribution of the time to third recidivism differs (though usually less radically) from the fourth, etc.

Our own interest in this area began with the analysis of a large data set consisting of prisoners released from Western Australian gaols over the years 1975–1984 (see Broadhurst *et al.* 1988). The Weibull mixture model provided quite a good fit to the times to first recidivism following release from prison, and clearly displayed differences between major groups, especially sexes and races (aboriginal prisoners form a large component of the Western Australian prison population).

In later papers we extended the method by examining times to return other than the first (Broadhurst and Maller 1991), demonstrating the differences referred to in the previous paragraph, and applied a covariate model to test for differences in recidivist proportions, and in rates of recidivism, between groups (Broadhurst and Maller 1992; Maller 1993). One of the features of recent work in this area is the size of the data files that are routinely being processed — one file contains around 250 000 individual arrest records — and are now easily handled with modern computing power, providing detailed information on recidivist behaviour that has never been available before (e.g. Broadhurst and Loh 1995). Aspects of some of these analyses, and those mentioned previously, are discussed later in the book.

Engineering reliability

Survival data is of extreme importance in the engineering sciences, since a knowledge of the distribution of the lifetime of a structure, a component of a structure, a motor, etc., is crucial to assessing its reliability.

In a life-testing experiment a number of components are put on test at time 0, say, and subjected to various regimes of stress or wear. Nelson (1982, pp. 123–125) describes a trial in which 10 electric motorettes were run at each of four temperatures (150, 170, 190 and 220 °C), with the time till breakdown of the insulation being recorded. At low temperatures the motors lasted almost indefinitely, at least by comparison with the durations of interest in the study, whereas at higher temperatures breakdowns began to occur earlier and earlier.

One might not expect a priori that there would be an immune component in populations such as these, since it is presumed that all motorettes will eventually break down even under benign operating conditions. Nevertheless the lifetimes under such conditions may be extremely long by comparison with others operating under more severe conditions, and give the impression of a proportion of the population which 'lives forever'. Nelson (1982, p. 52) mentions such models which 'degenerate at infinity', along with two other applications.

A related point is that the lifetime of a motorette, for example, could well end as a result of some 'competing' cause, a cause other than the breakdown of the insulation which the experiment was designed to study. Such competing risk models are closely related to immune proportion models; they are discussed in Section 9.3.

Meeker and LuValle (1995) described a 'humidity accelerated' life-test experiment in which printed circuit boards were subjected to various levels of humidity stress. At low levels of humidity, a proportion of the boards gave the appearance of lasting indefinitely, as noted by Meeker and LuValle. Meeker (1987) applies similar ideas to tests for integrated circuit reliability. See also Lawless (1982, Ch. 10) and Ansell and Phillips (1994, Fig. 2.8, p. 40) for similar data sets. Time-to-event data also occurs in engineering applications such as predicting the timing of warranty claims on a new product, perhaps a car. For many cars this type of claim may never occur, and the manufacturer is surely interested in estimating the proportion. Another application is in computer science: the time taken by a computer or a network of computers to process a certain task. Under heavy operating conditions, the computer may approach the state where the process effectively never receives sufficient priority to be performed.

Market penetration

A new product is launched on the market and is bought at varying times by consumers. At a certain date we survey consumers and ask for the date at which each individual first bought the product. All have the potential to do so, but not all will have done so at the date of sampling, and those who have not contribute censored observations to the data set. The 'immunes', if any, to be discerned from an analysis of data, are those who will never buy the product. Estimation of the distribution of the time taken to buy the product, and of the size of the immune component, is clearly important to market researchers.

To analyse such data, Anscombe (1961) suggested the exponential mixture model and a version of a logistic law.

Fisheries research

Gulland (1955) suggested the exponential mixture model for recapture dates of fish which have been marked and released back to the wild. The 'immune' proportion represents fish which are never recaptured, possibly due to predation or other misadventure, or perhaps as a result of the experience of their initial capture.

Education theory

A mathematical puzzle is presented to a group of students and the time taken for each to solve it is recorded, up till a cutoff time — the end of the experiment — at which remaining observations are censored. Those who will not solve the problem, even given indefinitely large amounts of time, constitute the immune component. Regal and Larntz (1978) fit a gamma mixture model and other models to such a set of data. As other examples, think of the time taken for a postgraduate student to complete a PhD degree, having commenced one — not all do so; or the time taken for the first, or second, . . ., paper from the thesis to appear.

Employment studies

Dunsmuir *et al.* (1989) analysed data collected from interviews of Australian youths aged 16–25 years as to their labour force experience prior to 1986. They were interested in individuals' transitions between the states of being employed, unemployed, or not in the labour force. Their analysis estimated the probabilities of transitions between the states, and also the parameters of the Weibull distributions of the 'sojourn' times (the times spent in the states). The sojourn times are possibly censored by the cutoff nature of the sampling.

Dunsmuir *et al.* fitted Weibull mixture models to the sojourn times, identifying the immune components as 'nonterminating employment' (for employment durations) or as 'hardcore unemployment' (for unemployment durations). They obtained good or adequate fits of the model to employment duration data and to the not-in-the-workforce durations, but the fit to durations of unemployment was not so good. They also tested for the effects of various covariates (sex, age, educational level) on the parameters of the models.

Yamaguchi (1992) gives an application of a mixture model to an analysis of 'permanent employment' in Japan.

1.5 SUMMING UP, SO FAR

We have tried to suggest above that there are many kinds of survival or time-to-event data for which the idea of an immune component of the population may be appropriate. The remainder of the book is devoted to studying the ramifications of this idea. To introduce in an informal way some of the issues this will raise, we conclude with a final example which serves to bring them out in a colourful way.

Sexing a species of turtle

The following example may seem whimsical but is based on an actual situation. For a certain species of turtle it is difficult to tell the sex of the turtle, at least without surgery, until (or if) it lays eggs, at which time it may safely be declared female. A number of the turtles are followed from an early age (let us say, for simplicity, from their hatching) and those that produce eggs are removed from the group or at least marked as female. The date of the first egg laying of a female is also recorded. As time goes by, more and more females are identified and we grow more and more confident that the remaining turtles are males without ever being sure of that fact.

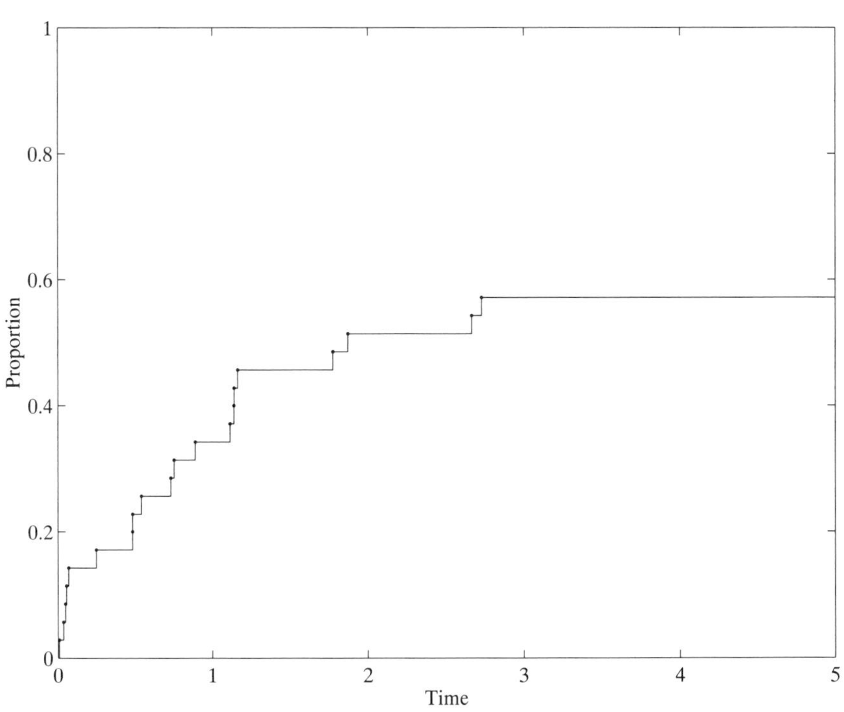

Figure 1.5 Time to egg laying for a species of turtle (simulated data).

The times at which eggs are laid constitute the 'times to event-of-interest' for this example. The males, if any, constitute the 'immune' component of the population. We are never certain that a given animal is male, even if it has not produced eggs up till and including times far in excess of the longest times at which females have been observed to produce them, but the likelihood grows as the time of observation increases. If we plot, as in Figure 1.5, the cumulative proportion of turtles laying eggs, as a function of the time they are laid, we obtain a representation of the distribution of the turtle age at which eggs are laid. If all turtles are eventually observed to lay eggs, then of course they all were and are females. But more generally the function in Figure 1.5 will level off below 1, suggesting that a component of males is present. How far below 1 the curve levels, obviously depends on the proportion of males present.

The above example suggests graphically the sorts of problems we will discuss in the following chapters. How do we estimate the proportion of males, if there are really any present? And how do we test if they are 'really present'? How do we assess whether a curve such as that in Figure 1.5 has 'levelled sufficiently', or been 'followed up' for long enough, for us to be 'reasonably confident' that no more egg laying will occur? If we are only able to observe the turtles' egg-laying experiences for a limited time, can we fit a function to the distribution which allows us to extrapolate, and can we assess how much precision is to be accorded such extrapolations? What is the probability that a turtle which has not produced eggs so far is male, and how accurately and precisely can we estimate this probability? (An agronomist might prefer to replace the above thought experiment with one in which a number of seeds are sown, and the germination times of each are noted as they occur. Sterile seeds may be present as 'long-term nongerminators', though we are never sure of their identity, given only finite follow-up time.)

All of the aforementioned are statistical questions, and we will in the remainder of the book attempt to provide statistical answers to them which are phrased in terms of the estimation of certain quantities. These may be the parameters of an assumed distribution for the survival times under consideration, and tests of hypotheses about those parameters may be used to encapsulate the questions of interest. In the next chapter we turn to this task.

CHAPTER 2

Formulating Tests for the Presence of Immunes, and Sufficient Follow-up

2.1 MODELS FOR CENSORED DATA WITH IMMUNES

Recall that our observed data consists of survival times $t_1, t_2, \ldots, t_n$, with associated censor indicators $c_1, c_2, \ldots, c_n$, where $c_i = 1$ if t_i is uncensored, and $c_i = 0$ otherwise. (Later, in Chapter 6, we will also have associated with t_i a covariate vector $\boldsymbol{x}_i$ of concomitant information, but for the present we disregard this extra information, if any.) We need to discuss how the censoring arises.

Types of censoring

A *censored* observation t_i with $c_i = 0$ arises when we do not in fact observe the event under study — the death of the individual, the recurrence of the disease, the failure of the component, etc. — we know only that the event of interest has not occurred for individual i by time t_i. How does this happen? There are various scenarios, all of which lead to essentially the same mathematical model, which can produce a censored observation.

In *Type I* censoring, n components, say, are put on test at time 0, and all are followed up for a *fixed* amount of time, u_0, say. We observe the failure times of those components that fail; the rest are still operating at time u_0 and their failure times are said to be censored at that time. Equivalent data arises when n individuals are released from hospital or prison at a given date, time 0, and at time u_0 we record the times at which those who returned to hospital or prison did so. There may be r, say, of these, and they constitute the 'survival times' t_i for those individuals. The remainder, who are still out of hospital or prison at time u_0, are recorded as censored individuals with 'survival' times u_0. All we know of these survival times, i.e. the potential times of these individuals' returns to hospital or prison if they ever occur, is that they must exceed u_0. In this scheme, all uncensored times are smaller than u_0, and all censored times equal u_0.

In *Type II* censorship, n items are again placed on test at time 0, but with this scheme we observe for the *random* time necessary until a specified number, r, say, of the n items have failed. We record the (ordered) survival times $t_{(1)} \leq t_{(2)} \leq \cdots \leq t_{(r)}$ of the failed items, and the remaining $n - r$ items are censored at time $t_{(r)}$.

Finally, we can allow a general censorship model in which censoring occurs at times which differ for each individual or item, but which will always, in our development, be statistically *independent* of the mechanism which causes death, failure or the event of interest. So we can think of a censoring time u_i associated with each individual or item i, which is all that we observe if in fact that individual is censored, whereas we observe the 'true' survival time for those not censored.

For further discussion on censoring, its causes and effects, and the possibly restrictive nature of the independent censoring assumption, see for example Gill (1980, pp. 21–25), Kalbfleisch and Prentice (1980, pp. 39–41), and Lawless (1982, pp. 31–44).

The independent and i.i.d. censoring models

All of the censoring schemes mentioned above can be subsumed under a general *independent censoring model* in which we postulate a 'true' survival time t_i^* for each individual i, which is only observed if it does not exceed individual i's censoring time u_i; otherwise, we observe u_i. We also know *which* individuals are censored; this is recorded in a *censor indicator* c_i, which takes the value 1 if t_i is an actual survival time and the value 0 if t_i is censored. The (possibly censored) survival times t_i, $1 \leq i \leq n$, which *are* observed, are assumed, under the model, to be the smaller of t_i^* and u_i, $1 \leq i \leq n$. We can write this as

$$t_i = \min(t_i^*, u_i), \qquad 1 \leq i \leq n. \tag{2.1}$$

The censor indicators can be written as

$$c_i = \begin{cases} 1 & \text{if } t_i^* \leq u_i \\ 0 & \text{if } t_i^* > u_i. \end{cases} \tag{2.2}$$

In general, both the t_i^* and u_i are random variables. Consequently the t_i and c_i are also random variables. We will furthermore assume throughout the book that for each i the random variables t_i^* and u_i are independent of each other, and also from (t_j^*, u_j), $1 \leq j \leq n$, $j \neq i$. In addition to this, it will be assumed until later notice (when we introduce covariates in Chapter 6) that the t_i^* all have the same cumulative distribution function (c.d.f.) F, and (with exceptions noted later) that the u_i all have the same cumulative (survival) distribution function G. We say that the t_i^* and u_i are independent and identically distributed (i.i.d.) and will refer to this as the *i.i.d. censoring model*. It is not hard to see then that the t_i and the c_i are i.i.d. sequences, $1 \leq i \leq n$ (but are not independent of each other), for the i.i.d. censoring model.

These basic assumptions put at our disposal all the usual results of probability and statistics relating to the behaviour of estimators constructed from independent and identically distributed random variables, such as the laws of large numbers, the central limit theorem and much of the well-developed theory of order statistics. Note that the general censoring model covers Type I, Type II, and general censoring schemes mentioned earlier. For Type I this is obvious (let G be degenerate at u_0 for Type I), while for Type II one needs to write down the joint distribution of the observations (i.e. their likelihood) and check that this is of the same form as for Type I, in essence. Expressions for the likelihoods are given in Chapter 5.

A simple consequence of the i.i.d. censoring model is that the cumulative distribution function of any one of the random variables t_i representing the observations, which we will call $H(t)$, is given by

$$H(t) = P\{t_i \leq t\} = 1 - (1 - F(t))(1 - G(t)), \qquad \text{for } t \geq 0. \tag{2.3}$$

Allowing for immunes and susceptibles

Note that we have not specified that the survival and censoring distributions be *proper*, i.e. have total mass 1. A survival distribution which is improper allows, formally, infinite values of the random variable to occur. We will always assume, throughout the book, that the *censoring* distribution is indeed proper, i.e. that $G(\infty) = \lim_{t\to\infty} G(t) = 1$. Allowing infinite censoring times would mean that some observations were never censored, and together with the possibility that some 'true' lifetimes are infinite, which we *will* permit—they represent immune individuals—would allow infinite *observed* lifetimes, which of course cannot occur.

So G will be proper, but F need not be, in our setup. In other words, the quantity $p = F(\infty) = \lim_{t\to\infty} F(t)$, which is always less than or equal to 1, may be *strictly* less than 1. (We will always assume that $F(\infty) > 0$, otherwise nonzero survival times could not occur.) The case $p < 1$ corresponds to the presence of immunes in the population. An immediate consequence of the fact that G is proper is that $H(t)$ as defined by (2.3) is always proper (whether or not F is proper).

Now we describe a probabilistic mechanism by which we can allow the 'true' survival times t_i^* to be, in effect, infinite. This is useful because it allows us to introduce a proper c.d.f. to describe the lifetimes of the nonimmunes (who we will also call *susceptibles*). The particular formulation used here was given by Farewell (1977a). To specify it, assume that associated with individual i is a Bernoulli random variable B_i which takes the value 1 with probability p, and the value 0 with probability $1 - p$. When $B_i = 1$, individual i is taken to be susceptible, or subject to death, failure or the event of interest, whereas $B_i = 0$ signifies an immune individual. We do not know whether a given individual is immune or not, so B_i is not observed.

Susceptibles, i.e. individuals with $B_i = 1$, are assumed to have cumulative survival distribution $F_0(t)$, with $F_0(\infty) = 1$, i.e. F_0 is a proper c.d.f. This is the distribution of t_i^* for these individuals. Individuals with $B_i = 0$ have, formally, $t_i^* = \infty$. We can write these relationships mathematically as

$$P\{t_i^* \le t | B_i = 1\} = F_0(t)$$

and

$$P\{t_i^* \le t | B_i = 0\} = 0$$

for all $t \ge 0$. They imply that, for all $t \ge 0$, the c.d.f. of the true survival times t_i^* is

$$\begin{aligned} F(t) = P\{t_i^* \le t\} &= P\{t_i^* \le t | B_i = 1\}P\{B_i = 1\} + P\{t_i^* \le t | B_i = 0\}P\{B_i = 0\} \\ &= pF_0(t) + 0 = pF_0(t). \end{aligned}$$

Consequently we can write, for all $t \ge 0$,

$$F_0(t) = \frac{F(t)}{p} = \frac{F(t)}{F(\infty)}. \tag{2.4}$$

This is, effectively, a rescaling of the possibly improper distribution F to a new distribution F_0 with total mass 1.

2.2 THE RIGHT EXTREMES OF F, G, H AND F_0

We need to consider the largest survival times that can possibly occur. These are related to the largest possible values of random variables with distributions F and G. If $F(t) < 1$ for all $t \ge 0$ then arbitrarily large survival times can occur, and we define the *right extreme*, τ_F, of F to be ∞ in this case. (This does not preclude the possibility that F is proper, i.e. $\lim_{t\to\infty} F(t) = 1$.) If on the other hand $F(t) = 1$ is achieved for some finite value of t, then no larger survival times than this can occur, and we let τ_F be the smallest of such values t_i. Technically, we define

$$\tau_F = \inf\{t \ge 0 : F(t) = 1\} \tag{2.5}$$

with $\tau_F = \infty$ if $F(t) < 1$ for all $t \ge 0$. If t^* is a random variable with distribution F then we can say

$$P\{t^* \le \tau_F\} = 1$$

(and this is not true if τ_F is replaced by any value smaller than τ_F). Simple examples are the exponential distribution, for which $\tau_F = \infty$, and the uniform distribution on $[0, A]$, for which $\tau_F = A$. Note that any *improper* F automatically has $\tau_F = \infty$, since $F(t) \le F(\infty) < 1$ for all $t \ge 0$.

Now τ_F represents the largest survival time that we could *potentially* observe (and this may be ∞); but due to the censoring, the largest survival time we can *actually* observe is the smaller of τ_F and τ_G, where τ_G is the right extreme of G. The quantity τ_G is defined in an analogous way:

$$\tau_G = \inf\{t \geq 0 : G(t) = 1\} \tag{2.6}$$

with $\tau_G = \infty$ if $G(t) < 1$ for all $t \geq 0$. When $\tau_F \leq \tau_G$, we may (with probability approaching 1 in large samples) observe the largest possible survival times, i.e. as close to τ_F as we please, but when $\tau_G < \tau_F$, the censoring is so 'heavy' that we can never observe the largest possible survival times, those close to τ_F. Thus τ_G represents *the limit of observation* in the sample in the sense that no survival times larger than it will be observed. Another way of expressing this is to define

$$\tau_H = \min(\tau_F, \tau_G), \tag{2.7}$$

the smaller of F and G, and then the above logic tells us that we never observe a (censored) survival time t_i larger than τ_H. Under the independent censoring model discussed in the previous section, τ_H is the right extreme—in the sense of (2.5) and (2.6)—of the following distribution (see (2.3)):

$$H(t) = 1 - (1 - F(t))(1 - G(t)).$$

The condition $\tau_F \leq \tau_G$, or equivalently, $\tau_H = \tau_F$, appears to signal a desirable situation in our formulation, since it allows the possibility of observing survival times up to the maximum possible. (If $\tau_F = \infty$ and $\tau_G = \infty$, as for example if F and G are both exponential distributions, we interpret $\tau_F \leq \tau_G$ as holding.) This is indeed so when F is proper, i.e. $F(\infty) = 1$. However, in general we wish also to allow for *improper* survival distributions F, that is, with $F(\infty) < 1$, and for these we always have $\tau_F = \infty$, by definition. Consequently when $\tau_G < \infty$, as is the case with the cutoff censoring (Type I) described in Section 2.1, for example, we always have $\tau_G < \tau_F$, and this signals the 'undesirable' case of too heavy censoring.

The problem here is that we have forgotten to cater for the existence of immunes in the population. To accommodate them, we introduce another version of the right extreme of F: the largest survival time potentially achievable by a *susceptible* (nonimmune) individual. This is best expressed in terms of the (proper) distribution F_0 of susceptibles introduced above. We have, from (2.4), $F_0(t) = F(t)/F(\infty)$, $t \geq 0$, and we define τ_{F_0} to be the right extreme of F_0, i.e.

$$\tau_{F_0} = \inf\{t \geq 0 : F_0(t) = 1\}, \tag{2.8}$$

with $\tau_{F_0} = \infty$ if $F_0(t) < 1$ for all $t \geq 0$. Survival times t_i^* of susceptibles, having the distribution F_0, may or may not be censored, according to whether or not it happens that $t_i^* \leq u_i$, where u_i is the censoring random variable for individual i, having distribution G. Thus for susceptibles we also have $t_i^* \leq \tau_G$ a.s. (almost surely). But now the condition $\tau_{F_0} \leq \tau_G$ indeed signals the desirable situation, that we can potentially observe the maximum lifetimes of *susceptibles* in the population.

Notice that

$$\tau_{F_0} = \inf\{t \geq 0 : F(t) = F(\infty)\}; \tag{2.9}$$

thus τ_{F_0} is indeed the time at which F achieves it maximum possible value, if this is in fact achieved. But since we permit $F(\infty) < 1$, so as to allow for the presence of immunes, we do not have $\tau_{F_0} = \tau_F$ in general. Figure 2.1 illustrates the possibilities. We see that when F is proper, as in Figure 2.1(a) and (b), then $\tau_{F_0} = \tau_F$, finite or infinite. When F is improper, as in Figure 2.1(c) and (d), we always have $\tau_F = \infty$, but we may have $\tau_{F_0} < \infty$, as in Figure 2.1(c). We summarise some useful properties of τ_{F_0} and τ_F as follows.

When $F(\infty) = 1$, we have

$$\tau_{F_0} = \tau_F; \tag{2.10}$$

we always have

$$\tau_{F_0} \leq \tau_F; \tag{2.11}$$

and we always have

$$F(\tau_{F_0}) = F(\tau_F) = F(\infty). \tag{2.12}$$

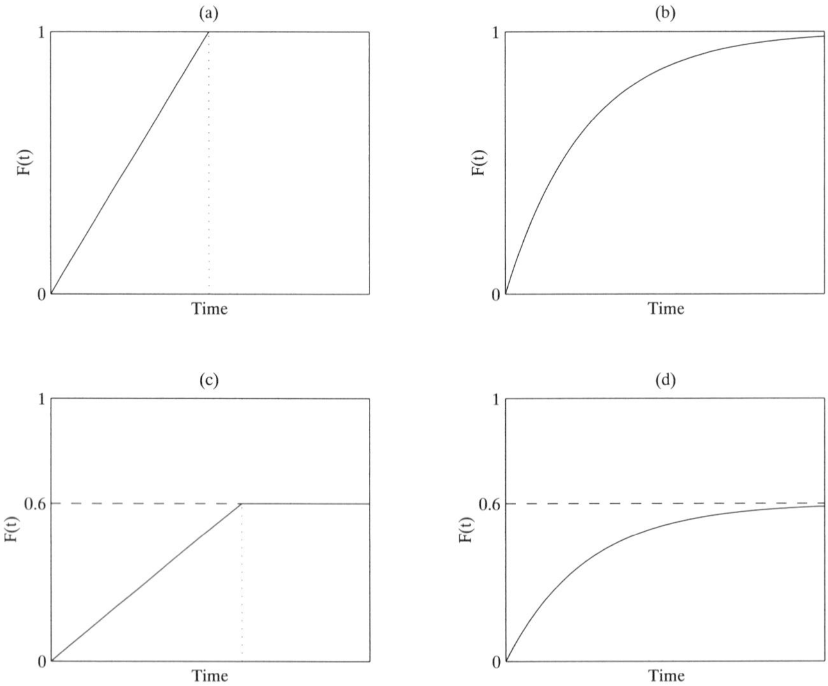

Figure 2.1 Schematics of possible c.d.f's: (a) Proper c.d.f. with $\tau_F < \infty$; (b) Proper c.d.f. with $\tau_F = \infty$; (c) Improper c.d.f. with $\tau_{F_0} < \infty$; (d) Improper c.d.f. with $\tau_{F_0} = \infty$.

For proof, just note that equation (2.10) is obvious because $F = F_0$ when $F(\infty) = 1$. Inequality (2.11) follows because the set $\{t \geq 0 : F(t) = 1\}$ is contained in the set $\{t \geq 0 : F(t) = F(\infty)\}$, so the infimum of the first set, which is τ_F, can be no smaller than the infimum of the second set, which is τ_{F_0}. Inequality (2.11) implies, since F is nondecreasing, that

$$F(\tau_{F_0}) \leq F(\tau_F) \leq F(\infty) \tag{2.13}$$

while, by the definition of the infimum in (2.9), we must have $F(t) = F(\infty)$ for any $t > \tau_{F_0}$. Thus $F(\tau_{F_0}) = F(\infty)$ since F is right continuous, and (2.13) then proves (2.12).

The situation now is as follows. We allow a general survival distribution F with total mass $p = F(\infty) \leq 1$; there may or may not be immunes present. Censoring is by the proper distribution G. If immunes are present, thus $F(\infty) < 1$, we may think of their 'true' survival times t_i^* as being ∞, corresponding to the 'improper' component $1 - p = 1 - F(\infty) > 0$. Their *actual* survival times, as observed in a sample, will *always* be censored, possibly at τ_G, the limit of follow-up, or possibly before. Survival times of susceptibles are distributed according to the 'proper' component of F, i.e. F_0. They may or may not be censored, possibly at τ_G, or possibly before.

2.3 FORMULATING TESTS FOR THE PRESENCE OF IMMUNES AND SUFFICIENT FOLLOW-UP

If we want to apply the ideas set out above, we need a clear and unequivocal procedure for disentangling two aspects of the population (and of the sample) which are confounded together. It is obviously worth asking (and answering!) the question: are there really immunes in the population from which the sample is drawn? If there are not, we can proceed with an 'ordinary' survival analysis of the kind described in Collett (1994), Aitkin *et al.* (1989), Cox and Oakes (1984), Kalbfleisch and Prentice (1980), etc., in which all individuals are assumed to fail. If on the other hand we are reasonably confident that there are immunes present, we will want to apply methods which are quite different, as we will see later. Furthermore, our conclusions from the analysis of the data set may well differ in the two cases. So the decision as to whether or not immunes are present is a critical one. The complicating factor, however, is whether there is sufficient follow-up in the sample to allow us to detect immunes, if present, with any confidence.

In this section we will apply some of the results already known in the theory of survival analysis — and some new ones as well — to suggest a way of addressing the issues raised above. Our methods here will be essentially nonparametric (parametric methods will be discussed in Chapters 5 and 6), and are motivated by considering the properties of the statistic $\hat{F}_n(t_{(n)})$. We will use throughout the notation of equations (1.6) and (1.7), whereby $t_{(n)}$ is the largest observation, censored or uncensored, in the sample, and $\hat{F}_n(t_{(n)})$ is the value of the Kaplan–Meier estimator (KME) at its right endpoint, which is also its largest value. Thus $\hat{F}_n(t_{(n)}) = 1$ if and only if all the largest tied observations are uncensored.

The questions we are addressing, concerning the existence of immunes in the population, and whether or not we have sufficient follow-up, can be put in a statistical framework by positing them as tests of hypotheses concerning certain features of the population. These features of the population can be described by certain constants defined in terms of the survival and censoring distributions, namely the right extremes of F_0 and G. They are not the kinds of parameters we are used to in statistics, which describe location or scale, etc., but they are still amenable to estimation and hypothesis testing in certain ways. The discussion in the previous two sections provides the clues as to how we can proceed.

What do we mean by 'immunes'?

First of all, let us specify exactly what we mean by saying that there are immunes in the population. We will take this to mean simply that *the survival distribution F is improper*, that is, has total mass less than 1. In other words, *we will say that immunes are present in the population* if and only if

$$p = F(\infty) < 1. \tag{2.14}$$

Also we will call $1 - F(\infty)$ the *immune proportion* of the population (p is the proportion of susceptibles). When (2.14) holds, $1 - F(\infty) > 0$, and there is a component of individuals in the population who are not subject to death, failure or the event of interest: they 'survive forever'. The remaining proportion, $F(\infty)$, are susceptible to death or failure at a rate specified by the proper component, $F_0(t)$, of $F(t)$. The interesting aspect of our data, and our problem, is that immunes or susceptibles are not identified—we can only assume or hypothesise their presence from indications in the sample such as we discussed earlier.

Now we wish to consider the general question of testing for the presence of immunes. This will lead us to some useful and important aspects of the analysis of any set of survival data, whether or not we are interested in the question of immunes. Recall from the previous section that τ_F is the right extreme of the survival distribution F, and τ_{F_0} is the right extreme of the distribution F_0 of the survival times of the susceptibles. In view of (2.14) we could formulate the null hypothesis of interest, that there are no immunes in the population, as the hypothesis that $F(\infty) = 1$. However, in order to phrase this hypothesis in terms of things we can estimate, it turns out to be more useful to consider the value of F at τ_G. So we will begin with the following step.

Step 1: test the hypothesis $H_{01} : F(\tau_G) = 1$
A method for doing this will be described in Section 4.2. It will be based on the fact that the value of the KME at its right extreme, $\hat{F}_n(t_{(n)})$, is a consistent

estimator of $F(\tau_G)$ when F is continuous at τ_H; these sorts of properties will be investigated in detail in Chapter 3. We leave further discussion of methods till Chapter 3, and go on with our formulation.

The idea of sufficient follow-up

Now consider the situation when H_{01} is true. One consequence then is that we have not been able to detect any immunes in the sample, because $F(\tau_G) = 1$ certainly implies that F is a proper distribution, so (2.14) cannot hold. Also when $F(\tau_G) = 1$, we have by the definition of τ_F that $\tau_F \leq \tau_G$, and by (2.11) we know that $\tau_{F_0} \leq \tau_F$. Thus a second consequence of $F(\tau_G) = 1$ is that $\tau_{F_0} \leq \tau_G$. This expresses the fact that *follow-up in the sample has been sufficient*, since it states that all lifetimes which can in fact be observed (the lifetimes of the susceptibles) occur within the limit of observation as represented by τ_G. In other words, the censoring allows us to observe the largest possible survival time of a susceptible individual.

Thus we see that *acceptance of* H_{01} *has two implications*:

1. We decide that there are no immunes in the population.
2. We decide that follow-up has been sufficient.

(We will discuss later the possibility that immunes may be present in the population even when $\hat{F}_n(t_{(n)}) = 1$, but except for some extreme circumstances they will be undetectable in the sample at hand.) These conclusions lead us back to doing an 'ordinary' analysis of the data, with no necessity to consider the presence of immunes.

Next suppose we *reject* H_{01} and decide that $F(\tau_G) < 1$. Then there are two conclusions we could come to:

1. There are immunes in the population and they have been detected by sufficient follow-up in the sample.
2. There may or may not be immunes on the population, but the follow-up in the sample has been insufficient to be decisive.

In order to distinguish between these possibilities we must develop a test for 'sufficient follow-up' in the sample. We mentioned in Step 1 that 'sufficient follow-up' may be formulated as '$\tau_{F_0} \leq \tau_G$'. So if we reject H_{01} we move on to the next step.

Step 2: test the hypothesis $H_{02} : \tau_{F_0} \leq \tau_G$ (knowing that $F(\tau_G) < 1$)

Now consider the consequences of accepting or rejecting H_{02}. If we *accept* that $\tau_{F_0} \leq \tau_G$, then since we have already decided that $F(\tau_G) < 1$, we have $F(\tau_{F_0}) < 1$. By (2.12) this means $F(\infty) < 1$, and according to our definition (2.14) there is evidence of a proportion $1 - F(\tau_{F_0}) > 0$ of immunes in the population. If on the other hand we *reject* H_{02} then follow-up has been insufficient; at best we can only estimate $F(\tau_G)$ within the limitations of the sample, and $F(\tau_G) < F(\tau_{F_0})$—since if $F(\tau_G) = F(\tau_{F_0}) = F(\infty)$ then $\tau_G \geq \tau_{F_0}$ by the definition of τ_{F_0}. Thus, under this last scenario, the fact that the KME is improper, or appears to level off below 1, may simply be due to the effect of the censoring, which has been too heavy to allow survival times near τ_{F_0} to occur.

Applying the procedures

The above gives us a clear prescription to follow, and will always lead to a conclusion, with due regard to the fact that the statistical hypothesis tests involved will have their associated Type I and Type II error rates. One of the conclusions to which we are led may be that follow-up has not been sufficient to confirm or rule out the possibility of the existence of immunes; but this also we must accept as part of the statistical procedure.

Of course we have not discussed *how* to do the tests, and that is not a trivial matter—we postpone discussion of this to the next section. But at this stage let us consider in a general way how the procedure can be applied. In Figure 2.2 we

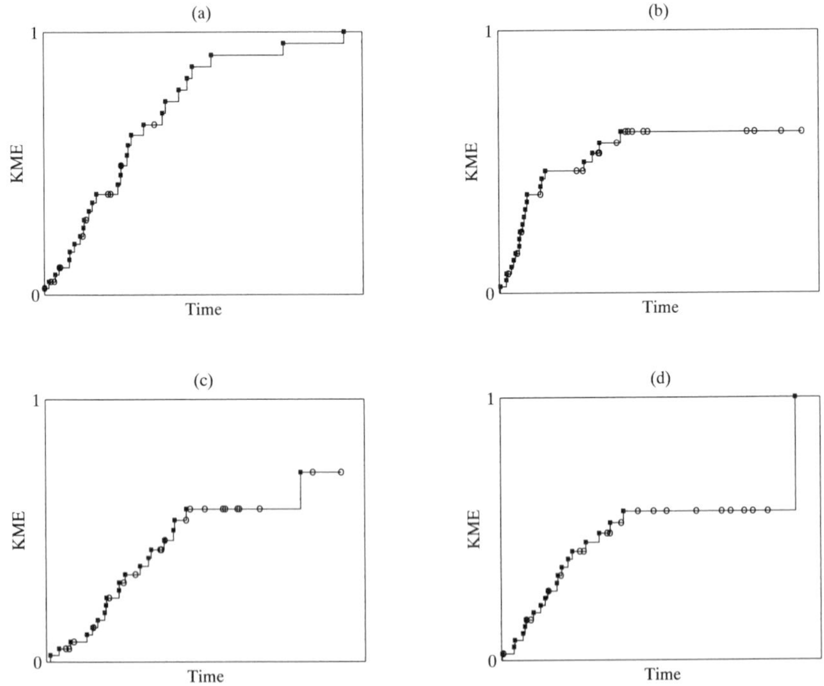

Figure 2.2 Various data sets exhibiting immunes or their absence: (a) 'ordinary' survival data with 'proper' KME; (b) KME with evidence of immunes; (c), (d) some evidence of immunes, but with possibly anomalous values.

have plotted the KMEs of some synthetic (simulated) data sets. Let us conjecture classifications for them in the light of the above discussion.

Figure 2.2(a) has $\hat{F}_n(t_{(n)}) = 1$, and we would not expect to reject H_{01} on the basis of a test which uses the discrepancy of $\hat{F}_n(t_{(n)})$ from 1. On the basis of the test to be introduced in Chapter 4, we will indeed decide that Figure 2.2(a) represents 'ordinary' survival data with no immunes. Even if the largest observation in Figure 2.2(a) had been censored, so that $\hat{F}_n(t_{(n)}) < 1$, still $\hat{F}_n(t_{(n)})$ would clearly be close to 1, and we would still be unlikely to reject H_{01}. We will see in the next chapter that it is not at all unlikely, in some situations, to observe $\hat{F}_n(t_{(n)}) < 1$ even when F is proper. (See Theorem 3.11 and the discussion preceding it.)

The KME in Figure 2.2(b) appears to level off well below 1 and we would guess that there is an immune component present. To substantiate this conclusion, we must reject H_{01} and accept H_{02}. As a matter of fact, the nonparametric tests that we propose later for these purposes do precisely that (at the 5% significance level) for this data set.

The data in Figure 2.2(c) is more problematic. There appears to be a levelling of the KME at its right end but this is spoiled by a single late failure. For this data we guess that we might reject H_{02}, leaving us inconclusive as to the existence of the immunes. However, as in any statistical analysis, robustness considerations might lead us to investigate further the high influence of the single late failure. In fact, omitting this point does lead to acceptance of H_{02} for this data according to the test we propose in Chapter 4. In Section 5.5 we discuss how to formalise a procedure for dealing with survival data with outlying observations.

Figure 2.2(d) is similar to (c) but here the possibly atypical large failure is the largest one and causes the KME to be proper, i.e. $\hat{F}_n(t_{(n)}) = 1$, even though the observations just prior to this suggest a levelling off and the possibility of immunes. When $\hat{F}_n(t_{(n)}) = 1$ we will not be able to reject H_{01}, at least with the test we propose, so if we suspect the presence of immunes we should again ask (and test) whether the single late failure is an outlier.

Testing H_{02} then H_{01}

Before leaving the formal framework of the hypothesis tests we propose, it is worthwhile considering the logic of first testing H_{02} rather than H_{01}. The hypothesis

$$\boldsymbol{H_{02}} : \tau_{F_0} \leq \tau_G$$

may be of interest in the case $F(\infty) = 1$, i.e. even if there are no immunes in the population. For example, in Appendix Note 2 to Chapter 3 we discuss the fact that when F is proper, the 'Kaplan–Meier' mean $\hat{\mu}_n$ defined by

$$\hat{\mu}_n = \int_0^{t_{(n)}} (1 - \hat{F}_n(t))\,\mathrm{d}t$$

is consistent for the mean survival time

$$\mu = \int_{[0,\infty)} t\,\mathrm{d}F(t) = \int_0^\infty (1 - F(t))\,\mathrm{d}t$$

if and only if H_{02} holds, i.e. according to our interpretation, if and only if there is sufficient follow-up in the sample.

Here $\hat{\mu}_n$ is a natural estimator of μ. We can similarly show that, under some reasonable conditions, $\tau_{F_0} \leq \tau_G$ is necessary and sufficient for the consistent estimation of other functionals of the survival time distribution by the corresponding functionals of the KME. Thus we see that the condition $\tau_{F_0} \leq \tau_G$ has diagnostic importance whether or not we are interested in the existence of immunes. Of course, when there *are* immunes in the population, a test for H_{02} is also important in deciding if the KME of the survival distribution has levelled off sufficiently, as we discussed above. In fact a test for H_{02} is useful even if we have a parametric representation for the underlying survival distribution, as we will illustrate in Chapters 5 and 6.

Let us now consider the interpretations possible after testing the two hypotheses H_{01} and H_{02} in the reverse order.

Step 1: test H_{02}: $\tau_{F_0} \leq \tau_G$

If H_{02} is rejected we have insufficient follow-up in the sense that a functional of the KME such as $\hat{\mu}_n$ is not a consistent estimator of μ, in the case of no immunes, or to be confident in the presence of immunes, if that is our interest. If we accept H_{02} then there is 'sufficient follow-up' for either purpose and we may or may not wish to move on to the next step.

Step 2: test H_{01}: $F(\tau_{F_0}) = 1$ (knowing that $\tau_{F_0} \leq \tau_G$)

Acceptance of H_{01} means that we are now confident that there are no immunes, rejection means the opposite.

From the above we see that it makes sense alternatively to test first H_{02}, then H_{01}. However, although this is so in principle, the tests we will propose rely for their good properties on our first testing H_{01}, then H_{02}. We turn in the next chapter to the means of doing this.

CHAPTER 3

Properties of the Kaplan–Meier Estimator

3.1 CONSISTENCY OF THE KAPLAN–MEIER ESTIMATOR

At the end of Chapter 2 we *formulated* tests for the presence of immunes, and for sufficient follow-up in our data. It would be logical next to discuss some methods of *performing* those tests on the data at hand. However, in order to motivate the forms of the tests we will propose, and to understand their properties, we need to consider some theoretical properties of the Kaplan–Meier estimator constructed from a sample of data drawn according to the i.i.d. censoring model of Section 2.1.

Some of this material is unavoidably technical, so first-time readers wanting mainly to get the flavour of the applications may wish to jump ahead to Chapter 4, or beyond. But it is important for us to set out certain results in some detail. The fact that the maximum value of the KME, $\hat{F}_n(t_{(n)})$, is consistent for $F(\tau_G)$ under a mild continuity condition (see Theorem 3.5) will suggest a test for the presence of immunes (i.e. a test of the hypothesis H_{01} formulated in Section 2.3); and an analysis of situations when the largest observation is censored or not will help us assess whether or not follow-up has been sufficient, and how to test for this. Accordingly, we take a large detour to study in detail the properties of the KME that will be needed in the sequel.

General concepts of consistency

An estimator $\hat{\theta}_n$ calculated from a sample of size n is *consistent* for a parameter θ, if, informally, $\hat{\theta}_n$ becomes close to θ as the sample size n increases. The idea here is that, as more and more sample information becomes available, a consistent estimator $\hat{\theta}_n$ of θ becomes more and more accurate. Formally, we can quantify this in terms of some mode of stochastic convergence of $\hat{\theta}_n$ to θ. We will be concerned mainly with *consistency in probability* of an estimator, which we denote by

$$\hat{\theta}_n \xrightarrow{P} \theta.$$

(We will often omit the qualifier '$n \to \infty$' which applies to convergences of this type, when it is obvious.) By this we mean that $\hat{\theta}_n$ converges in probability to θ, or formally

$$\lim_{n\to\infty} P\{|\hat{\theta}_n - \theta| > \varepsilon\} = 0 \tag{3.1}$$

for each $\varepsilon > 0$. Condition (3.1) clearly expresses the fact that $\hat{\theta}_n$ is within a distance ε of θ if n is large enough, except for a set whose probability measure converges to 0 as $n \to \infty$. Another way of expressing this is to say that $\hat{\theta}_n$ *converges to* θ *with probability approaching* 1, in the sense that the sequence of events

$$A_n(\varepsilon) = \{|\hat{\theta}_n - \theta| \leq \varepsilon\}$$

satisfies

$$\lim_{n\to\infty} P(A_n(\varepsilon)) = 1$$

for each $\varepsilon > 0$ if (3.1) holds. We will often use the concept of a sequence of events A_n occurring with probability approaching 1 as n diverges to ∞, meaning that $\lim_{n\to\infty} P(A_n) = 1$, and abbreviate it to '$A_n$ occurs WPA1'.

The other mode of stochastic convergence we will occasionally consider is *almost sure* (sometimes abbreviated to 'a.s.') and we will say that $\hat{\theta}_n$ is *almost surely consistent* for θ and write $\hat{\theta}_n \to \theta$ a.s. if

$$P\left(\lim_{n\to\infty} \hat{\theta}_n = \theta\right) = 1.$$

For a discussion of these kinds of convergence see, for example, Grimmett and Stirzaker (1992) and Chow and Teicher (1988). Most of the emphasis in this book will be placed on convergence in probability, rather than on almost sure convergence, since this seems adequate for most statistical applications. (We will also need the notion of *convergence in distribution* so as to talk about the asymptotic distributions of estimators, but we defer this to Section 3.4.)

Now consistency of an estimator is clearly a large-sample property, which strictly speaking says nothing about the properties of the estimator in any finite sample—and of course any sample we have to hand must be finite! That consistency is a desirable property of an estimator, however, may perhaps be elucidated by considering that *inconsistency* is clearly undesirable; inconsistency of $\hat{\theta}_n$ would mean, in effect, that even with an *infinite* amount of information, we would not estimate θ correctly! Even so, by itself consistency of $\hat{\theta}_n$ only tells us about the *accuracy of estimation* of θ by $\hat{\theta}_n$; it must be supplemented by an assessment of the *precision of estimation* of θ by $\hat{\theta}_n$—how widely $\hat{\theta}_n$ varies around θ—to be really useful.

We will leave a discussion of precision of estimation till later, and now turn to the question of the consistency of the KME. An important precursor of this for us is to study the behaviour of $t_{(n)}$, the largest observed survival or censored time. This we do in the next section.

Large-sample behaviour of the largest survival time

Throughout the remainder of this chapter we will assume the i.i.d. censoring model outlined in Section 2.1. Recall that this model specifies that the observed survival times t_i are the minima of 'true' survival times t_i^* and censoring times u_i. The t_i^* and u_i are i.i.d. and independent of each other. We also have given the values of censor indicators: $c_i = 1$ if t_i is uncensored and $c_i = 0$ otherwise. The t_i^* have c.d.f. $F(t)$, possibly improper, and the u_i have proper c.d.f. $G(t)$. Thus the t_i are i.i.d. with c.d.f.

$$H(t) = 1 - (1 - F(t))(1 - G(t)), \qquad t \geq 0.$$

Recall also the notations τ_F, τ_G, and τ_H for the right extremes of F, G, and H (see Section 2.2). Denote the ordered observed survival times by $t_{(i)}$, $1 \leq i \leq n$, so that

$$t_{(1)} \leq t_{(2)} \leq \cdots \leq t_{(n)},$$

and let $c_{(i)}$ denote the censoring indicator of $t_{(i)}$. We do not exclude the possibility that there may be ties among the $t_{(i)}$, and if this occurs, uncensored observations tied at $t_{(i)}$ will precede those censored, according to our convention.

In this section we obtain formulae for the c.d.f. of $t_{(n)}$, the largest of the observations. One such formula is obtained from the fact that the maximum of n i.i.d. random variables with c.d.f. $H(t)$ is easily shown to have c.d.f. $H^n(t)$. The second representation, which effectively shows us how to differentiate the c.d.f. of $t_{(n)}$, is more complicated, but will be needed in the proof of Theorem 3.10 below. To obtain this formula, we need to allow for the fact that there may be $1, 2, \ldots, n$ observations tied for the largest value.

Theorem 3.1 *Suppose that* $t_{(n)} = \max_{1\leq i\leq n} t_i$, *where the* t_i *are i.i.d. with c.d.f.* $H(t)$. *Then we have for all* $t \geq 0$

$$P\{t_{(n)} \leq t\} = H^n(t) = \sum_{m=1}^{n} \int_{[0,t]} H^{n-m}(y)H^{m-1}(y-)\,\mathrm{d}H(y). \tag{3.2}$$

Proof The first equality in (3.2) follows from

$$P\{t_{(n)} \leq t\} = P\{t_i \leq t, 1 \leq i \leq n\} = H^n(t).$$

For the second, take $t \geq 0$ and write

$$\begin{aligned} P\{t_{(n)} \leq t\} &= \sum_{m=1}^{n} P\{t_{(n)} \leq t, \text{ exactly } m \text{ survival times are tied at } t_{(n)}\} \\ &= \sum_{m=1}^{n} \binom{n}{m} P\{t \geq t_1 = \cdots = t_m > \max_{m+1\leq i\leq n} t_i\} \end{aligned}$$

$$= \sum_{m=1}^{n} \binom{n}{m} \int_{[0,t]} P\{t_2 = \cdots = t_m = y\} P\{y > \max_{m+1 \le i \le n} t_i\} \, dH(y)$$

$$= \int_{[0,t]} \sum_{m=1}^{n} \binom{n}{m} (\Delta H(y))^{m-1} H^{n-m}(y-) \, dH(y). \tag{3.3}$$

Here we interpret $\max_{n+1 \le i \le n}$ as $-\infty$, and use the notation

$$\Delta H(t) = H(t) - H(t-) = P(t_1 = t). \tag{3.4}$$

Note that when $\Delta H(y) > 0$

$$\sum_{m=1}^{n} \binom{n}{m} (\Delta H(y))^{m-1} H^{n-m}(y-) = \frac{(H(y-) + \Delta H(y))^n - H^n(y-)}{\Delta H(y)}$$

$$= \frac{H^n(y) - H^n(y-)}{\Delta H(y)} = \sum_{m=1}^{n} H^{n-m}(y) H^{m-1}(y-). \tag{3.5}$$

Thus (3.3) agrees with (3.2) in this case. When $\Delta H(y) = 0$ the series in (3.3) reduces to $nH^{n-1}(y)$, so again (3.3) agrees with (3.2). This completes the proof of Theorem 3.1. ■

Note that we must have $P\{t_{(n)} \le \tau_H\} = 1$, where τ_H denotes the right extreme of H, because no observation from H can exceed its right extreme a.s. Also of course $t_{(n)}$ can only increase as n increases. Thus as an application of Theorem 3.1 we have Theorem 3.2.

Theorem 3.2 *Under the conditions of Theorem 3.1,*

$$t_{(n)} \uparrow \tau_H \text{ a.s.} \tag{3.6}$$

Proof By (3.2) and the formula for the sum of a geometric series we have

$$\sum_{n=1}^{\infty} P\{t_{(n)} \le t\} = \sum_{n=1}^{\infty} H^n(t) = \frac{H(t)}{(1 - H(t))}.$$

When $t < \tau_H$ this is finite, because $H(t) < 1$ when $y < \tau_H$ by the definition of τ_H. So we see that $\Sigma P\{t_{(n)} \le t\}$ converges for $t < \tau_H$, and since $t_{(n)} \le \tau_H$ a.s., we obtain from the Borel–Cantelli lemma that $t_{(n)} \to \tau_H$ almost surely. Since $t_{(n)}$ is nondecreasing in n, this proves (3.6). ■

Equation (3.2) can of course also be used to calculate the probability that $t_{(n)}$ takes on a particular value. This is only nonzero if H has a jump at that value, i.e. if $\Delta H(t) > 0$. We have by (3.2) that, in general,

$$P\{t_{(n)} = t\} = H^n(t) - H^n(t-).$$

In particular, the probability that $t_{(n)}$ takes on its maximum possible value, τ_H, is

$$P\{t_{(n)} = \tau_H\} = 1 - H^n(\tau_H -),$$

and this is 0 unless H has a jump at τ_H, so that $H(\tau_H -) < 1$. It follows that

$$\lim_{n\to\infty} P\{t_{(n)} = \tau_H\} = \begin{cases} 1 & \text{if } H(\tau_H -) < 1 \\ 0 & \text{if } H(\tau_H -) = 1. \end{cases} \tag{3.7}$$

In particular, *the limit in (3.7) always exists*, and is 0 if H is continuous at τ_H, and 1 otherwise. Moreover, if $H(\tau_H -) < 1$, then

$$\sum_{n\geq 1} P\{t_{(n)} < \tau_H\} = \sum_{n\geq 1} H^n(\tau_H -) \leq \frac{1}{(1 - H(\tau_H -))} < \infty,$$

so we obtain from the Borel–Cantelli lemma again that $t_{(n)} = \tau_H$ for sufficiently large n, almost surely.

Consistency of the KME and the limiting behaviour of its maximum value

Now we turn to the consistency of the KME for the underlying survival distribution. Since we are talking about the convergence of (random) *functions*, we must consider the values of t for which convergence takes place. Most useful are theorems which yield that $\hat{F}_n(t)$ converges to $F(t)$ *uniformly* for a range of values of t. Such theorems have been proved in great generality by Gill (1980, 1983) and Wang (1987), and we shall apply their results to obtain a very general result of the kind we need to analyse survival data which may contain immunes. Gill's and Wang's results apply to possibly improper c.d.f.'s F, and so are well adapted to our purpose. (Some related results are mentioned in Appendix Note 3.)

We begin by stating Wang's consistency result.

Theorem 3.3 (Wang 1987) *Under the i.i.d. censoring model we always have*

$$\sup_{0\leq t\leq t_{(n)}} |\hat{F}_n(t) - F(t)| \xrightarrow{P} 0, \qquad \text{as } n \to \infty. \tag{3.8}$$

Condition (3.8) establishes the consistency of the KME $\hat{F}_n(t)$ for $F(t)$ for values of t in the range $[0, t_{(n)}]$. (See Winter (1987) for some other consistency results related to the KME.) We next examine the limiting behaviour of $F(t)$ evaluated at the largest observed survival or censored time, $t_{(n)}$. This will tell us, via (3.8), about the asymptotic behaviour of the maximum value of the KME, $\hat{F}_n(t_{(n)})$, and this statistic will turn out to be a consistent estimator of the proportion of susceptibles, when follow-up is sufficient (see Section 4.1).

Theorem 3.4 *Suppose the i.i.d. censoring model holds. If*

$$H(\tau_H -) < 1 \text{ then } \hat{F}_n(t_{(n)}) \xrightarrow{P} F(\tau_H) \qquad \text{as } n \to \infty, \tag{3.9}$$

while if

$$H(\tau_H -) = 1 \text{ then } \hat{F}_n(t_{(n)}) \xrightarrow{P} F(\tau_H -) \qquad \text{as } n \to \infty. \tag{3.10}$$

Proof According to (3.8) we always have

$$\hat{F}_n(t_{(n)}) - F(t_{(n)}) \xrightarrow{P} 0. \tag{3.11}$$

When $H(\tau_H -) < 1$, $P\{t_{(n)} = \tau_H\} \to 1$ as $n \to \infty$ by (3.7). Thus

$$P\{F(t_{(n)}) = F(\tau_H)\} \to 1,$$

and the required result (3.9) follows from (3.11). Alternatively, if $H(\tau_H -) = 1$, then $t_{(n)} < \tau_H$ for all large n, a.s., while $t_{(n)} \xrightarrow{P} \tau_H$ by (3.6). Thus $F(t_{(n)}) \xrightarrow{P} F(\tau_H -)$, so (3.10) follows from (3.11) again. ∎

This theorem shows that the limit in probability of $\hat{F}_n(t_{(n)})$ *always exists* under the i.i.d. censoring model, and also that, *if* F *is continuous at* τ_H, *then we have*

$$\hat{F}_n(t_{(n)}) \xrightarrow{P} F(\tau_H), \tag{3.12}$$

since the right-hand sides of (3.9) and (3.10) are equal when F is continuous at τ_H. We allow the possibility that $\tau_H = \infty$ in all of these, of course, and in this case we automatically have F continuous at τ_H, that is, $F(\tau_H -) = F(\tau_H)$, so (3.12) holds then. For our later purposes we will restate (3.12) as Theorem 3.5.

Theorem 3.5 *Suppose the i.i.d. censoring model holds and* F *is continuous at* τ_H *in case* $\tau_H < \infty$. *Then*

$$\hat{F}_n(t_{(n)}) \xrightarrow{P} F(\tau_G) \qquad \text{as } n \to \infty. \tag{3.13}$$

Proof If $\tau_G \le \tau_F$ then $\tau_H = \tau_G$ and (3.13) follows from (3.12). If $\tau_F < \tau_G$ then (3.12) gives

$$\hat{F}_n(t_{(n)}) \xrightarrow{P} F(\tau_F) = 1 = F(\tau_G)$$

so (3.13) holds again. ∎

Our next result considers the estimation of $F(\infty)$. Recall from Section 2.2 that τ_{F_0} is the point (perhaps ∞) at which F achieves its maximum value $F(\infty)$; see (2.9).

Theorem 3.6 *Suppose the i.i.d. censoring model holds and* F *is continuous at* τ_H *in case* $\tau_H < \infty$. *Then*

$$\hat{F}_n(t_{(n)}) \xrightarrow{P} F(\infty) \qquad \text{as } n \to \infty \text{ if and only if } \tau_{F_0} \le \tau_G. \tag{3.14}$$

Proof By Theorem 3.5 we have

$$\hat{F}_n(t_{(n)}) \xrightarrow{P} F(\tau_G)$$

so (3.14) follows from the fact that

$$F(\tau_G) = F(\infty) \text{ if and only if } \tau_{F_0} \leq \tau_G. \qquad \blacksquare$$

The requirement that F be continuous at τ_H in case $\tau_H < \infty$ in Theorems 3.5 and 3.6 (and also in some later theorems) is a technicality which is a nuisance from a practical point of view since it is uncheckable in reality. Commonly, however, we do not expect, or postulate, that F is likely to be discontinuous at its right endpoint (or, even less likely, at the right endpoint of G, if $\tau_G < \tau_F$) so the continuity assumption hardly restricts us, in practice. However, continuity is a necessary condition for (3.14) to hold, as can easily be seen from (3.10).

More consistency theorems for the KME

We give two more theorems which establish consistency of the KME for the true survival distribution $F(t)$, uniform in intervals of t. Condition (3.8) established the consistency of $\hat{F}_n$ for F for the only values of t that matter in the sample: there are no survival or censored times larger than $t_{(n)}$. However, we did define $\hat{F}_n(t)$ for values of $t > t_{(n)}$ by

$$\hat{F}_n(t) = \hat{F}_n(t_{(n)}), \qquad t > t_{(n)},$$

(see (1.13) of Section 1.2) and for this extended definition it will be useful to have the following results.

Theorem 3.7 *Suppose the i.i.d. censoring model holds and F is continuous at τ_H in case $\tau_H < \infty$. Then*

$$\sup_{0 \leq t \leq \tau_H} |\hat{F}_n(t) - F(t)| \xrightarrow{P} 0 \qquad \text{as } n \to \infty. \tag{3.15}$$

Proof This follows immediately from the theorem and the first corollary of Wang (1987), which show that, with no further assumptions,

$$\sup_{0 \leq s \leq t} |\hat{F}_n(s) - F(s)| \xrightarrow{P} 0, \qquad \text{as } n \to \infty, \tag{3.16}$$

for any $t < \tau_H$, and this remains true with $t = \tau_H$ provided $G(\tau_H -) < 1$ or $F(\tau_H -) = F(\tau_H)$. (See also Gill 1994, Section 8, p. 151, for further discussion and an alternative proof of Wang's result). ■

The final result in this section relates the uniform consistency of $\hat{F}_n(t)$ on the whole line to the idea of sufficient follow-up.

Theorem 3.8 *Suppose the i.i.d. censoring model holds and F is continuous at τ_H in case $\tau_H < \infty$. Then*

$$\sup_{t \geq 0} |\hat{F}_n(t) - F(t)| \xrightarrow{P} 0, \qquad \text{as } n \to \infty, \tag{3.17}$$

if and only if $\tau_{F_0} \leq \tau_G$.

Proof Let (3.17) hold and suppose $\tau_{F_0} > \tau_G$. Take $t = \tau_{F_0}$ in (3.17) to get

$$\hat{F}_n(\tau_{F_0}) \xrightarrow{P} F(\tau_{F_0}) = F(\infty).$$

Also, since $\tau_{F_0} \geq t_{(n)}$ a.s., by the extended definition of the KME we have $\hat{F}_n(\tau_{F_0}) = \hat{F}_n(t_{(n)})$. Thus $\hat{F}_n(t_{(n)}) \xrightarrow{P} F(\infty)$. However, by (3.9) and (3.10),

$$\hat{F}_n(t_{(n)}) \xrightarrow{P} F(\tau_H) = F(\tau_G) < F(\infty)$$

or

$$\hat{F}_n(t_{(n)}) \xrightarrow{P} F(\tau_H -) = F(\tau_G -) < F(\infty).$$

So we obtain a contradiction, and it must be that $\tau_{F_0} \leq \tau_G$.

Conversely let $\tau_{F_0} \leq \tau_G$. To prove (3.17), by (3.15) we need only consider values of $t > \tau_H = \tau_{F_0}$. For these, $t > t_{(n)}$ for large n, a.s., so by (3.14) and the extended definition of the KME, we have

$$\hat{F}_n(t) = \hat{F}_n(t_{(n)}) \xrightarrow{P} F(\infty) = F(\tau_{F_0}) = F(t).$$

Thus (3.17) holds. ■

We stress that Theorems 3.7 and 3.8 hold only when the KME is extended by defining $\hat{F}_n(t) = \hat{F}_n(t_{(n)})$ for values of t exceeding $t_{(n)}$. Theorem 3.8 serves as a reminder that the value of the KME, projected beyond the largest survival or censored time observed, is consistent for the underlying survival distribution if and only if follow-up in the sample is sufficient in the sense that we defined it in Section 2.3. A similar result holds for the asymptotic bias of the KME (see Section 3.3).

3.2 WHEN DOES THE KME HAVE TOTAL MASS 1?

It is clearly highly relevant to our concerns to ask when the KME equals 1 at its right-hand endpoint, because if this occurs there are probably no immunes in the data. But conversely, if the largest value of the KME is less than 1, we cannot conclude that there are immunes in the sample, because even when all individuals fail or die eventually, the censoring mechanism may be such that the largest survival time is censored, and this will produce a KME from the sample which does not reach 1. Consequently we need to consider the possibility that this may occur even when all individuals fail eventually. We will see that this may happen with nonzero probability even with (ordinary) survival models with no immunes.

The probability that the largest observation is censored—or not

In this section it will be convenient to let

$$\overline{F}(t) = 1 - F(t), \overline{G}(t) = 1 - G(t) \text{ and } \overline{H}(t) = 1 - H(t)$$

denote the 'tails' or 'survivor functions' of the distributions F, G, and H. With the notation and conventions setup in the previous section, the largest observation $t_{(n)}$ is censored if and only if $c_{(n)} = 0$, and we wish to calculate the probability of this event in terms of F and G. This is done in the next theorem, where we also obtain some bounds on that probability.

Theorem 3.9 *Under the i.i.d. censoring model, with $p = F(\infty) \leq 1$, we have the formula*

$$P\{c_{(n)} = 1\} = p \int_{[0,\tau_{F_0} \wedge \tau_G]} f_n(t)\, \mathrm{d}F_0(t) = p \int_{[0,\tau_H]} f_n(t)\, \mathrm{d}F_0(t) \qquad (3.18)$$

where for all $t \geq 0$

$$f_n(t) = \overline{G}(t-) \sum_{j=1}^{n} H^{n-j}(t-)[1 - \overline{F}(t)\overline{G}(t-)]^{j-1}. \qquad (3.19)$$

The function $f_n(t)$ satisfies the inequalities

$$n\overline{G}(t-)[1 - \overline{F}(t-)\overline{G}(t-)]^{n-1} \leq f_n(t) \leq n\overline{G}(t-)[1 - \overline{F}(t)\overline{G}(t-)]^{n-1}, t \geq 0. \qquad (3.20)$$

When F is continuous for all $t \leq \tau_H$ we have

$$\begin{aligned} P\{c_{(n)} = 1\} &= np \int_{[0,\tau_H]} \overline{G}(t-)H^{n-1}(t-)\, \mathrm{d}F_0(t) \\ &= np \int_{[0,\tau_H]} \overline{G}(t-)H^{n-1}(t)\, \mathrm{d}F_0(t). \end{aligned} \qquad (3.21)$$

Proof The largest observation is uncensored, or equivalently, $c_{(n)} = 1$, if and only if all observations tied at the largest time are uncensored. Recall that, under the i.i.d. censoring model, the random variables $t_1^*, \ldots, t_n^*$ are i.i.d. with possibly improper c.d.f. $F(t) = pF_0(t)$. Now there may be $m = 1, 2, \ldots, n$ observations tied at $t_{(n)}$, but we can allow for them by an extension of the method we used in Theorem 3.1. Thus we calculate

$$P\{c_{(n)} = 1\} = \sum_{m=1}^{n} P\left\{t_{i_1} = \cdots = t_{i_m} > \max_{i \neq i_1,\ldots,i_m} t_i,\, t^*_{i_j} \le u_{i_j},\, 1 \le j \le m, \right.$$

$$\left. \text{for some } i_1, \ldots, i_m \in \{1, 2, \ldots, n\}\right\}$$

$$= \sum_{m=1}^{n} \binom{n}{m} P\{t^*_1 = \cdots = t^*_m > \max_{m+1 \le i \le n} t_i,\, \min_{1 \le i \le m} u_i \ge t^*_1\}. \quad (3.22)$$

Since the u_i are independent of the t^*_i, conditioning on t^*_1 gives

$$P\{c_{(n)} = 1\}$$

$$= \sum_{m=1}^{n} \binom{n}{m} \int_{[0,\tau_{F_0}]} P\{t = t^*_2 = \cdots = t^*_m > \max_{m+1 \le i \le n} t_i\} P\{\min_{1 \le i \le m} u_i \ge t\}\,\mathrm{d}F(t)$$

$$= \sum_{m=1}^{n} \binom{n}{m} \int_{[0,\tau_{F_0}]} (\Delta F(t))^{m-1} H^{n-m}(t-) \overline{G}^m(t-)\,\mathrm{d}F(t). \quad (3.23)$$

Here $\Delta F(t) = F(t) - F(t-) = P\{t^*_1 = t\}$, and we need only extend the range of integration over the interval $[0, \tau_{F_0}]$, since F attributes no mass to values of $t > \tau_{F_0}$. Consider the function

$$\sum_{m=1}^{n} \binom{n}{m} (\Delta F(t))^{m-1} H^{n-m}(t-) \overline{G}^m(t-). \quad (3.24)$$

When $\Delta F(t) = 0$, this equals

$$n\overline{G}(t-)H^{n-1}(t-). \quad (3.25)$$

This is the same as (3.19) in this case because $\Delta F(t) = 0$ implies

$$1 - \overline{F}(t)\overline{G}(t-) = 1 - \overline{F}(t-)\overline{G}(t-) = 1 - \overline{H}(t-) = H(t-)$$

so that (3.19) becomes

$$f_n(t) = \overline{G}(t-) \sum_{j=1}^{n} H^{n-j}(t-) H^{j-1}(t-) = n\overline{G}(t-)H^{n-1}(t-).$$

When $\Delta F(t) > 0$ the function in (3.24) equals

$$\sum_{m=1}^{n} \binom{n}{m} (\overline{G}(t-)\Delta F(t))^m H^{n-m}(t-)/\Delta F(t)$$

$$= \left((H(t-) + \overline{G}(t-)\Delta F(t))^n - H^n(t-)\right)/\Delta F(t) \quad (3.26)$$

$$= \overline{G}(t-) \sum_{j=1}^{n} H^{n-j}(t-)[H(t-) + \overline{G}(t-)\Delta F(t)]^{j-1}. \quad (3.27)$$

Since

$$H(t-) + \overline{G}(t-)\Delta F(t) = 1 - \overline{F}(t-)\overline{G}(t-) + \overline{G}(t-)[\overline{F}(t-) - \overline{F}(t)]$$
$$= 1 - \overline{F}(t)\overline{G}(t-) \tag{3.28}$$

the function in (3.27) equals

$$\overline{G}(t-)\sum_{j=1}^{n} H^{n-j}(t-)[1 - \overline{F}(t)\overline{G}(t-)]^{j-1}.$$

This again is the same as (3.19).

Note that $f_n(t) = 0$ for $t > \tau_G$, so we can write the upper limit to the integral in (3.23) as $\tau_{F_0} \wedge \tau_G$. Thus the first equality in (3.18) follows from (3.23). For the second equality in (3.18), note that if $p = 1$, $\tau_{F_0} \wedge \tau_G$ equals τ_H, while if $p < 1$, $\tau_H = \tau_G$, and the integral over $[0, \tau_H] = [0, \tau_G]$ is the same as that over $[0, \tau_{F_0} \wedge \tau_G]$ because $F_0(t) = 1$ on the interval $(\tau_{F_0} \wedge \tau_G, \tau_G]$.

The inequalities in (3.20) follow from (3.19) and the inequalities

$$[1 - \overline{F}(t-)\overline{G}(t-)]^{n-1} \le H^{n-j}(t-)[1 - \overline{F}(t)\overline{G}(t-)]^{j-1} \le [1 - \overline{F}(t)\overline{G}(t-)]^{n-1}$$

which are a result of

$$H(t-) = 1 - \overline{F}(t-)\overline{G}(t-) \le 1 - \overline{F}(t)\overline{G}(t-). \tag{3.29}$$

When F is continuous for all $t \in [0, \tau_H]$, we have $\Delta F(t) \equiv 0$, so by (3.25) and (3.18),

$$P\{c_{(n)} = 1\} = n \int_{[0,\tau_{F_0} \wedge \tau_G]} \overline{G}(t-)H^{n-1}(t-)\,\mathrm{d}F(t).$$

This is the same as

$$n \int_{[0,\tau_{F_0} \wedge \tau_G]} \overline{G}(t-)H^{n-1}(t)\,\mathrm{d}F(t),$$

since when F is continuous it assigns no mass to a single point, so that

$$\int_{[0,\tau_{F_0} \wedge \tau_G]} \overline{G}(t-)[H^{n-1}(t-) - H^{n-1}(t)]\,\mathrm{d}F(t) = 0.$$

Thus (3.21) holds. ■

We next obtain an analogous formula for the complementary probability to that in (3.18).

Theorem 3.10 *Under the i.i.d. censoring model, with $p = F(\infty) \le 1$, we have the formula*

$$P\{c_{(n)} = 0\} = \int_{[0,\tau_H]} g_n(t)\,\mathrm{d}G(t) \tag{3.30}$$

where

$$g_n(t) = \overline{F}(t) \sum_{j=1}^{n} H^{n-j}(t)[1 - \overline{F}(t)\overline{G}(t-)]^{j-1}. \tag{3.31}$$

The function $g_n(t)$ satisfies the inequalities

$$n\overline{F}(t)[1 - \overline{F}(t)\overline{G}(t-)]^{n-1} \leq g_n(t) \leq n\overline{F}(t)[1 - \overline{F}(t)\overline{G}(t))]^{n-1}, \qquad t \geq 0. \tag{3.32}$$

When G is continuous for all $t \leq \tau_H$ we have

$$P\{c_{(n)} = 0\} = n \int_{[0,\tau_H]} \overline{F}(t)H^{n-1}(t)\,\mathrm{d}G(t) = n \int_{[0,\tau_H]} \overline{F}(t-)H^{n-1}(t)\,\mathrm{d}G(t). \tag{3.33}$$

Proof We could derive (3.30) and (3.31) directly by a probabilistic argument as for Theorem 3.9, but the following analytic method using the representation for the distribution of $t_{(n)}$ from Theorem 3.1 is slightly shorter.

We proceed by showing that the right-hand sides of (3.18) and (3.30) add to 1, and this of course will prove (3.30) since the event $\{c_{(n)} = 0\}$ is the complementary event to $\{c_{(n)} = 1\}$. In other words, we will show that, when $f_n(t)$ and $g_n(t)$ are defined by (3.19) and (3.31),

$$\int_{[0,\tau_H]} g_n(t)\,\mathrm{d}G(t) + \int_{[0,\tau_H]} f_n(t)\,\mathrm{d}F(t) = 1. \tag{3.34}$$

This is really an integration by parts formula, but we have to take into account that F and G are not necessarily continuous. To prove (3.34), note that when $\Delta G(t) > 0$, (3.31) gives

$$\begin{aligned} H^n(t) &- [1 - \overline{F}(t)\overline{G}(t-)]^n \\ &= [H(t) - 1 + \overline{F}(t)\overline{G}(t-)] \sum_{j=1}^{n} H^{n-j}(t)[1 - \overline{F}(t)\overline{G}(t-)]^{j-1} \\ &= \Delta G(t) g_n(t) \end{aligned} \tag{3.35}$$

because

$$H(t) - 1 + \overline{F}(t)\overline{G}(t-) = -\overline{F}(t)\overline{G}(t) + \overline{F}(t)\overline{G}(t-) = \Delta G(t)\overline{F}(t). \tag{3.36}$$

This together with

$$1 - \overline{F}(t)\overline{G}(t-) = H(t-) + (\overline{F}(t-) - \overline{F}(t))\overline{G}(t-) = H(t-) + \Delta F(t)\overline{G}(t-) \tag{3.37}$$

shows that

$$\Delta G(t) g_n(t) = H^n(t) - [H(t-) + \Delta F(t)\overline{G}(t-)]^n. \tag{3.38}$$

Also when $\Delta F(t) > 0$, (3.19) gives by a similar calculation

$$\Delta F(t) f_n(t) = [H(t-) + \Delta F(t)\overline{G}(t-)]^n - H^n(t-). \tag{3.39}$$

Adding (3.38) and (3.39) gives

$$\Delta G(t) g_n(t) + \Delta F(t) f_n(t) = H^n(t) - H^n(t-). \tag{3.40}$$

When $\Delta F(t) = 0$ but $\Delta G(t) > 0$, (3.38) also gives

$$\Delta G(t) g_n(t) = H^n(t) - H^n(t-)$$

so (3.40) remains true in this case. When $\Delta G(t) = 0$ but $\Delta F(t) > 0$, (3.37) and (3.39) give

$$\Delta F(t) f_n(t) = [H(t-) + \Delta H(t)]^n - H^n(t-) = H^n(t) - H^n(t-)$$

so again (3.40) remains true.

Now suppose $\Delta H(t) > 0$. Then either $\Delta F(t) > 0$ or $\Delta G(t) > 0$ so by (3.40)

$$\begin{aligned} \int_{\{t:\Delta H(t)>0\}} g_n(t)\,\mathrm{d}G(t) &+ \int_{\{t:\Delta H(t)>0\}} f_n(t)\,\mathrm{d}F(t) \\ &= \int_{\{t:\Delta H(t)>0\}} \left\{ \frac{\Delta G(t)}{\Delta H(t)} g_n(t) + \frac{\Delta F(t)}{\Delta H(t)} f_n(t) \right\} \mathrm{d}H(t) \\ &= \int_{\{t:\Delta H(t)>0\}} \left\{ \frac{H^n(t) - H^n(t-)}{\Delta H(t)} \right\} \mathrm{d}H(t). \end{aligned} \tag{3.41}$$

Alternatively, when $\Delta H(t) = 0$, then $\Delta F(t) = \Delta G(t) = 0$, so by (3.31) and (3.19),

$$g_n(t) = n\overline{F}(t)H^{n-1}(t) \text{ and } f_n(t) = n\overline{G}(t)H^{n-1}(t).$$

It follows that, at points of continuity of $H(t)$,

$$\begin{aligned} g_n(t)\,\mathrm{d}G(t) + f_n(t)\,\mathrm{d}F(t) &= n\{\overline{F}(t)\,\mathrm{d}G(t) + \overline{G}(t)\,\mathrm{d}F(t)\}H^{n-1}(t) \\ &= nH^{n-1}(t)\,\mathrm{d}H(t). \end{aligned} \tag{3.42}$$

Integrating in (3.42) then adding the expression in (3.41) gives

$$\int_{[0,\tau_H]} g_n(t)\,\mathrm{d}G(t) + \int_{[0,\tau_H]} f_n(t)\,\mathrm{d}F(t) = \int_{[0,\tau_H]} h_n(t)\,\mathrm{d}H(t) \tag{3.43}$$

where

$$h_n(t) = \sum_{m=1}^{n} H^{n-m}(t)H^{m-1}(t-) = \begin{cases} \dfrac{H^n(t) - H^n(t-)}{\Delta H(t)} & \text{if } \Delta H(t) > 0 \\ nH^{n-1}(t) & \text{if } \Delta H(t) = 0. \end{cases}$$

Consequently we see from (3.2) or (3.5), and the fact that $P(t_{(n)} \le \tau_H) = 1$, that (3.34) does indeed hold.

This proves that (3.30) holds with $g_n(t)$ defined by (3.31). We obtain bounds for $g_n(t)$ from (3.31) and $1 - \overline{F}(t)\overline{G}(t-) \leq H(t)$. This proves (3.32). When G is continuous, (3.31) reduces to

$$g_n(t) = n\overline{F}(t)H^{n-1}(t),$$

and so (3.33) follows from (3.30) in this case. ∎

Limiting values of the probabilities

There is a special case in which we can evaluate $P\{c_{(n)} = 1\}$ and $P\{c_{(n)} = 0\}$ given by Theorems 3.9 and 3.10 explicitly—when F and G are exponential distributions. Suppose in fact that

$$F(t) = 1 - e^{-t/\phi} \text{ and } G(t) = 1 - e^{-t/\mu}$$

for some $\phi > 0$, $\mu > 0$. In other words, F and G are exponential with means ϕ and μ, respectively. These are continuous distributions so we can apply (3.33) to get

$$P\{c_{(n)} = 0\} = n\int_0^\infty e^{-t/\phi}[1 - e^{-(1/\phi+1/\mu)t}]^{n-1}\frac{1}{\mu}e^{-t/\mu}\,dt = \frac{\phi}{\phi+\mu}. \tag{3.44a}$$

Hence we also have

$$P\{c_{(n)} = 1\} = \frac{\mu}{\phi+\mu}. \tag{3.44b}$$

Equation (3.44b) displays clearly how the probability that the largest observation is censored depends on the mean survival times ϕ and μ of F and G. When the mean censoring time μ is very small compared with the mean survival time ϕ, there is a tendency for more survival times to be censored, and the censoring is 'heavy'. Correspondingly, (3.44b) shows that the probability that the largest observation is uncensored is small. The opposite is true when μ is large compared with ϕ; fewer observations will be censored, on average, and this corresponds to 'light' censoring, and a higher probability that the largest observation is uncensored. But note that $P\{c_{(n)} = 0\}$ is never 0; even for this quite 'ordinary' survival model, there is a nonzero chance of finding that the largest observation is censored—or equivalently, that the Kaplan–Meier estimator calculated from the sample is improper.

The exponential distribution is one of the few for which we can explicitly calculate $P\{c_{(n)} = 0\}$. For other distributions we must calculate the integral numerically. Alternatively, we can use a large-sample analysis to gain some intuition about the behaviour of $P\{c_{(n)} = 0\}$ as the sample size becomes large. In other words, we let $n \to \infty$ and use the expressions in Theorems 3.9 and 3.10 to find the limiting values of $P\{c_{(n)} = 1\}$ and $P\{c_{(n)} = 0\}$, when they exist. These will be of interest to us in deciding when we expect the KME to be proper.

First consider what we expect to happen. From the discussion above for the exponential distributions, when censoring is light we expect $P\{c_{(n)} = 0\}$ to be close to 0, and we expect the opposite when censoring is heavy. Now the condition $\tau_G < \tau_F$ certainly is 'heavy' censoring, since it means that some survival times will always be censored in large samples — no observed time can exceed the censoring limit τ_G. Likewise when there are immunes present, we expect that the largest observation will be censored with probability approaching 1 in large samples. Theorem 3.11 below shows that these expected results are true, and likewise, Theorem 3.12 shows that when $\tau_F < \tau_G$, i.e. censoring is light, the largest observation will be uncensored, with probability approaching 1 in large samples.

Theorem 3.11 *Assume the i.i.d. censoring model. If* $F(\infty) = 1$ *and* $\tau_G < \tau_F$, *or if* $F(\infty) < 1$, *then*

$$\lim_{n\to\infty} P\{c_{(n)} = 0\} = 1. \tag{3.45}$$

Proof Suppose $\tau_G < \tau_F$ in case $F(\infty) = 1$. Then $F(\tau_G) < 1$ in either case. Since $\tau_F = \infty$ when $F(\infty) < 1$, we also have $\tau_H = \tau_G$ in either case. Now assume that $G(\tau_G-) < 1$, so G is discontinuous at τ_G. When $t \le \tau_G$ use (3.20) together with $\overline{F}(t)\overline{G}(t-) \ge \overline{F}(\tau_G)\overline{G}(\tau_G-)$ to obtain

$$f_n(t) \le n(1 - \overline{F}(t)\overline{G}(t-))^{n-1} \le n(1 - \overline{F}(\tau_G)\overline{G}(\tau_G-))^{n-1} = n(1-a)^{n-1}, \text{ say.}$$

Here $a = \overline{F}(\tau_G)\overline{G}(\tau_G-) > 0$ since $F(\tau_G) < 1$ and $G(\tau_G-) < 1$. Thus from (3.18) we have

$$P\{c_{(n)} = 1\} = p\int_{[0,\tau_G]} f_n(t)\,\mathrm{d}F_0(t) \le n(1-a)^{n-1}$$

and this converges to 0 as $n \to \infty$, since $a > 0$. This implies (3.45).

Alternatively, assume $G(\tau_G-) = 1$, so G is continuous at τ_G. Then we can find a $\delta > 0$ so that G is continuous on $(\tau_G - \delta, \tau_G]$. When $\tau_G - \delta < t < \tau_G$ we have

$$\overline{F}(\tau_G-) \le \overline{F}(t) \le \overline{F}(\tau_G - \delta),$$

and also $\overline{G}(t-) = \overline{G}(t)$. Thus from (3.30) and (3.32) we have

$$\begin{aligned} P\{c_{(n)} = 0\} &\ge n\int_{(\tau_G-\delta,\tau_G)} \overline{F}(\tau_G-)[1 - \overline{F}(\tau_G - \delta)\overline{G}(t)]^{n-1}\,\mathrm{d}G(t) \\ &= \frac{\overline{F}(\tau_G-)}{\overline{F}(\tau_G-\delta)}\{1 - [1 - \overline{F}(\tau_G - \delta)\overline{G}(\tau_G - \delta)]^n\} \\ &\to \frac{\overline{F}(\tau_G-)}{\overline{F}(\tau_G-\delta)} \qquad (\text{as } n \to \infty). \end{aligned} \tag{3.46}$$

Here we were able to evaluate the integral since G is continuous on $(\tau_G - \delta, \tau_G]$. Now let $\delta \downarrow 0$ in (3.46) to obtain (3.45) again. ■

Theorem 3.12 *Assume the i.i.d. censoring model and suppose $\tau_F < \tau_G$ (so that $F(\infty) = 1$). Then*

$$\lim_{n\to\infty} P\{c_{(n)} = 1\} = 1. \tag{3.47}$$

Proof This is similar to the proof of Theorem 3.11. Here we have $\tau_H = \tau_F$. Suppose $F(\tau_F-) < 1$. We have from (3.30) that

$$P\{c_{(n)} = 0\} = \int_{[0,\tau_F]} g_n(t)\,\mathrm{d}G(t) = \int_{[0,\tau_F)} g_n(t)\,\mathrm{d}G(t) \tag{3.48}$$

since $g_n(\tau_F) = 0$ (because $F(\tau_F) = 1$). Now from (3.32) we get

$$P\{c_{(n)} = 0\} \le n[1 - \overline{F}(\tau_F-)\overline{G}(\tau_F-)]^{n-1} \to 0 \tag{3.49}$$

since $F(\tau_F-) < 1$ and $G(\tau_F-) \le G(\tau_G-) < 1$. Thus (3.47) holds in this case.

If on the other hand $F(\tau_F-) = 1$ choose $\delta > 0$ so that F is continuous on $(\tau_F - \delta, \tau_F]$, and use (3.18) and (3.20) to write

$$\begin{aligned}
P\{c_{(n)} = 1\} &\ge n\int_{(\tau_F-\delta,\tau_F)} \overline{G}(\tau_F-)[1-\overline{F}(t)\overline{G}(\tau_F-\delta)]^{n-1}\,\mathrm{d}F(t)\\
&= \frac{\overline{G}(\tau_F-)}{\overline{G}(\tau_F-\delta)}\{1-[1-\overline{F}(\tau_F-\delta)\overline{G}(\tau_F-\delta)]^n\}\\
&\to \frac{\overline{G}(\tau_F-)}{\overline{G}(\tau_F-\delta)} \qquad (n\to\infty)\\
&\to 1 \qquad (\delta \downarrow 0).
\end{aligned}$$

Thus again (3.47) holds. ■

Together Theorems 3.11 and 3.12 show that

(i) $F(\infty) = 1$ and $\tau_G < \tau_F$, or $F(\infty) < 1$, imply $P\{c_{(n)} = 0\} \to 1$.
(ii) $\tau_F < \tau_G$ (so that $F(\infty) = 1$) implies $P\{c_{(n)} = 0\} \to 0$.

For information on what happens when $\tau_G = \tau_F$, in which case the limit of $P\{c_{(n)} = 0\}$ may or may not exist, depending on a certain balance condition of the tails of F and G, see Appendix Note 4.

3.3 THE BIAS OF THE KAPLAN-MEIER ESTIMATOR

An estimator $\hat{\theta}_n$ based on a sample of size n is said to be *unbiased* for a parameter θ if

$$E(\hat{\theta}_n) = \theta. \tag{3.50}$$

Operationally, this means that if we were to repeat the same experiment under the same conditions many times (or drew many times a sample of the same size from the population), and calculated a version of $\hat{\theta}_n$ for each 'realisation' of the experiment, then these various values of $\hat{\theta}_n$ would *average* close to θ, with the closeness improving as the experiment was repeated more and more often. This is clearly a desirable property, capturing as it does the 'repeatability' of an experimental situation and of the corresponding estimate of the parameter, although it must be augmented by some estimate of the precision of the estimator to be useful in practice.

In general, therefore, biasedness in an estimator is undesirable, especially if the amount of bias is unknown, since it means that repeated experimentation cannot produce accurate estimation of the parameter of interest. It may, and often does, occur, however, that the bias of a biased estimator can be proved to converge to 0 for large sample sizes n. Then the estimator is said to be *asymptotically unbiased*, and this property might be taken as some consolation for the fact that an estimator is biased, since it means that in 'large' samples we can be confident of accurate estimation.

We need to distinguish this situation from the consistency of an estimator $\hat{\theta}_n$, as discussed in Section 3.1, where the concept is that, as more and more information in the form of experimental units or individuals is added, i.e. as sample size n becomes large, $\hat{\theta}_n$ converges (in some mode of stochastic convergence) to the parameter θ. This is also a desirable property, but by itself it says little about the behaviour of $\hat{\theta}_n$ in a sample of given (finite) size.

The concepts of consistency and asymptotic unbiasedness are closely connected, however. If we strengthen asymptotic unbiasedness a little to the requirement that

$$E|\hat{\theta}_n - \theta| \to 0 \qquad (\text{as } n \to \infty) \tag{3.51}$$

(called 'convergence in mean of $\hat{\theta}_n$ to θ') then a simple application of Markov's inequality:

$$P\{|\hat{\theta}_n - \theta| > \varepsilon\} \leq \frac{E|\hat{\theta}_n - \theta|}{\varepsilon} \qquad (\text{for all } \varepsilon > 0)$$

shows that $\hat{\theta}_n \xrightarrow{P} \theta$, so $\hat{\theta}_n$ is consistent for θ. Thus (3.51) implies consistency in probability of $\hat{\theta}_n$. On the other hand, one can deduce from $\hat{\theta}_n \xrightarrow{P} \theta$ that $E(\hat{\theta}_n) \to \theta$, i.e. asymptotic unbiasedness of $\hat{\theta}_n$ holds, under some other conditions. (Some form of uniform integrability is needed; see for example Chow and Teicher 1988, Theorem 3, p. 100.) Note that we have implicitly assumed that $E|\hat{\theta}_n|$ exists in formulating the idea of unbiasedness, and also when formulating (3.51). Under some conditions, then, we have the hierarchies:

$$\textit{convergence in mean} \Rightarrow \textit{consistency} \Rightarrow \textit{asymptotic unbiasedness}$$

while also of course

unbiasedness $\Rightarrow$ *asymptotic unbiasedness.*

With this discussion in mind, let us turn to the question of the bias of the KME. In general, the KME $\hat{F}_n(t)$ is a biased estimator of $F(t)$, the distribution of survival times, and the bias is negative, so $\hat{F}_n(t)$ underestimates $F(t)$. The bias is computable and converges to 0 for large sample size n, so the KME is asymptotically unbiased for $F(t)$, provided follow-up is sufficient.

These follow from Theorem 3.13, which we will prove below. Theorem 3.13 in turn relies on a result of Gill (1980, pp. 34–38). Gill extended some results of Meier (1975) to derive the following formula. Assume the i.i.d. censoring model, let τ_F be the right extreme of F and let $t_{(n)}$ be the largest observed survival time. Then if $0 \le t < \tau_F$

$$E(\hat{F}_n(t)) = F(t) - E\left\{\frac{I(t_{(n)} \le t)(1 - \hat{F}_n(t_{(n)}))(F(t) - F(t_{(n)}))}{1 - F(t_{(n)})}\right\}. \tag{3.52}$$

(See Gill 1980, 3.2.16, p. 38.) Equation (3.52) clearly demonstrates the negative bias of $\hat{F}_n(t)$, and now we investigate when this bias converges to 0 for large sample sizes.

It will be useful to define the random function

$$B_n(t) = \begin{cases} \dfrac{I(t_{(n)} \le t)(1 - \hat{F}_n(t_{(n)}))(F(t) - F(t_{(n)}))}{1 - F(t_{(n)})} & \text{if } t < \tau_F \quad (3.53\text{a}) \\ F(t) - \hat{F}_n(t) = 1 - \hat{F}_n(t) = 1 - \hat{F}_n(t_{(n)}) & \text{if } t \ge \tau_F. \quad (3.53\text{b}) \end{cases}$$

Then the bias in $\hat{F}_n(t)$ is $-E[B_n(t)]$. Notice that if $t < \tau_H$ then

$$E[B_n(t)] \le E[I(t_{(n)} \le t)] = P(t_{(n)} \le t) \to 0 \tag{3.54}$$

as $n \to \infty$, since $t_{(n)} \xrightarrow{P} \tau_H$ by Theorem 3.2. Consequently $\hat{F}_n(t)$ is asymptotically unbiased for $F(t)$ for any $t < \tau_H$ without any further assumptions. We can extend this to general t as follows.

Theorem 3.13 *Assume the i.i.d. censoring model, suppose $\tau_{F_0} \le \tau_G$, and suppose F is continuous at τ_H in case $\tau_H < \infty$. Then for each $t \ge 0$*

$$\lim_{n\to\infty} E[B_n(t)] = 0. \tag{3.55}$$

Conversely if (3.55) holds for each $t \ge 0$ and F is continuous at τ_H in case $\tau_H < \infty$, then $\tau_{F_0} \le \tau_G$.

Proof Suppose first that $\tau_{F_0} \le \tau_G$, and consider two cases. We may have $\tau_F \le \tau_G$. If so, take $t < \tau_F$. Then since $\tau_H = \tau_F$, (3.55) follows from (3.54).

Next take $t \geq \tau_F$. Then $t_{(n)} \leq \tau_F \leq t$ a.s., so by Theorem 3.5

$$\hat{F}_n(t) \geq \hat{F}_n(t_{(n)}) \xrightarrow{P} F(\tau_G) = 1 \qquad (\text{since } \tau_F \leq \tau_G).$$

Thus $\hat{F}_n(t) \xrightarrow{P} 1$ and so by (3.53b)

$$B_n(t) = 1 - \hat{F}_n(t) \xrightarrow{P} 0.$$

Now (3.53a) and (3.53b) show that $B_n(t) \leq 1$ for all t. Thus if $\varepsilon > 0$

$$\begin{aligned} E[B_n(t)] &= \int_0^1 P(B_n(t) > x)\,\mathrm{d}x \leq \varepsilon + P(B_n(t) > \varepsilon) \\ &\to \varepsilon \quad (\text{as } n \to \infty) \\ &\to 0 \quad (\text{as } \varepsilon \to 0) \end{aligned}$$

so we indeed have $E[B_n(t)] \to 0$ as $n \to \infty$.

The alternative case is that $\tau_G < \tau_F$. Since $\tau_{F_0} \leq \tau_G$, this forces $\tau_F = \infty$ and $p = F(\tau_{F_0}) < 1$. Also $\tau_H = \tau_G$, and if $t < \tau_H$, (3.55) follows from (3.54). So take $t \geq \tau_H = \tau_G$. Then $t_{(n)} \leq \tau_H = \tau_G \leq t$ a.s., and $F(t) \geq F(\tau_G) \geq F(\tau_{F_0}) = p$, while $F(t_{(n)}) \xrightarrow{P} F(\tau_G) = p$ by Theorem 3.5 since F is assumed continuous at τ_G. Thus from (3.53a) (note that $t < \tau_F = \infty$)

$$B_n(t) = \frac{(1 - \hat{F}_n(t_{(n)}))(p - F(t_{(n)}))}{1 - F(t_{(n)})} \leq \frac{p - F(t_{(n)})}{1 - F(t_{(n)})} \xrightarrow{P} 0$$

so (3.55) holds just as before.

Conversely suppose $E[B_n(t)] \to 0$ for all $t \geq \tau_H$. Suppose by way of contradiction that $\tau_G < \tau_{F_0}$, so $\tau_G < \tau_F$ and $\tau_H = \tau_G$. Suppose $\tau_F < \infty$ and take $t \geq \tau_F$. Then $t_{(n)} \leq t$ a.s., $F(t_{(n)}) \to F(\tau_G)$ a.s. by Theorem 3.5, and $F(\tau_G) < 1 = F(\tau_F) = F(t)$. So by (3.53a), if $t < \tau_F$,

$$B_n(t) = \frac{(1 - \hat{F}_n(t_{(n)}))(F(t) - F(t_{(n)}))}{1 - F(t_{(n)})} = 1 - \hat{F}_n(t_{(n)}), \tag{3.56}$$

and this is also true by (3.53b) if $t \geq \tau_F$. Since $E[B_n(t)] \to 0$ we have by Markov's inequality that $B_n(t) \xrightarrow{P} 0$, so (3.56) shows that $1 - \hat{F}_n(t_{(n)}) \xrightarrow{P} 0$. In other words, in the terminology of the previous section, $P(c_{(n)} = 1) \to 1$, or equivalently, $P(c_{(n)} = 0) \to 0$. From Theorem 3.11 this implies $F(\infty) = 1$ and $\tau_F \leq \tau_G$, a contradiction. The other case is $\tau_F = \infty$. Since $\tau_G < \infty$, then, F must have points of increase at times greater than τ_G, so choose t larger than one of these. Then again $t_{(n)} \leq t$ a.s., $F(t_{(n)}) \xrightarrow{P} F(\tau_G)$, and $F(\tau_G) < F(t) \leq 1$. So by (3.53a),

$$B_n(t) = \left(\frac{(F(t) - F(\tau_G))}{1 - F(\tau_G)} + o_P(1) \right) (1 - \hat{F}_n(t_{(n)}))$$

which again forces $1 - \hat{F}_n(t_{(n)}) \xrightarrow{P} 0$, and this gives a contradiction just as before. ■

Notice that when $\tau_{F_0} > \tau_G$ we have $\tau_H = \tau_G$, so $t_{(n)} \xrightarrow{P} \tau_G$ and $F(\tau_G) < 1$. Then when $t \geq \tau_H = \tau_G$, (3.53a) and Theorem 3.4 tell us that

$$B_n(t) \xrightarrow{P} \frac{(1 - F(\tau_G))(F(t) - F(\tau_G))}{1 - F(\tau_G)} = F(t) - F(\tau_G) > 0 \qquad (3.57)$$

(if F is continuous at τ_G), so in this case the bias is not asymptotically negligible. The case $\tau_{F_0} > \tau_G$ is of course the case when follow-up in the sample is not sufficient. Similar comments apply here as we made at the end of Section 3.1.

3.4 THE LARGE-SAMPLE DISTRIBUTION OF THE KAPLAN-MEIER ESTIMATOR

To supplement the consistency results of Section 3.1, we also need some results on the large-sample distribution of the KME. The theorem we give here is a special case of some theorems of Gill (1980), who developed his large-sample theory of the KME allowing for a possibly improper survival c.d.f. F, thus making possible our applications. Earlier versions, such as that of Breslow and Crowley (1974), are not general enough for our use.

The asymptotic distribution of the KME, is, under the independent censoring model and some other reasonable restrictions, a normal distribution, and furthermore one has 'weak convergence', in subsets of the real line, of the Kaplan-Meier empirical process to a Gaussian process. We will not need in this book any of the technical apparatus related to results like these, and our applications in the end will come down to inferring only results such as the asymptotic normality of $\hat{F}_n(t_{(n)})$ from them (just as we inferred consistency properties of $\hat{F}_n(t_{(n)})$ from the more general stochastic process results in Section 3.1). But we do need to state these advanced results here since they will be needed for some of the proofs in the next chapter. Readers with no interest in such technicalities may with little loss skip the remainder of this section and jump to Chapter 4.

The first result we need is a special case of Theorem 4.2.3 of Gill (1980), simply translated into our notation. Define a real-valued increasing function $v(t)$ on $[0, \tau_H)$ by

$$v(t) = \int_{[0,t]} \frac{\mathrm{d}F(s)}{(1 - F(s))(1 - F(s-))(1 - G(s-))}. \qquad (3.58)$$

Let $Z(t)$ be a stochastic process on $[0, \tau_H]$ such that for each $t \leq \tau_H$, $Z(t)$ is normally distributed with mean 0 and variance $v(t)$; also $Z(0) = 0$, and the increment $Z(t_2) - Z(t_1)$ is independent of $Z(t_1)$ for each $t_2 > t_1 > 0$. Also let

$$N_n(t) = \sum_{i=1}^{n} I\{t_i^* = t_i \le t\} \quad (t \ge 0)$$

be the number of uncensored survival times observed up till time t, and let $\Delta N_n(t) = N_n(t) - N_n(t-)$ be the jump in $N_n(t)$ at time t. Suppose

$$Y_n(t) = \sum_{i=1}^{n} I(t_i \ge t)$$

denotes the number of individuals still at risk at time t.

Define a stochastic process $X_n(t)$ on $[0, \infty)$ by

$$X_n(t) = \sqrt{n}\frac{(1 - F(t))}{(1 - F(t \wedge t_{(n)}))}\{\hat{F}_n(t \wedge t_{(n)}) - F(t \wedge t_{(n)})\} \tag{3.59}$$

(where, as usual, $t_{(n)} = \max_{1 \le i \le n} t_i$). Notice that $X_n(t) = \sqrt{n}(\hat{F}_n(t) - F(t))$ when $t \le t_{(n)}$. Knowledge of the limiting behaviour of $X_n(t)$ will tell us about the limiting behaviour of the KME $\hat{F}_n$. The main result we need is Theorem 3.14.

Theorem 3.14 *Assume the i.i.d. censoring model and suppose F is continuous at τ_H in case $\tau_H < \infty$. Assume also that*

$$\lim_{t \uparrow \tau_H} (F(\tau_H) - F(t))^2 v(t) = 0 \tag{3.60}$$

and

$$\lim_{t \uparrow \tau_H} \int_{(t, \tau_H)} \frac{I\{0 \le G(s-) < 1\}(1 - F(s))}{(1 - G(s-))(1 - F(s-))} \mathrm{d}F(s) = 0. \tag{3.61}$$

Then the random variable $\lim_{t \uparrow \tau_H}(1 - F(t))Z(t)$ exists and is finite a.s., and, as $n \to \infty$, $X_n(t)$ converges weakly in $D[0, \tau_H]$ to the process whose value at $t \ge 0$ is

$$I\{[0, \infty)\}(1 - F(t))Z(t) + I\{(\infty)\}R \tag{3.62}$$

where

$$R = \begin{cases} (1 - F(\tau_H))Z(\tau_H) & \text{if } H(\tau_H -) < 1 \\ \lim_{t \uparrow \tau_H}(1 - F(t))Z(t) & \text{if } H(\tau_H -) = 1. \end{cases} \tag{3.63}$$

Proof We need only note the correspondence between Gill's notation (Theorem 4.2.3) and ours. In our situation, his $L_j^n(t)$ are all equal to $G(t)$, our censoring distribution, so we take $L(t) = G(t)$ in his theorem, and $L_-(t) = L(t-) = G(t-)$. Likewise his $F_-(t) = F(t-)$ in our case. His $y(t)$ is $1 - H(t-)$ in our notation, so his u is our τ_H, and he supposes $y(u) > 0$, equivalently

$H(\tau_H-) < 1$, or $\Delta F(u) = 0$, equivalently $F(\tau_H) = F(\tau_H-)$, as we do. Our condition (3.60) is his (4.2.2), and our (3.61) implies his (4.2.3) (we have neglected the hazard jump term, ΔG, in the notation of Gill's (4.2.3) which of course is not negative). Gill's conclusions are that $\lim_{t\uparrow\tau_H}(1-F(t))Z(t)$ exists and is finite a.s. (note that we did not define $Z(t)$ at τ_H, nor did we assume $v(\tau_H)$ finite), and that (3.62) holds, with R defined by (3.63). ■

APPENDIX NOTES TO CHAPTER 3

1. Notes on the KME

The KME must be one of the most interesting of all statistical estimators, and few such simply defined functions can have exerted so great a fascination on statisticians and probabilists alike—or proved to have been so interesting to analyse. For an account of its conception by Kaplan and Meier (1958) as a generalisation of earlier actuarial estimators, see Breslow (1992). A vast literature has grown up concerning its theoretical and practical properties, and similarly for related processes and estimators; see for example Andersen *et al.* (1993) and their references.

As a random variable, the KME is a remarkable object; survival and censoring distributions combine to produce (possibly) censored observations, then ordering takes place before the 'product-limit' form, (1.7) or (1.10), of the KME emerges.

Kaplan and Meier themselves gave some large-sample analyses of the KME, mainly concerning asymptotic unbiasedness and consistency, by noticing that (1.10) lends itself to an analysis via conditional binomial random variables. It was not until some 16 years later that Breslow and Crowley (1974) obtained the asymptotic distribution of $\hat{F}_n$, after norming and centring, as a form of Brownian bridge, using quite complex arguments involving empirical processes.

The modern theory, though, involves sophisticated counting process and martingale techniques, and, at least concerning the results we have been applying, was developed in large part by Gill (1980, 1983) (see also Andersen *et al.* 1993). To our present good fortune, Gill (and others) were able to develop the large sample theory of the KME *allowing a sub- or improper survival distribution.* The main reason, apparently, was the fact that the KME is itself improper, in general (as we observed), since Gill (1980) does not mention the possibility of data analyses allowing for immunes. But it is precisely this generalised theory of Gill's which finds application in our Chapters 3 and 4, where we develop tests for the presence of immunes and/or for sufficient follow-up in the data, and their properties.

We have only briefly surveyed some of the main results in this area, and we have not attempted to relate them, or Gill's results (mentioned in Section 3.4), to possible precursors; nor have we attempted to survey work in the field more widely. For this we refer the reader to Gill's papers mentioned above, and to Andersen *et al.* (1993) and their references. See also Gill (1994) for an up-to-date treatment of 'product integration' and the large-sample theory of the KME.

2. Redistribution of mass rules for the KME

The KME places probability mass only on the (ordered) uncensored observations, as we saw in Section 1.2, unlike the EDF (if we were to calculate it) of (all) the observations, which would place mass $1/n$ at each censored and uncensored point. The latter would produce an estimate whose bias would be unacceptable in general, but we can think of beginning with this unacceptable estimator and 'redistributing' the mass initially associated with the censored observations among the uncensored observations. We can do this in such a way as to end up with the KME.

The 'redistribution from the right' rule of Dinse (1985) accomplishes this, and in so doing provides us with some nice intuition as to why the KME 'works'. Dinse's construction is as follows. Think of the ordered censored and uncensored survival times arranged from left (smallest) to right (largest). Initially, associate mass $1/n$ with each observation. Now begin at the far right, move to the left and distribute the mass $1/n$ of the first censored observation encountered to all the uncensored times to its right, in proportion to the masses already accumulated at those points. (If the largest time is censored, treat it as uncensored for the purposes of this exercise.) Keep moving to the left and repeat the procedure until the mass of all censored observations has been distributed. The resulting distribution of masses, as shown by Dinse (1985), is precisely the KME.

The intuitive idea here, useful for pedagogical purposes, is that a censored survival time contributes information only to larger, uncensored, survival times, so in a sense, we are distributing the information regarding the probability of failing carried by a censored observation to those uncensored observations larger than it. A similar procedure involving 'redistribution from the left' was formulated by Efron (1967); see also Miller (1981).

3. Bias and consistency of functionals of the KME

Suppose that F is proper, so that no immunes are present in the population. In Section 2.3 we mentioned that the 'sufficient follow-up' condition $\tau_{F_0} \le \tau_G$ is necessary and sufficient (provided F is continuous at τ_H in case $\tau_H < \infty$) for the 'Kaplan–Meier mean'

$$\hat{\mu}_n = \int_0^{t_{(n)}} (1 - \hat{F}_n(t))\,\mathrm{d}t$$

to be consistent for the mean survival time

$$\mu = \int_{[0,\infty)} t\,\mathrm{d}F(t) = \int_0^{\infty} (1 - F(t))\,\mathrm{d}t.$$

This is proved in Theorem 4.1 of Maller and Zhou (1993), who also prove the result of Theorem 3.6. (The mean μ need not be finite in the Maller–Zhou result.) Their methods are to apply the results of Gill and of Stute and Wang (1993), as we did in Sections 3.2 and 3.3. See also Matthews (1984) for another application.

Stute and Wang (1993) prove a much more general theorem to the effect that a *functional* of the Kaplan–Meier distribution of the form

$$S_n = \int_{[0,\infty)} \varphi(t)\, \mathrm{d}\hat{F}_n(t)$$

is, under some mild conditions, *almost surely* consistent for the corresponding population functional

$$\int_{[0,\infty)} \varphi(t)\, \mathrm{d}F(t).$$

Specialising to indicator functions for $\varphi(t)$ gives an almost sure version of Gill's consistency result (see Stute and Wang 1993, Corollary 1.2.) This important and useful result was reproved in Gill (1994) using martingale methods, having held out against them for some time, before succumbing to the more 'order statistic' oriented methods of Stute and Wang (as Gill remarks). Gill (1994) also gives an alternative proof of Wang's (1987) result (our Theorem 3.3).

Stute (1993) extends the Stute–Wang result to survival data with covariates of a certain kind. Prior to these, Stute (1992a, b) had shown the almost sure consistency of Kaplan–Meier functionals, under the i.i.d. censoring model, when the survival and censoring c.d.f.'s F and G satisfy the 'Koziol–Green proportional hazards model'. This specifies that the censoring and survival distributions are related by

$$1 - G(t) = (1 - F(t))^{\beta}, \qquad t \geq 0,$$

for some $\beta > 0$. This assumption is often useful for examples and to develop intuition regarding the effect of censoring on estimators. Incidentally, when F is continuous, we can evaluate (3.33) for the Koziol–Green model, obtaining $\beta/(\beta+1)$ for the probability that the largest observation is censored.

On the subject of bias, Stute (1994) derives an expression for the bias of a Kaplan–Meier functional, generalising that of Gill (1980) (see Section 3.3), and obtains bounds for the rate at which the bias converges to 0 as $n \to \infty$. Stute and Wang (1994) derive an explicit formula for the Quenouille 'jackknife' estimate of a Kaplan–Meier functional, and discuss its effect on reducing the bias of the estimator.

We stress that the above results are for *proper* survival c.d.f.'s F. For improper F, functionals such as the mean

$$\mu = \int_{[0,\infty)} (1 - F(t))\, \mathrm{d}t$$

are of course infinite, but we might expect Kaplan–Meier functionals defined in terms of the proper c.d.f. F_0 of the survival times of the susceptibles to be consistent for the corresponding population functionals. We conjecture that a result like

$$\int_{[0,t_{(n)}]} \varphi(t)\, \mathrm{d}\hat{F}_{0n}(t) \xrightarrow{P} \int_{[0,\infty)} \varphi(t)\, \mathrm{d}F_0(t)$$

(or even for convergence almost sure) should hold under mild conditions on φ if and only if $\tau_{F_0} \leq \tau_G$, but we know of no investigations along these lines. Here $\hat{F}_{0n}(t)$ is the estimator of $F_0(t)$ defined in equation (4.2) of the next chapter.

4. The probability that the largest observation is censored

Some of the results in Theorems 3.7 to 3.10 appeared in Maller and Zhou (1993), but we have generalised the original versions, which only applied when the c.d.f. F of the survival times is proper, and in some proofs required continuity of F and G. The lower bound for $P\{c_{(n)} = 0\}$ which results from (3.30) and (3.32), namely

$$P\{c_{(n)} = 0\} \geq n \int_{[0,\tau_H]} \overline{F}(t)[1 - \overline{F}(t)\overline{G}(t-)]^{n-1}\, \mathrm{d}G(t),$$

is a little better than that given in Lemma 2.2 of Maller and Zhou (1993), since $\overline{F}(t) \leq \overline{F}(t-)$, but they are the same when F is continuous for all t. Maller and Zhou (1993, Theorem 2.1) also cover the case (not included in Theorems 3.11 and 3.12 above) when $\tau_F = \tau_G$, showing then that $P(c_{(n)} = 0) \to 0$ if and only if

$$\lim_{t \uparrow \tau_H} \frac{\int_t^{\tau_H} \overline{F}(y)\, \mathrm{d}G(y)}{\overline{F}(t)\overline{G}(t)} = 0.$$

They also show in Theorem 3.1 of their paper that, if F is continuous at τ_H, then $P\{c_{(n)} = 1 \text{ infinitely often}\} = 1$ if and only if $\tau_F \leq \tau_G$, so the 'sufficient follow-up' idea is important here too (but remember that they deal with the case of F proper, so no immunes are present). Maller and Zhou (1993) also discuss the calculation of the probability that the r largest observations are censored, for $r \geq 2$.

CHAPTER 4

Nonparametric Estimation and Testing

4.1 NONPARAMETRIC ESTIMATION OF p

Consistent estimation of p and of the distribution of the susceptibles

In this section we begin to apply the results derived in Chapter 3, on the large-sample behaviour of the Kaplan–Meier Estimator (KME), to investigate the questions we raised in Chapters 1 and 2 regarding the estimation of an immune proportion, testing for its presence and also testing to see whether we have sufficient follow-up in the sample, in the sense discussed in Section 2.3. In the present chapter we concentrate on *nonparametric* methods, meaning that we make no assumptions on the type or shape of the underlying survival or censoring distributions, although we assume that the i.i.d. censoring model of Section 2.1 holds.

The results of Section 3.1 suggest that the nonparametric estimator $\hat{p}_n$ defined by

$$\hat{p}_n = \hat{F}_n(t_{(n)}), \tag{4.1}$$

where as usual $\hat{F}_n(t)$ denotes the KME, may be used to estimate the parameter p, the proportion of susceptibles in the population. Note that $\hat{p}_n$ is the right extreme of the KME, or equivalently, its maximum value. Our first theorem shows that $\hat{p}_n$ is consistent for p whenever $0 < p \leq 1$, provided follow-up is sufficient, and only in that case. It was first proved by Maller and Zhou (1992).

Theorem 4.1 *Assume the i.i.d. censoring model with* $0 < p \leq 1$, *and suppose that* F *is continuous at* τ_H *in case* $\tau_H < \infty$. *Then*

$$\hat{p}_n \xrightarrow{P} p \text{ as } n \to \infty \text{ if and only if } \tau_{F_0} \leq \tau_G.$$

Proof This follows immediately from Theorem 3.6. ■

Next we show how to consistently estimate the distribution of the susceptibles. This is $F_0(t) = F(t)/p$ (see Section 2.1). A natural estimator of $F_0(t)$ is the nondecreasing, right continuous statistic given by

$$\hat{F}_{0n}(t) = \hat{F}_n(t)/\hat{p}_n, \qquad t \geq 0, \tag{4.2}$$

where $\hat{p}_n$ is defined in (4.1).

Theorem 4.2 *Assume the i.i.d. censoring model with $0 < p \leq 1$, and suppose that F is continuous at τ_H in case $\tau_H < \infty$. Then*

$$\sup_{t\geq 0} |\hat{F}_{0n}(t) - F_0(t)| \xrightarrow{P} 0 \qquad \text{as } n \to \infty$$

if and only if $\tau_{F_0} \leq \tau_G$.

Proof Let $\tau_{F_0} \leq \tau_G$. Then by Theorem 4.1 $\hat{p}_n \xrightarrow{P} p$, and by Theorem 3.8 $\hat{F}_n(t) \xrightarrow{P} F(t)$ uniformly in $t \geq 0$. (Recall our convention that $\hat{F}_n(t) = \hat{F}_n(t_{(n)})$ when $t > t_{(n)}$.) Thus

$$\begin{aligned} |\hat{F}_{0n}(t) - F_0(t)| &= \left| \frac{\hat{F}_n(t)}{\hat{p}_n} - \frac{F(t)}{p} \right| \\ &\leq \left| \frac{1}{\hat{p}_n} - \frac{1}{p} \right| + \frac{1}{p} |\hat{F}_n(t) - F(t)| \xrightarrow{P} 0 \end{aligned}$$

uniformly in $t \geq 0$.

Conversely, suppose that $\hat{F}_{0n}(t) \xrightarrow{P} F_0(t)$ for each $t > 0$, but that $\tau_{F_0} > \tau_G$. Then $\tau_H = \tau_G < \infty$. By Theorem 3.4 and the continuity of F at τ_H we have

$$\hat{p}_n \xrightarrow{P} F(\tau_H) = F(\tau_G) < F(\tau_{F_0}) = p.$$

By Theorem 3.7 $\hat{F}_n(t) \xrightarrow{P} F(t)$ for each $t \leq \tau_H$. Thus for every t in $(0, \tau_H)$ such that $F_0(t) > 0$,

$$\hat{F}_{0n}(t) = \frac{\hat{F}_n(t)}{\hat{p}_n} \xrightarrow{P} \frac{F(t)}{F(\tau_H)} = \frac{pF_0(t)}{F(\tau_H)} > F_0(t).$$

This gives a contradiction and proves the theorem. ■

As we mentioned in Section 3.1, consistency of an estimator is a valuable property to establish, but it is just one of a number of criteria by which we can judge it. We can also discuss the bias of $\hat{p}_n$ using the results of Section 3.3. We mentioned there that the KME is always biased *downwards* for $F(t)$, and gave an expression for the bias, namely $-B_n(t)$, where $B_n(t)$ is given in equation (3.53). Consequently we have

$$E(\hat{p}_n) = E(\hat{F}_n(t_{(n)})) \leq E(\hat{F}_n(\tau_H)) \leq F(\tau_H) \leq F(\infty) = p,$$

showing that $\hat{p}_n$ is also biased downwards for p, in general. However, $\hat{p}_n$ is asymptotically unbiased if follow-up is sufficient (supposing that F is continuous at τ_H in case $\tau_H < \infty$). This is a consequence of Theorem 4.1, as follows. When $0 < p \leq 1$, $\hat{p}_n \xrightarrow{P} p$ as $n \to \infty$ implies $E(\hat{p}_n) \to p$ as $n \to \infty$, since $0 \leq \hat{p}_n \leq 1$. So by Theorem 4.1 this holds if follow-up is sufficient, that is, if $\tau_{F_0} \leq \tau_G$.

Asymptotic normal distribution for $\hat{p}_n$ when $0 < p < 1$

The consistency and asymptotic unbiasedness of the estimator $\hat{p}_n$ show that, at least in large samples, $\hat{p}_n$ can be expected to be close to p, and hence $1 - \hat{p}_n$ will, with high probability, be close to the immune proportion $1 - p$ which we wish to estimate. Thus $\hat{p}_n$ is an *accurate* estimator of p, at least in large samples. To assess the *precision* of estimation of $\hat{p}_n$, we must measure the variation of $\hat{p}_n$ around p in some way. Best of all would be to know the distribution of $\hat{p}_n$ in some calculable form. This is too much to hope for in general, however, since that distribution will depend on the unknown survival and censoring distributions F and G in some (probably complicated) way. Second best, then, is to show that, in large samples, the distribution of $\hat{p}_n$ can be approximated by a known distribution — the normal distribution in this case — for a wide variety of survival and censoring distributions. In the next theorem we are able to show that $\hat{p}_n$ is asymptotically normally distributed around p, when $0 < p < 1$, under quite general conditions.

Theorem 4.3 *Assume the i.i.d. censoring model with* $p = F(\infty) \leq 1$. *Suppose that*

$$\int_{[0,\tau_{F_0}]} \frac{\mathrm{d}F_0(t)}{1 - G(t-)} < \infty \tag{4.3}$$

(which implies $\tau_{F_0} \leq \tau_G$) and that F is continuous at τ_{F_0} in case $\tau_{F_0} < \infty$. Then as $n \to \infty$

$$\sqrt{n}(\hat{p}_n - p) \xrightarrow{D} N(0, (1 - p)^2 v_0) \tag{4.4}$$

for some $v_0 < \infty$. When $p = 1$, the convergence in (4.4) should be interpreted as saying that $\sqrt{n}(\hat{p}_n - 1) \xrightarrow{P} 0$.

Proof Suppose that (4.3) holds and F is continuous at τ_{F_0}. Inequality (4.3) obviously implies $\tau_{F_0} \leq \tau_G$. Suppose first that $0 < p < 1$. Then $\tau_F = \infty$, $\tau_H = \tau_G$, and $F(\tau_H) = F(\tau_G) = pF_0(\tau_G) = p$. Recall Gill's process $X_n(t)$ defined in Section 3.4. From (3.59) we have, taking $t = \tau_G$,

$$\sqrt{n}(\hat{p}_n - F(t_{(n)})) = \frac{(1 - F(t_{(n)}))}{1 - p} X_n(\tau_G). \tag{4.5}$$

Also recall the function $v(t)$ defined in (3.58). We have from (3.58) and (4.3) that, when $t \le \tau_{F_0}$,

$$v(t) \le \frac{1}{(1-p)^2} \int_{[0,t]} \frac{\mathrm{d}F(s)}{1-G(s-)} \le \frac{p}{(1-p)^2} \int_{[0,\tau_{F_0}]} \frac{\mathrm{d}F_0(s)}{1-G(s-)} < \infty.$$

Thus in particular $v(\tau_{F_0}) < \infty$. Since F_0 attributes no mass to the interval $(\tau_{F_0}, \tau_H]$, this means that $v(\tau_H) = v(\tau_{F_0}) < \infty$. We assumed that F is continuous at τ_{F_0} in case $\tau_{F_0} < \infty$, so because $\tau_{F_0} \le \tau_G$, F is continuous at $\tau_G = \tau_H$ in case $\tau_H < \infty$. Thus (3.60) of Theorem 3.14 holds. Now (3.61) of Theorem 3.14 also holds under (4.3), because $1 - F(s) \le 1 - F(s-)$, and for $t < \tau_G$

$$\int_{(t,\tau_G)} \frac{\mathrm{d}F(s)}{1-G(s-)} = \max\left(p \int_{(t,\tau_{F_0})} \frac{\mathrm{d}F_0(s)}{1-G(s-)}, 0\right) \to 0, \qquad \text{as } t \uparrow \tau_G.$$

Furthermore, since we are in the case $0 < p < 1$ and F is continuous at τ_{F_0}, we have $(1 - F(t_{(n)}))/(1-p) \xrightarrow{P} (1 - F(\tau_{F_0}))/(1-p) = 1$. We now have by (4.5) and Theorem 3.14 that

$$\sqrt{n}(\hat{p}_n - F(t_{(n)})) \xrightarrow{D} \begin{cases} (1 - F(\tau_G))Z(\tau_G) & \text{if } H(\tau_G -) < 1 \\ \lim_{\tau \uparrow \tau_G} (1 - F(t))Z(t) & \text{if } H(\tau_G -) = 1 \end{cases} \tag{4.6}$$

where $Z(t)$ is the process defined in Section 3.4.

Next we show that

$$\sqrt{n}(p - F(t_{(n)})) \xrightarrow{P} 0. \tag{4.7}$$

This holds for $0 < p \le 1$ under (4.3). To see this, note that for $t < \tau_{F_0}$

$$\frac{1 - F_0(t)}{1 - G(t)} \le \int_{(t,\tau_{F_0}]} \frac{\mathrm{d}F_0(s)}{1-G(s-)},$$

and the right-hand side converges to 0 as $t \uparrow \tau_{F_0}$ (since F and hence F_0 is continuous at τ_{F_0}). Thus, given an arbitrary $\delta > 0$, we have $\delta(1-G(t)) \ge 1-F_0(t)$ for all t near τ_{F_0}. For an arbitrary $\varepsilon > 0$, define $a_n = a_n(p, \varepsilon)$ by

$$a_n = \inf\left\{t > 0 : 1 - F_0(t) \le \frac{\varepsilon}{p\sqrt{n}}\right\}. \tag{4.8}$$

Then $1-F_0(a_n-) \ge \varepsilon/(p\sqrt{n})$ and $a_n \uparrow \tau_{F_0}$ as $n \to \infty$. Thus for n large enough

$$1 - G(a_n-) \ge (1 - F_0(a_n-))/\delta \ge \varepsilon/(\delta p \sqrt{n}).$$

Also

$$1 - F(a_n-) = 1 - pF_0(a_n-) \ge 1 - p + \varepsilon/\sqrt{n}.$$

Consequently

$$\begin{aligned} P\{\sqrt{n}(p - F(t_{(n)})) > \varepsilon\} &= P\left\{1 - F_0(t_{(n)}) > \frac{\varepsilon}{p\sqrt{n}}\right\} = P\{t_{(n)} < a_n\} \\ &= H^n(a_n-) = [1 - (1 - F(a_n-))(1 - G(a_n-))]^n \\ &\le \left[1 - \left(1 - p + \frac{\varepsilon}{\sqrt{n}}\right)\frac{\varepsilon}{\delta p\sqrt{n}}\right]^n. \end{aligned}$$

When $p < 1$, this converges to 0 as $n \to \infty$. When $p = 1$, it converges to $e^{-\varepsilon^2/\delta}$, and this can be made arbitrarily small by choosing δ small. Thus indeed (4.7) holds for $0 < p \le 1$.

Combining (4.6) and (4.7) gives, for $0 < p < 1$,

$$\sqrt{n}(\hat{p}_n - p) \xrightarrow{D} \begin{cases} (1 - F(\tau_G))Z(\tau_G) & \text{if } H(\tau_G -) < 1 \\ \lim\limits_{t \uparrow \tau_G} (1 - F(t))Z(t) & \text{if } H(\tau_G) = 1. \end{cases} \tag{4.9}$$

However, both the limits in (4.9) are the same when $\tau_{F_0} \le \tau_G$ and F is continuous at τ_{F_0}, since then $1 - F(\tau_G) = 1 - p$, while also $\lim_{t \uparrow \tau_G}(1 - F(t)) = 1 - p$ since F continuous at τ_{F_0} implies F is continuous at τ_G. In addition, $Z(\tau_G)$ is distributed as $N(0, v(\tau_G))$, where $v(\tau_G) = v(\tau_{F_0})$. Letting $v_0 = v(\tau_{F_0})$ proves (4.4) when $p < 1$.

Next consider the case when $p = 1$. This time we use the inequality

$$P\{1 - \hat{F}_n(t_{(n)}) > \beta^{-1}(1 - F(t_{(n)})\} \le \beta \tag{4.10}$$

which holds for every $\beta \in (0, 1)$. This follows from Theorem 3.2.1 of Gill (1980, p. 39). We also have $\sqrt{n}(1 - F(t_{(n)})) \xrightarrow{P} 0$, by (4.7). Consequently, if $\varepsilon > 0$,

$$\begin{aligned} P\{\sqrt{n}(1 - \hat{F}_n(t_{(n)})) > \varepsilon\} &\le P\{1 - \hat{F}_n(t_{(n)}) > \beta^{-1}(1 - F(t_{(n)}))\} \\ &\quad + P\{\sqrt{n}(1 - F(t_{(n)}) > \varepsilon\beta\} \\ &\le \beta + o(1) \end{aligned}$$

and this can be made arbitrarily small by choosing β small. So (4.4) holds also when $p = 1$. ∎

In applying Theorem 4.1, we need to distinguish the cases when immunes are present ($p < 1$) or not. When $p = F(\infty) < 1$, (4.4) gives a nondegenerate normal limit for $\sqrt{n}(\hat{p}_n - p)$. When $p = F(\infty) = 1$, so there are no immunes in the population, (4.3) merely says that $\hat{p}_n - p$ converges to 0 faster than $1/\sqrt{n}$; it does not specify an asymptotic distribution, which could be used to set confidence intervals, for example, for p. We do not know how to scale $\hat{p}_n - p$ to get a nondegenerate proper limiting distribution when $p = 1$. For this important boundary case, as we will discuss in the next section, we have to rely on simulations to get some idea of the distribution of $\hat{p}_n$.

When immunes are judged to be present, on the other hand, we do have from Theorem 4.1 that $\hat{p}_n$ is approximately normally distributed in large samples, but still (4.4) is not usable as it stands because of the unknown quantities p and $v(\tau_{F_0})$ on the right-hand side of (4.4). However, this problem is easily overcome by 'Studentising' the result, that is, by scaling $\hat{p}_n - p$ by statistics — quantities that can be calculated from the sample — so as to obtain a standard normal distribution, in the limit. Now p of course is consistently estimated by $\hat{p}_n = \hat{F}_n(t_{(n)})$, under the conditions of Theorem 4.3, so we need only find a consistent estimator for $v_0 = v(\tau_{F_0})$. To do this we again refer to Gill's results. As in Section 3.4, let

$$N_n(t) = \sum_{i=1}^{n} I(t_i^* = t_i \le t) \qquad (\text{for } t \ge 0)$$

be the number of uncensored survival times up till time t, and let

$$\Delta N_n(t) = N_n(t) - N_n(t-)$$

be the jump in N_n at time t. Also let

$$Y_n(t) = \sum_{i=1}^{n} I(t_i \ge t) \qquad (\text{for } t \ge 0)$$

be the number of individuals still at risk at time t.

Note that $N_n(t)$ is a step function with a nonzero jump at $t \ge 0$ if and only if there is at least one *uncensored* survival time at t, and then $\Delta N_n(t)$ equals the number of uncensored survival times tied at t. In the notation that we used for the KME itself (see Section 1.2), if

$$t_{(1)}^{(d)} < t_{(2)}^{(d)} < \cdots t_{(m)}^{(d)}$$

are the *distinct* ordered (censored or uncensored) failure times, d_i is the number of individuals failing at time $t_{(i)}^{(d)}$, and a_i is the number censored at time $t_{(i)}^{(d)}$, $1 \le i \le m$, then

$$\Delta N_n(t) = \begin{cases} d_i & \text{if } t = t_{(i)}^{(d)} \text{ for some } i \le m \\ 0 & \text{otherwise.} \end{cases}$$

(Note that we may also have $d_i = 0$.) $Y_n(t)$, on the other hand, is a step function with jumps at *each* survival time t_i, the size of the jump being equal to the number of observations tied at t_i. The function $Y_n(t)$ is nonzero only for values of t in $[0, t_{(n)}]$, and $Y_n(t) > \Delta N_n(t)$ for all $t < t_{(n)}$, but we may have $Y_n(t_{(n)}) = \Delta N_n(t_{(n)})$, if all the tied, largest, observations are uncensored.

With this in mind, define for t in $[0, t_{(n)}]$ the estimator (see Gill 1980, p. 79)

$$v_n(t) = \int_{[0,t]} \frac{nI(\Delta N_n(s) < Y_n(s))\, \mathrm{d}N_n(s)}{(Y_n(s) - \Delta N_n(s)) \quad Y_n(s)}. \tag{4.11}$$

The denominator is positive, and the indicator function equals 1, except possibly at $t = t_{(n)}$. If $Y_n(t_{(n)}) = \Delta N_n(t_{(n)})$, we take the value of the integrand in (4.11) as 0. Gill (1980, p. 80) shows that $v_n(t)$ so defined is uniformly consistent for the variance function $v(t)$ of $Z(t)$ for bounded t, in the sense that

$$\sup_{0\le s\le t} |v_n(s) - v(s)| \xrightarrow{P} 0 \qquad \text{as } n \to \infty \tag{4.12}$$

for each $t < \tau_H$. Again for our purposes we need a version of this which works on $[0, \tau_H]$, and again we can obtain this from Gill's working. We have

Theorem 4.4 *Assume the independent censoring model with $p = F(\infty) < 1$, and suppose*

$$\int_{[0,\tau_{F_0}]} \frac{\mathrm{d}F_0(t)}{1 - G(t-)} < \infty, \tag{4.13}$$

so that $\tau_{F_0} \le \tau_G$. Suppose also that F is continuous at τ_{F_0} if $\tau_{F_0} < \infty$, or there is an $M > 0$ such that F is continuous on (M, ∞) if $\tau_{F_0} = \infty$. Then $v(\tau_{F_0})$ is finite,

$$v_n(t_{(n)}) \xrightarrow{P} v_0 = v(\tau_{F_0}) \qquad \text{as } n \to \infty, \tag{4.14}$$

and

$$\frac{\sqrt{n}(\hat{p}_n - p)}{(1 - \hat{p}_n)\sqrt{v_n(t_{(n)})}} \xrightarrow{D} N(0, 1). \tag{4.15}$$

Proof Suppose first that $\tau_{F_0} < \tau_G = \tau_H$. Since $t_{(n)} \xrightarrow{P} \tau_H = \tau_G$, we have $t_{(n)} > \tau_{F_0}$ on a set whose probability approaches 1, and $v_n(t_{(n)}) = v_n(\tau_{F_0})$ on this set, since $N_n(t)$ is constant on $[\tau_{F_0}, \infty)$. But $v_n(\tau_{F_0}) \to v(\tau_{F_0}) = v_0$ by (4.12), so (4.14) is true in this case. Next suppose that $\tau_{F_0} = \tau_G = \tau_H$. Then $t_{(n)} \uparrow \tau_{F_0}$ in probability. Choose $t < \tau_{F_0} = \tau_H$. Then, on a set whose probability approaches 1, $t_{(n)} \in (t, \tau_{F_0}]$. Write

$$v(t_{(n)}) = v_n(t) + \int_{(t,t_{(n)})} \frac{nI(\Delta_n(s) < Y_n(s))\,\mathrm{d}N_n(s)}{(Y_n(s) - \Delta N_n(s))\;\; Y_n(s)}.$$

Since $v_n(t) \xrightarrow{P} v(t)$ by (4.12), it suffices to show that, for every $\varepsilon > 0$,

$$\lim_{t\uparrow\tau_{F_0}} \limsup_{n\to\infty} P\left\{\int_{(t,t_{(n)})} \frac{nI(\Delta N_n(s) < Y_n(s))\,\mathrm{d}N_n(s)}{(Y_n(s) - \Delta N_n(s))\;\; Y_n(s)} > \varepsilon\right\} = 0. \tag{4.16}$$

(In fact since $N_n(s)$ is constant on $(t_{(n)}, \tau_{F_0}]$, we can extend the interval of integration in (4.16) to $(t, \tau_{F_0}]$ and the result remains true.)

To prove (4.16), write the integral in (4.16) as

$$\sum_{i:t_i>t} \frac{nI(d_i < Y_n(t_i))}{(Y_n(t_i) - d_i)} \frac{d_i}{Y_n(t_i)} \le \sum_{i:t_i>t} \frac{nI(1 < Y_n(t_i))}{(Y_n(t_i) - d_i)} \frac{d_i}{Y_n(t_i)} \tag{4.17}$$

where $d_i = \Delta N_n(t_i)$ is the jump in $N_n(t)$ at t_i. Since we assumed F continuous at τ_{F_0} if $\tau_{F_0} < \infty$, or on large enough intervals if $\tau_{F_0} = \infty$, $N_n(t)$ can have jumps of size at most 1, when t is sufficiently close to τ_{F_0}. In that case $d_i \le 1$ in (4.17), and the right-hand side of that inequality is no larger than

$$\sum_{i:t_i>t} \frac{nI(1 < Y_n(t_i))}{(Y_n(t_i)-1)} \frac{d_i}{Y_n(t_i)} = \int_{(t,\tau_{F_0}]} \frac{nI(1 < Y_n(s))}{(Y_n(s)-1)} \frac{\mathrm{d}N_n(s)}{Y_n(s)}. \tag{4.18}$$

By Markov's inequality, if $\varepsilon > 0$

$$\begin{aligned} P&\left\{\int_{(t,\tau_{F_0}]} \frac{nI(1 < Y_n(s))}{(Y_n(s)-1)} \frac{\mathrm{d}N_n(s)}{Y_n(s)} > \varepsilon\right\} \\ &\le \frac{1}{\varepsilon} E\left\{\int_{(t,\tau_{F_0}]} \frac{nI(1 < Y_n(s))}{(Y_n(s)-1)} \frac{\mathrm{d}N_n(s)}{Y_n(s)}\right\} \\ &= \frac{1}{\varepsilon} E\left\{\int_{(t,\tau_{F_0}]} \frac{nI(1 < Y_n(s))}{(Y_n(s)-1)} \frac{\mathrm{d}F(s)}{1-F(s-)}\right\} \end{aligned} \tag{4.19}$$

where the last equality follows from the equation in the middle of p. 86 of Gill (1980). On the same page, Gill shows that

$$E\left\{\frac{nI(1 < Y_n(s))}{(Y_n(s)-1)}\right\} \le \frac{3}{(1-F(s-))(1-G(s-))} = \frac{3}{1-H(s-)}$$

so we obtain from (4.19) that

$$\begin{aligned} P&\left\{\int_{(t,t_{(n)})} \frac{nI(\Delta N_n(s) < Y_n(s))}{(Y_n(s)-\Delta N_n(s))} \frac{\mathrm{d}N_n(s)}{Y_n(s)} > \varepsilon\right\} \\ &\le \frac{1}{\varepsilon} \int_{(t,\tau_{F_0}]} \left(\frac{3}{1-H(s-)}\right) \frac{\mathrm{d}F(s)}{1-F(s-)} \\ &\le \frac{3p}{\varepsilon(1-p)^2} \int_{(t,\tau_{F_0}]} \frac{\mathrm{d}F_0(s)}{1-G(s-)}. \end{aligned}$$

This converges to 0 as $t \uparrow \tau_{F_0}$ by (4.13), since F and hence F_0 are continuous at τ_{F_0}. The continuity also means that $v(t) \to v(\tau_{F_0}) = v_0$ as $t \uparrow \tau_{F_0}$. This proves (4.16) and thus (4.14), and (4.15) then follows from (4.4) and the fact that $\hat{p}_n \xrightarrow{P} p < 1$. ■

Notice that we do not allow the case $p = 1$ in Theorem 4.4, since we do not know how to normalise $\hat{p}_n - p$ to get a nondegenerate limit in this case. When $p < 1$, however, Theorem 4.4 tells us to normalise by a quantity which includes

$v_n(t_{(n)})$. Equation (4.11) looks rather complicated, but we can reduce it to a simple expression which we now derive. As usual, assume the t_i have been condensed into distinct ordered failure or censor times $t_i^{(d)}$, of which there are m, in total, and d_i and a_i are the associated numbers failing or censored at these times. Also let

$$n_i = (d_i + \cdots + d_m) + (a_i + \cdots + a_m)$$

be the number at risk just prior to time $t_i^{(d)}$ (see Section 1.2, where we introduce this notation).

Just as in (4.17) we can write

$$v_n(t_{(n)}) = \sum_{i=1}^{m} \frac{nI(d_i < Y_n(t_{(i)}^{(d)}))}{(Y_n(t_{(i)}^{(d)}) - d_i)} \frac{d_i}{Y_n(t_{(i)})}.$$

Now $Y_n(t_{(i)}^d)$ is the number at risk just prior to $t_{(i)}^{(d)}$, so it equals n_i. Also $I(d_i < Y_n(t_{(i)}^{(d)})) = I(d_i < n_i) = 1$ except possibly when $t = t_{(n)}$. So

$$v_n(t_{(n)}) = \sum_{i=1}^{m-1} \frac{nd_i}{n_i(n_i - d_i)} + \frac{nI(d_m < n_m)d_m}{n_m(n_m - d_m)}. \tag{4.20}$$

This simple formula is closely related to Greenwood's formula for the (approximate or asymptotic) variance of the KME, which can be constructed by elementary arguments involving the binomial distribution (see Section 1.2). It makes calculation of $v_n(t_{(n)})$ quite easy. Note that the last term in (4.20) should be taken as 0 when $n_m = d_m$, i.e. when all the largest, tied, survival times are uncensored, but this will most likely not be the case in the data we are analysing, with its suspected immune individuals.

Formula (4.20) also simplifies when F and G are continuous functions, in which case there will be zero probability of a tied censor or survival time, and so we do not need to condense the t_i into $t_i^{(d)}$. When F and G are continuous we also have

$$n_i = n - i + 1, \qquad d_i = c_{(i)},$$

where $c_{(i)}$ is the censor indicator associated with $t_{(i)}$. The number at risk at time $t_{(n)}$ is 1, and $I(c_{(n)} < 1)c_{(n)} = 0$, so (4.20) reduces for continuous F and G to

$$v_n(t_{(n)}) = \sum_{i=1}^{n-1} \frac{nc_{(i)}}{(n - i + 1)(n - i + 1 - c_{(i)})}. \tag{4.21}$$

This is valid for data with no ties present, and is easily calculated.

4.2 USING $\hat{p}_n$ TO TEST FOR THE PRESENCE OF IMMUNES

A nonparametric method of testing H_{01}

Our basic paradigm as set out in Section 2.3 is first to test $H_{01} : F(\tau_G) = 1$. If H_{01} is accepted, there is no evidence of immunes and we judge follow-up to be sufficient. If H_{01} is rejected, we move on to test H_{02}. But how can we test H_{01}?

We showed in Theorem 3.6 that $\hat{p}_n = \hat{F}_n(t_{(n)})$ is a consistent estimator of $F(\tau_G)$ (unless both F and G are discontinuous at τ_H; let us rule out this possibility for the moment). But $F(\tau_G) = 1$ under H_{01}, so $\hat{p}_n$ should be close to 1 if H_{01} is true. Thus we can base a nonparametric test of H_{01} on the closeness of $\hat{p}_n$ to 1.

Now it is quite possible that

$$\hat{p}_n = \hat{F}_n(t_{(n)}) = 1,$$

in fact this occurs precisely when the largest observation is uncensored, or in the notation of Section 1.2, when $c_{(n)} = 1$. When $\hat{p}_n = 1$ we immediately accept H_{01}, that there are no immunes in the population, as seems reasonable from the fact that the KME has not levelled off below 1. (However, we will consider in Section 5.4 the possibility that the largest observation may be outlying—see Figure 2.2 for an illustration—or the possibility that H_{01} is accepted as a result of late failures which we may wish to treat as atypical.)

Now suppose $\hat{p}_n < 1$ occurs in our sample, so the largest observation is censored. We must decide if this is an improbable event under H_{01}, keeping in mind that, even when H_{01} is true, we may observe $\hat{p}_n < 1$, or equivalently $c_{(n)} = 0$, with positive probability for a fixed n or as $n \to \infty$, as we showed in Chapter 3. So we must specify that $\hat{p}_n$ be sufficiently far below 1. Our proposed test for H_{01} is of the following form:

$$\text{Reject } H_{01} \text{ if } \hat{p}_n < c_{0.05} \tag{4.22}$$

where $c_{0.05}$ is the 5th percentile of the distribution of $\hat{p}_n$ calculated under H_{01}, that is, the largest value such that

$$P(\hat{p}_n < c_{0.05}) \le 0.05.$$

If we observe $\hat{p}_n < c_{0.05}$ we conclude with 95% confidence that H_{01} is false and so immunes are likely to be present. But how do we obtain the value of $c_{0.05}$? For this, we need the distribution of $\hat{p}_n$ when $p = 1$, and unfortunately so far this distribution is not known in theoretical form, even for large samples, and in any case it will depend on the unknown distributions F and G of the censoring and survival times. Consequently we resort to simulations in order to obtain some guidance on the percentage points for the distribution of $\hat{p}_n$.

Simulation Results for $\hat{p}_n$

To simulate the distribution of $\hat{p}_n = \hat{F}_n(t_{(n)})$, we used the software program Matlab (MathWorks 1994). The idea is to generate many sets of survival data (we used 10 000 'replicate' sets) which 'simulate' a specified number of i.i.d. censored observations, and for each set of data, have the computer calculate the KME and then the value of $\hat{p}_n$. From the 10 000 observations on $\hat{p}_n$ we can then construct a histogram which approximates its distribution, and read off simulated values of the percentiles at 1%, 5%, 10% and 20%.

To carry out this procedure we must decide on both a survival distribution for susceptibles and a censoring distribution. In this instance we are interested in the testing of H_{01}, which specifies that $p = 1$, so the data we simulate should satisfy this restriction, thus it will have no immunes present. We are therefore simulating some 'ordinary' survival data in this exercise.

As a simple but fairly general survival distribution of susceptibles, we selected the exponential distribution with mean equal to 1. (Any other exponential distribution can be transformed to this by a scalar multiple; the Weibull and gamma distributions can be similarly transformed.) We examine two families of censoring distributions: the exponential distribution and the uniform distribution. The idea is to cover a relatively wide range of censoring distributions, and this is the case with the exponential and uniform distributions because they are quite different from each other in their shapes. And they often appear, at least approximately, as censoring distributions in practical examples, as we will discuss in Section 9.4.

For the exponential censoring distributions G, we take the values of the means as $\mu = 1, 2, 3, 4, 5$; these distributions will be denoted in general by $\text{Exp}(1/\mu)$. In this way we cover a fairly wide range from heavy to light censoring. In order to match, the uniform censoring distributions are taken over intervals $[0, B]$ (denoted by $U[0, B]$), with $B = 2, 4, 6, 8, 10$, which correspond again to means of $1, 2, 3, 4, 5$ respectively. Given that the survival distribution is exponential with mean 1, a censoring distribution of Exp(1) or $U[0, 2]$ corresponds to rather heavy censoring, whereas Exp(1/5) and $U[0, 10]$ represent very light censoring.

Our final choice is to select n, the sample size. We choose n ranging from 20 to 1000, so as to cover a fairly realistic set of situations. The simulated percentiles are given in Tables A.1 and A.2. For those data which are far away from the range covered by Tables A.1 and A.2, either in sample size or in type of survival or censor distribution, simulated points can easily be obtained for the particular situations in the same way we do here (or consult with us if necessary). The Matlab program which carries out the task of obtaining the simulated values is quite simple.

To give some visual indication of the shape of the distribution of $\hat{p}_n$, we have plotted in Figures 4.1 and 4.2 eight histograms of the simulated values of $\hat{p}_n$. In each of the two figures, the histograms correspond to four different censoring distributions as specified on the graph. The sample size is $n = 50$ for Figure 4.1 and $n = 100$ for Figure 4.2.

It can be seen from Figures 4.1 and 4.2 that the distribution of $\hat{p}_n$ has a quite stable shape, regardless of the censoring distributions and sample sizes, at least for $n = 50$ and $n = 100$. (The same is true for larger sample sizes, as can be seen from Tables A.1 and A.2). The distributions are noticeably bimodal, with

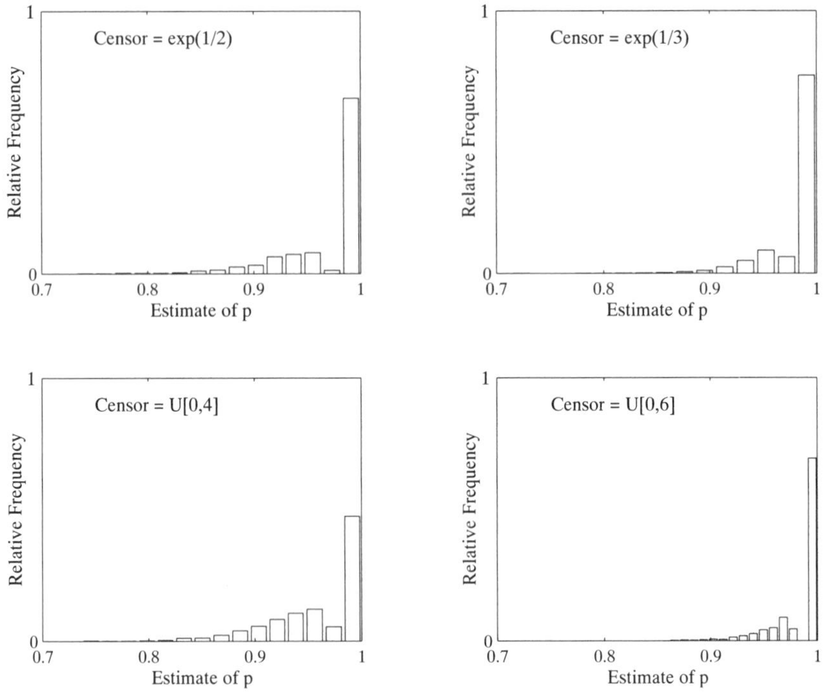

Figure 4.1 Histograms of simulated values of $\hat{p}_n = \hat{F}_n(t_{(n)})$ with $n = 50$.

a concentration of $\hat{p}_n$ values at 1 (as we would expect), but also with a modal value slightly below 1, which moves closer to 1 as n increases. This bimodality is due to the fact that $\hat{p}_n$ cannot take any value between 1 and 1 minus the last jump of the KME—which is at least $1/n$ but could be much larger if there are many censored observations.

Figures 4.1 and 4.2 also show that lighter censoring (Exp(1/3) or $U[0, 6]$) tends to produce $\hat{p}_n$ values which are more concentrated near 1, the null value of $F(\tau_G)$, than a heavier censoring (Exp(1/2) or $U[0, 4]$). The effect of this is to increase the values of the percentage points for lighter censoring or larger sample sizes as compared with heavier censoring or smaller samples.

Furthermore, we observe from Figures 4.1 and 4.2, and Tables A.1 and A.2, that the value of $\hat{p}_n$ tends to increase with the sample size, again as we expect, since $\hat{p}_n$ converges to 1 in probability under H_{01} by Theorem 4.1 when $\tau_{F_0} \leq \tau_G$. A comparison of Table A.1 with Table A.2 also reveals that the percentage points, for most values, do not differ very much between exponential and uniform censoring distributions with the same means except for small sample sizes ($n \leq 40$) and heavy censoring ($\mu = 1$ and $B = 2$).

These observations suggest, admittedly with fairly scant evidence (though if the demand arises we can easily add to the evidence by doing more extensive

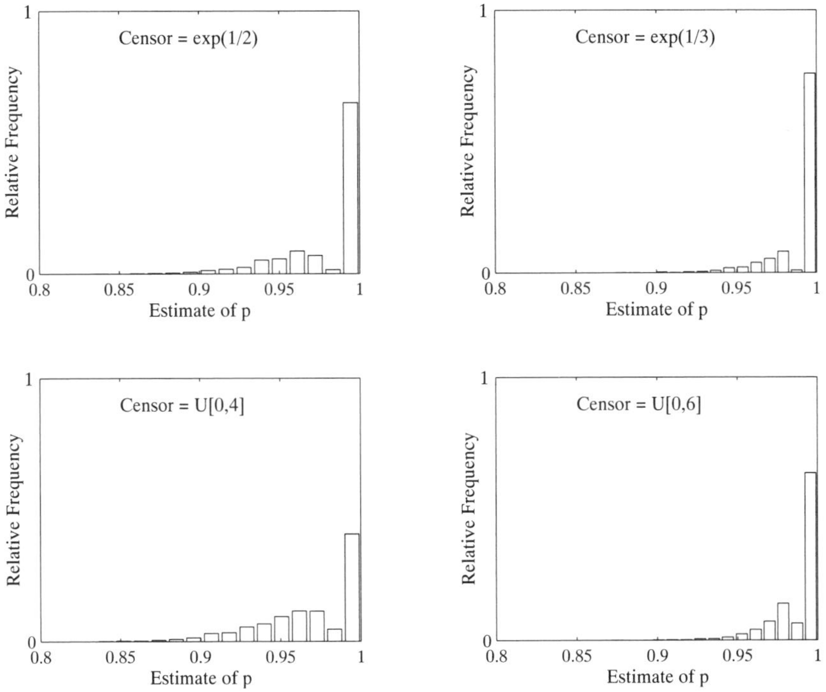

Figure 4.2 Histograms of simulated values of $\hat{p}_n = \hat{F}_n(t_{(n)})$ with $n = 100$.

simulations over a range of survival and censoring distributions), that heaviness of censoring and sample size are major factors determining the percentage points of $\hat{p}_n$, but that the shape of the censoring distribution is of lesser importance. As a rough guide to the heaviness of the censoring, we can compare the observed proportion of censored observations with the probability of obtaining a censored observation, assuming exponential survival and censoring distributions. This probability is easily seen to be $\phi/(\phi + \mu)$, where ϕ and μ are the survival and censoring means, respectively. Since Table A.1 was constructed with $\phi = 1$, this means that we look it up with a value of μ such that $1/(1 + \mu)$ approximately equals the observed proportion of censored observations. For Table A.2, we use the value of B which gives the same censoring distribution mean as for the exponential distribution.

Admittedly this procedure is very approximate but it provides some guidance in a situation for which we currently have little. We illustrate the procedure below.

Based on the intuition provided by the simulation results in Tables A.1 and A.2, and the above ideas, we propose to test H_{01} nonparametrically for a given data set of size n, as follows:

(i) Calculate the value of $\hat{p}_n$ from the data.

(ii) Calculate the proportion of the data points which are *censored*.

(iii) Look up in Table A.1 or A.2 the percentage point which corresponds to the closest value of n to that of the data, and also has the tabulated probability of an observation being censored closest to the proportion of censored observations in the data. For example, if about one-third of the data consists of censored observations, then look at the column with $\mu = 2$ in Table A.1 or with $B = 4$ in Table A.2. The rationale for this, as explained above, is that for an Exp(1) survival distribution with an Exp(1/2) censoring distribution, corresponding to a mean of $\mu = 2$, the probability of an observation being censored is equal to 1/3. The uniform with a mean of 2 has $B = 4$.

(iv) Having obtained all of this information, we then compare the value of $\hat{p}_n$ calculated from the data with the percentage points read from the tables. If the former is *smaller* than the latter (recall that $\hat{p}_n$ is the estimated *susceptible proportion*), then we reject H_{01}; otherwise we accept H_{01}.

We must also take into account the issue of sufficient follow-up, and some further exercise of judgement is required. We illustrate now with one set of data, and further examples are given at the end of the section, and in the next chapter.

Example: leukaemia data

Let us illustrate the procedure outlined for the leukaemia data analysed by A. Goldman and discussed in Section 1.4. The KMEs of the data are shown in Figure 4.3 and the data is reproduced in Table 4.1; see also Kersey *et al.* (1987, Fig. 4). Note that in Table 4.1 the superscript 2 in Group 1 represents two uncensored observations tied at the value $t_i = 0.2137$. In Group 2 there are two pairs of ties at times 0.2685 years (both uncensored) and 4.2055 years (both censored). Thus the numbers of observations are 46 for Group 1 and 44 for Group 2.

For allogeneic transplants (Group 1 data), we have $n = 46$ and we find that the estimated susceptible proportion is $\hat{p}_n = 0.7366$, simply by writing down the largest value of the KME. Since about one-third of the observations (13 out 46) are censored, we look up the percentiles in Table A.1 corresponding to $\mu = 2$ and $n = 50$, from which we find that the 1% point of the null distribution of $\hat{p}_n$ under exponential censoring to be about 0.8123; or from Table A.2, corresponding to $B = 4$ and $n = 50$, about 0.8718 under uniform censoring. In either case, we see that the observed value of $\hat{p}_n = 0.7366$ is decidedly *smaller* than the tabulated 1% point. Hence we *reject* H_{01} and provisionally decide that there is evidence that immunes are present in Group 1; but with the caveat that first we must convince ourselves that follow-up has been sufficient. We will illustrate this test (for H_{02}) in the next section.

For autologous transplants (Group 2), we have a sample size of $n = 44$ and we find in a similar way that $\hat{p}_n = 0.8011$. There are 9 censored observations—about one-fifth of the sample. So we look under $\mu = 4$ in

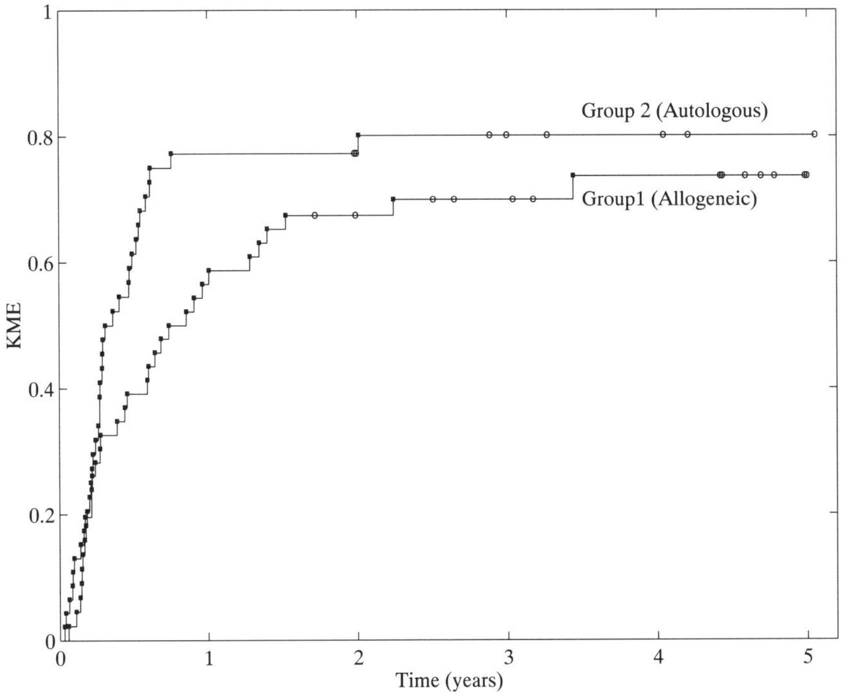

Figure 4.3 KMEs of leukaemia data (Kersey *et al.* 1987).

Table A.1 and under $B = 8$ in Table A.2, (with $n > 40$) which give us 1% points of 0.8926 and 0.9001 for exponential and uniform censoring respectively. Again the 1% points are considerably greater than the observed value of $\hat{p}_n = 0.8011$. Thus we reject H_{01} for autologous transplant data as well, and decide that immunes may be present in Group 2—but we also admit the possibility that there is not sufficient follow-up to be confident of this. In the next section we will test for sufficient follow-up in this data too.

4.3 TESTING FOR SUFFICIENT FOLLOW-UP

Heuristics of the q_n-test

Acceptance of H_{01} allows us to conclude that no immunes are present, but its rejection leaves us indecisive, as it does for both groups in the leukaemia data analysed in the previous section. We must next test the hypothesis

$$H_{02} : \tau_{F_0} \leq \tau_G \tag{4.23}$$

to decide whether or not we have sufficient follow-up to be confident that immunes are present. How can we devise a test for H_{02}? Looking back at a data

Table 4.1 Leukaemia data of Kersey *et al.* (1987): time to recurrence of leukaemia, in years; the superscript 2 denotes two tied observations.

Group 1				Group 2			
t_i	c_i	t_i	c_i	t_i	c_i	t_i	c_i
0.0301	1	0.9096	1	0.0575	1	0.3589	1
0.0384	1	0.9644	1	0.1096	1	0.4027	1
0.0630	1	1.0082	1	0.1370	1	0.4685	1
0.0849	1	1.2822	1	0.1452	1	0.4712	1
0.0877	1	1.3452	1	0.1479	1	0.4904	1
0.0959	1	1.4000	1	0.1534	1	0.5178	1
0.1397	1	1.5260	1	0.1671	1	0.5342	1
0.1616	1	1.7205	0	0.1753	1	0.5452	1
0.1699	1	1.9890	0	0.1836	1	0.5836	1
0.2137^2	1	2.2438	1	0.2000	1	0.6110	1
0.2164	1	2.5068	0	0.2082	1	0.6137	1
0.2384	1	2.6466	0	0.2164	1	0.7589	1
0.2712	1	3.0384	0	0.2219	1	1.9836	0
0.2740	1	3.1726	0	0.2411	1	1.9973	0
0.3863	1	3.4411	1	0.2603	1	2.0110	1
0.4384	1	4.4219	0	0.2685^2	1	2.8849	0
0.4548	1	4.4356	0	0.2712	1	2.9973	0
0.5918	1	4.5863	0	0.2849	1	3.2658	0
0.6000	1	4.6904	0	0.2877	1	4.0411	0
0.6438	1	4.7808	0	0.2904	1	4.2055^2	0
0.6849	1	4.9863	0	0.3068	1	5.0548	0
0.7397	1	5.0000	0				
0.8575	1						

set such as the one illustrated in Figure 4.3, we see that evidence of an immune proportion is contained in the fact that the KME levels off at its right-hand end; in other words, a number of the largest observations are censored. How many must be censored, or how much of a level stretch at the right-hand end of the KME should there be in order to provide convincing evidence that immunes are present? Let us write the alternative hypothesis to H_{02} as the complementary hypothesis

$$H_{02}^c : \tau_{F_0} > \tau_G . \tag{4.24}$$

We will be prepared to believe that the hypothesis H_{02} formulated in (4.23) is true if the data contradicts H_{02}^c, as in (4.24). In other words, to be confident that H_{02} is true, we want to *reject* H_{02}^c *in a convincing manner*. (Likewise, to be confident of H_{02}^c, we would have to reject H_{02} strongly.) To test H_{02}^c we must use the information in the sample regarding the magnitude of $\tau_G - \tau_{F_0}$, which is effectively a measure of how far the large censored lifetimes (the potential immunes) lie from the main mass of the susceptibles' lifetimes. This information is conveyed by the censored observations which exceed the largest uncensored failure time. Thus we need a measure of the distance between the *largest* failure time and the *largest uncensored* failure time.

To see the origin of the test statistic we will use, denote the largest failure time and the largest uncensored failure time respectively by

$$t_{(n)} = \max_{1\le i\le n} t_i \text{ and } t^*_{(n)} = \max\{t_i : t_i = t^*_i, 1 \le i \le n\}. \tag{4.25}$$

The distribution of $t^*_{(n)}$ may be calculated under the i.i.d. censoring model as follows. Note that $t^*_{(n)} > t$ if and only if there is an uncensored observation exceeding t. Thus

$$\begin{aligned} P\{t^*_{(n)} > t\} &= P\{u_i \ge t^*_i > t \text{ for some } i \in \{1, \ldots, n\}\} \\ &= 1 - (P\{u_1 \ge t^*_1 > t\}^c)^n. \end{aligned}$$

This gives the distribution of $t^*_{(n)}$ in the form

$$P\{t^*_{(n)} \le t\} = (1 - P\{u_1 \ge t^*_1 > t\})^n, \qquad t \ge 0. \tag{4.26}$$

To motivate the test we will use, suppose at first that $p < 1$, so there is in fact an immune proportion in the population. Thus $\tau_F = \infty$ and $\tau_H = \tau_G$. We showed in Theorem 3.2 that $t_{(n)} \to \tau_H = \tau_G$ almost surely as $n \to \infty$. Also for all $t \ge 0$ we have

$$P\{u_1 \ge t^*_1 > t\} = \int_{(t,\infty)} [F(u) - F(t)]\, \mathrm{d}G(u) = \Delta(t), \text{ say.}$$

Assume for the remainder of this section that for some $a_0 > 0$,

$$F \text{ and } G \text{ assign positive mass to } (t, \tau_{F_0} \wedge \tau_G) \text{ for all } t \in (a_0, \tau_{F_0} \wedge \tau_G). \tag{4.27}$$

Then $\Delta(t) > 0$ for such t. Hence by (4.26), for all $t < \tau_{F_0} \wedge \tau_G$ we have

$$\sum_{n\ge1} P(t^*_{(n)} \le t) = \sum_{n\ge1} [1 - \Delta(t)]^n \le 1/\Delta(t) < \infty.$$

So by the Borel–Cantelli lemma, $t^*_{(n)} > t$ almost surely for all but finitely many n. Also since $t^*_{(n)} \le \tau_{F_0} \wedge \tau_G$ almost surely, it must be the case that

$$t^*_{(n)} \to \tau_{F_0} \wedge \tau_G \text{ almost surely.} \tag{4.28}$$

It follows that, almost surely as $n \to \infty$,

$$t_{(n)} - t^*_{(n)} \to \begin{cases} \tau_G - \tau_{F_0} & \text{if } \tau_{F_0} \le \tau_G \\ 0 & \text{if } \tau_{F_0} > \tau_G. \end{cases} \tag{4.29}$$

This quantifies our intuition that the distance $t_{(n)} - t^*_{(n)}$ provides information on the difference between τ_{F_0} and τ_G at least when there is an immune proportion in the population. If $\tau_{F_0} > \tau_G$, then $t_{(n)} - t^*_{(n)}$ should be small, at least for large n. Conversely, if the observed value of $t_{(n)} - t^*_{(n)}$ is 'large', then we will accept that $\tau_{F_0} \le \tau_G$, and conclude that the follow-up has been sufficient.

However, we do not know the distribution or even the large sample distribution of $t_{(n)} - t^*_{(n)}$, and in any case this is likely to depend on the underlying survival and censoring distributions in a complicated way. So we look for a simpler nonparametric test statistic for H_{02}.

Consider the following heuristic argument which will be used to motivate the statistic we propose to analyse here. Let

$$\delta_n = \text{ observed value of } t_{(n)} - t^*_{(n)}$$

and

$$w_n = \text{ observed value of } t^*_{(n)}.$$

If $\tau_{F_0} > \tau_G$, so that H^c_{02} is true, then $t_{(n)} - t^*_{(n)} \xrightarrow{P} 0$ by (4.29). Thus we will reject H^c_{02} if δ_n is too large, if δ_n exceeds c_α, say. Here α is a prespecified level of significance ($\alpha = 0.05$, say), and c_α is the largest value such that $P(t_{(n)} - t^*_{(n)} > c_\alpha) \leq \alpha$.

Equivalently, this logic tells us to reject H^c_{02} if the p-value $P(t_{(n)} - t^*_{(n)} \geq \delta_n)$ does not exceed α. This last probability equals $P\{t^*_{(n)} \leq t_{(n)} - \delta_n\}$, and since $t_{(n)}$ is close to τ_G if the sample is large enough, we can take $P\{t^*_{(n)} \leq \tau_G - \delta_n\}$ as an approximation to it. But since $t^*_{(n)} \to \tau_G$ almost surely under H^c_{02} by (4.28), we replace the last probability by

$$P\{t^*_{(n)} \leq w_n - \delta_n\} = \{1 - P(u_1 \geq t^*_1 > w_n - \delta_n)\}^n. \tag{4.30}$$

(Recall that this is all simply heuristics, so that we can try to 'guess' a reasonable test statistic.) In (4.30) we used (4.26) to substitute for the distribution of t^*_n.

Of course we do not know F or G, as required to evaluate the right-hand side of (4.30), but we can estimate unbiasedly the probability in (4.30) by its sample empirical function estimator

$$\left\{1 - \frac{\text{number of uncensored } t_i \text{ exceeding } w_n - \delta_n}{n}\right\}^n. \tag{4.31}$$

So we let

$$N_n = \text{ number of uncensored } t_i \text{ in } (2t^*_{(n)} - t_{(n)}, t^*_{(n)}], \tag{4.32}$$

and normalise N_n by defining

$$q_n = \frac{N_n}{n} = \text{ proportion of uncensored } t_i \text{ in } (2t^*_{(n)} - t_{(n)}, t^*_{(n)}].$$

Then the expression in (4.31) is a monotone function of q_n, and we can equivalently test H^c_{02} in terms of N_n or q_n. (If $t^*_{(n)} = t_{(n)}$, so that the largest observation is uncensored, we take $N_n = 0$ and $q_n = 0$.) Note that small values of q_n correspond to large values of the expression in (4.31) so we test H^c_{02} by rejecting it if a large value of N_n or q_n is observed in the sample.

Our proposed test can now be formulated as follows:

> Reject H_{02}^c in favor of H_{02} if q_n exceeds a critical value; or
> Accept H_{02}^c otherwise.

The above argument was quite heuristic and nonrigorous, but we will show in Theorem 4.5 (Appendix Note 1) that if $N_n \xrightarrow{P} \infty$ as $n \to \infty$, then under some mild conditions, H_{02} is true, and conversely. Thus, indeed, if we observe a large enough value of N_n, or of q_n, then we can be confident that H_{02}^c is false, in which case we accept H_{02} and conclude that follow-up has been sufficient. But how large is large enough? This depends on the distribution of q_n, which is not yet known, and so we will again resort to simulations for some guidance.

Simulations of the critical points of the q_n-test

Our proposed test is to reject H_{02}^c for large values of $q_n = N_n/n$, and to reject H_{02} for small values of q_n. However, we do not know the distribution of N_n or of q_n, even in large samples, and anyway this in general will depend on the sample size and on the unknown distributions of the censoring and survival times from which the data were generated. Consequently, we again turn to simulations.

As we did for the test of H_{01}, we applied the software package Matlab to simulate values for q_n from simulated censored survival data. In order to see whether there is strong evidence for or against either H_{02} or H_{02}^c, we need the 5%, 10%, 90% and 95% percentiles of simulated q_n values, which we obtained based on 100 000 replicates. These are tabulated in Tables B.1 to B.8, at the end of the book, for a range of sample sizes and values of p, and as for the H_{01} test, for an exponential survival and a uniform censoring distribution.

For a given data set, we proceed as follows. If the observed value of q_n in our data set exceeds the tabulated 95th percentile, there is strong evidence against H_{02}^c, thus supporting H_{02}, whereas an observed value of q_n less than the 5% point provides strong evidence against H_{02}, thus supporting H_{02}^c. If the observed value of q_n lies between the tabulated 10% and 90% points, we do not have strong evidence against either of the two hypotheses. In this case the result of the test is inconclusive.

The simulation results in Tables B.1 to B.8 are for sample sizes ranging from 20 to 1000, with the survival distribution of susceptibles again taken as the exponential with mean 1. For the censoring distribution, we considered only the family of uniform distributions over the interval $[0, B]$ with $B = 2, 4, 6, 8, 10$. In addition, since immunes are permitted under H_{02} or H_{02}^c, we need also to take the value of the susceptible proportion, p, into account. So we took a range of values $p = 0.2, 0.3, \ldots, 0.9$ in Tables B.1 to B.8.

The exponential censoring distribution was not used for this situation because, strictly speaking, it is irrelevant to testing for sufficient follow-up: in principle, follow-up is always sufficient under an exponential censoring distribution, because it has $\tau_G = \infty$, hence the inequality $\tau_{F_0} \le \tau_G$ will hold regardless of the value of τ_{F_0}, finite or not. In other words, in a technical sense, we *always* have sufficient follow-up in such a model. However, the exponential distribution cannot, strictly speaking, be a model for censoring in a real situation since it would potentially allow infinite survival times to be observed, which cannot occur in reality. (Note that we did use the exponential distribution for the simulations of the statistic $\hat{p}_n$ which we suggested for use in testing H_{01}, but there the question of the endpoints of the distributions did not arise. In that situation we regarded it merely as a convenient mathematical abstraction or approximation to an observed exponential-like distribution.)

The reverse situation occurs with uniform censoring when we postulate a distribution such as the exponential for the survival distribution, which is the case in Tables B.1–B.8. In this case, we have $\tau_{F_0} = \infty$ but $\tau_G < \infty$, so once again, strictly speaking, the hypothesis $H_{02} : \tau_{F_0} \le \tau_G$ can never hold. In other words, we *never* have sufficient follow-up in such a model. Again, however, the exponential distribution is merely a convenient mathematical fiction (no real survival time can ever have such a distribution precisely) and we argue that we effectively have sufficient follow-up if the value of $F_0(\tau_G)$ is sufficiently close to 1. For example, if the censoring distribution is uniform over $[0, 8]$ and $F_0(t) = 1 - \mathrm{e}^{-t}$, then $F_0(\tau_G) = 1 - \mathrm{e}^{-8} = 0.9997$, which is very close to 1, and so in this case follow-up may be considered as sufficient for practical purposes.

In order to display the distribution of q_n, we have plotted in Figures 4.4 and 4.5 some histograms of the simulated values of q_n. These are based on 2000 replicates. The censoring distributions and the sample sizes are specified on the graphs.

From Figures 4.4 and 4.5 (and Tables B.1 to B.8) we can see that the values of q_n tend to concentrate closer to 0 when n increases, and this trend increases as the censoring becomes heavier. On the other hand, the impact of the immune proportion seems less important overall. The effect of the censoring is explained by Theorem 4.5 (Appendix Note 1), which shows that N_n is bounded in probability, so $q_n = N_n/n \xrightarrow{P} 0$ as $n \to \infty$, when $\tau_{F_0} > \tau_G$, i.e. when censoring is heavy.

Let us now consider how to use the percentage points in Tables B.1 to B.8 to carry out a significance test for sufficient follow-up in a given data set. Once again we have to exercise some judgement in making this assessment, because the tabulated points may apply either to H_{02} or H_{02}^c, and there is the possibility that the test may be inconclusive. The simulations for $B = 2$ (with a censoring distribution which is $U[0, 2]$) apply to the case where follow-up is not sufficient, since, for a survival distribution which is Exp(1), we have $F_0(\tau_G) = 1 - \mathrm{e}^{-2} = 0.8647$, which is substantially less than 1. Hence the results for $B = 2$ apply to the case when H_{02}^c holds. Conversely, with $B = 10$ or $B = 8$, for example, follow-up is quite sufficient, as $1 - \mathrm{e}^{-10} = 1.0000$ and $1 - \mathrm{e}^{-8} = 0.9997$. Thus the tabulated values for $B = 10$ and $B = 8$ apply to the case when H_{02} holds. Similarly, follow-up can be considered as reasonably sufficient with $B = 6$. The case $B = 4$ seems about on the borderline. We thus advocate the following approach to testing H_{02}^c. Compare the observed value of q_n, as calculated from

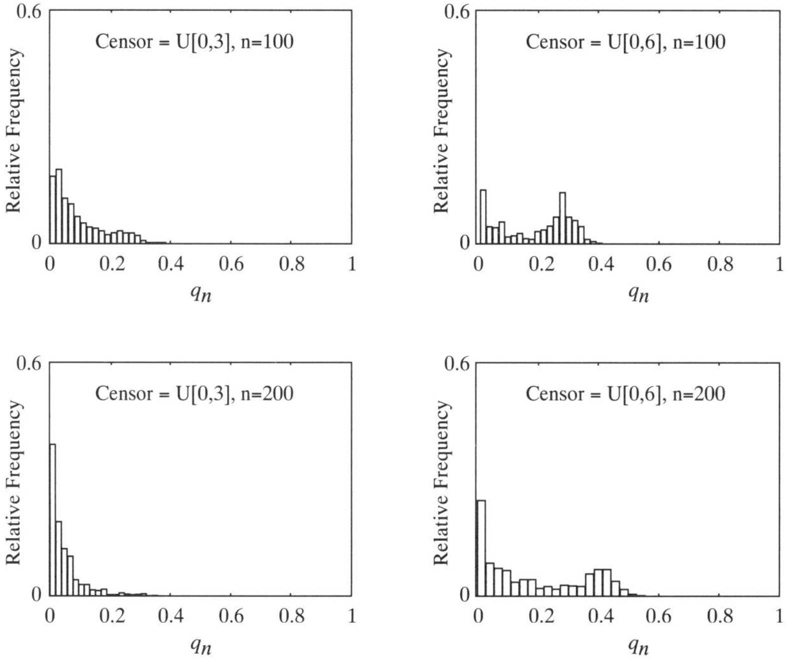

Figure 4.4 Histograms of simulated values of q_n with $p = 0.4$.

the data, with the percentage points tabulated in Tables B.1 to B.8 for each of the B values given there, and with a value of n close to the sample size of the data. Denote the $100\alpha\%$ percentage points thus obtained by $c_{\alpha,B}$. If the data indeed came from a population whose survival distribution was Exp(1) and whose censoring distribution was $U[0, B]$, and in which follow-up was sufficient, there would be only a 5% chance of observing a high value of q_n, that is, a value of q_n exceeding the tabulated value of $c_{0.95,B}$. If this occurs in our data, we can be 95% confident that the follow-up is at least as good as in the population represented by a censoring distribution of $U[0, B]$. Conversely, if a value of q_n smaller than the tabulated value of $c_{0.05,B}$ is observed in the sample, we can be 95% confident that the follow-up is no better than that in the population represented by a censoring distribution of $U[0, B]$.

More specifically, suppose that we find $q_n > c_{0.95,8}$ for our particular data set. Then we are confident that follow-up in the sample is sufficient, since $B = 8$ corresponds to sufficient follow-up in the population. If, on the other hand, we find that $q_n < c_{0.05,2}$ for our data, there is good evidence that follow-up is not sufficient. There still remains the possibility that we observe a value of q_n between, say, $c_{0.10,6}$ and $c_{0.90,4}$. This is the inconclusive case. It is doubtful that the data has levelled off sufficiently, but neither is there strong evidence that the follow-up is insufficient.

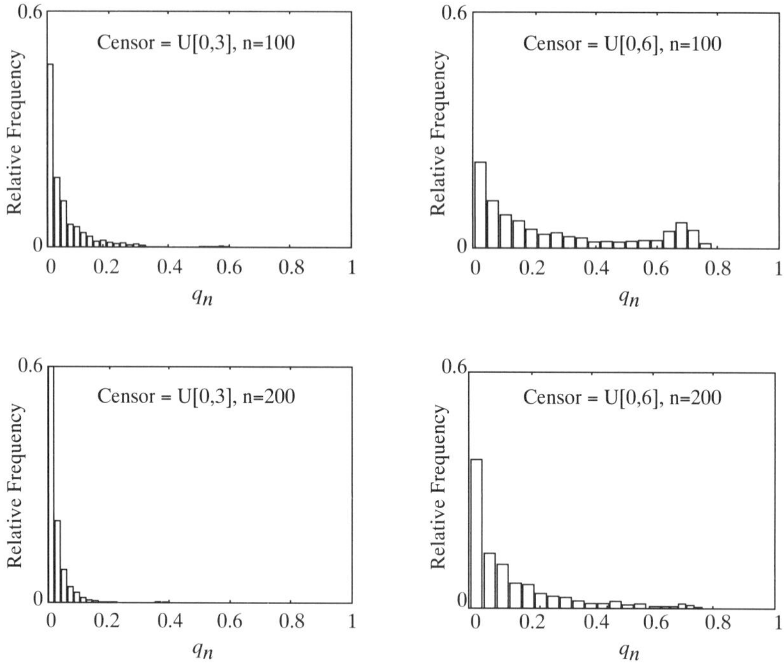

Figure 4.5 Histograms of simulated values of q_n with $p = 0.8$.

The logic set out here may seem a little complicated, but this is a necessary consequence of the fact that we are trying to make assessments at the extreme right-hand endpoint of the distribution, a notoriously difficult situation. Further study is indicated, but the procedure outlined above gives reasonable guidance as to how to proceed. In practice, as we will demonstrate, the test seems to give results which are intuitively reasonable: when the test is inconclusive, it is as a result of small samples and (visually) a lack of levelling of the KME at its right extreme. Remember also that our simulated results refer to survival times which are exponentially distributed, are close to an exponential distribution or can be transformed so they are close to an exponential distribution.

Examples

1. A medical example Consider again the data of Kersey *et al.* (1987) in Figure 4.3. For both types of transplants we decided in Section 4.2 to reject H_{01}, so we have evidence of immunes providing we can convince ourselves that follow-up is sufficient. Let us apply the q_n-test to see if this is so.

For allogeneic transplants, for which the sample size is $n = 46$, the interval of length $t_{(n)} - t^*_{(n)}$ to the left of $t^*_{(n)}$ contains two uncensored observations. Thus q_n has value

$$\frac{2}{46} = 0.0435$$

in this data. We also found in Section 4.2 that $\hat{p}_n$ is around 0.74, as can also be estimated from Figure 4.3. Looking up Table B.6, we see that the 90% point corresponding to $p = 0.7$ and $B = 2$ is 0.325 for $n = 40$ and 0.30 for $n = 50$. Since the observed value of $q_n = 0.0435$ is much smaller even than 0.30, we cannot reject H_{02}^c (follow-up is insufficient) even at the 10% level of significance. Thus we cannot be confident that $\tau_{F_0} \leq \tau_G$. In fact, 0.0435 is between the 10% point with $p = 0.7$, $B = 6$ and $n = 40$, which is 0.050, and the 10% point with $p = 0.7$, $B = 6$ and $n = 50$, which is 0.040, so there is reasonable evidence for rejecting H_{02} at the 10% significance level and concluding that follow-up is not sufficient. Thus for this data set the odds lean towards insufficient follow-up.

For autologous transplants (Group 2), with $n = 44$, however, an interval of length $t_{(n)} - t^*_{(n)}$ to the left of $t^*_{(n)}$ includes 0, so the observed value of N_n is the number of uncensored observations, 35. This gives

$$q_n = \frac{N_n}{n} = \frac{35}{44} = 0.7955.$$

For this data, we found that $\hat{p}_n = 0.80$. From Table B.7, we see that the 95% point corresponding to $p = 0.8$, $n = 40$ and $B = 6$ is 0.75. Since the observed value 0.7955 of q_n well exceeds 0.75, and follow-up can be considered as reasonably sufficient for $B = 6$, we conclude that this data set has sufficiently levelled off according to the q_n-test. (In fact, 0.7955 is very close to the 95% point, 0.80, corresponding to $p = 0.8$, $n = 40$ and $B = 8$ in Table B.7.) So for autologous transplants, we are confident in concluding that an immune or cured proportion exists, and that the sample has sufficient follow-up to decide this.

2. A criminological example Now let us consider two criminological examples. They are drawn from a large data set described in Broadhurst and Maller (1990), and consist of two groups of prisoners convicted of serious sexual offences, either having no prior records (no-prior group) or with prior convictions (prior group) and released, following imprisonment for the serious offence, after 30 June 1975. 'Failure' for these individuals is 'recidivism' (a return to prison for any offence), and we observe their times to return, possibly censored by the necessity to cease observation at a predetermined 'cutoff' time (30 June 1987). Follow-up is different for each individual since their times of release from prison differed. The KMEs of the data are plotted in Figure 4.6.

Visual assessment of Figure 4.6 suggests fairly strong evidence of immune proportions (called in this context 'nonrecidivists', or, simply, 'successes') in the population, and indeed we find that $\hat{p}_n = 0.6951$ with $n = 121$ for the prior group and $\hat{p}_n = 0.4512$ with $n = 296$ for the no-prior group. In either case, the observed value of $\hat{p}_n$ is well below the 1% point for any case with comparable n in Tables A.1 and A.2. So H_{01} is strongly rejected for either group.

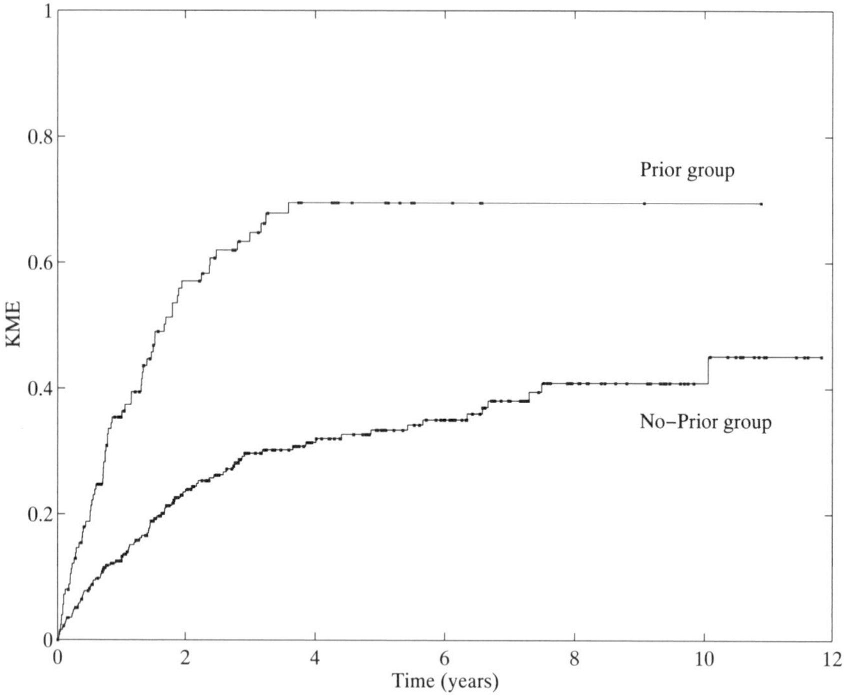

Figure 4.6 KMEs of the recidivism data.

Now let us consider whether follow-up is sufficient in these data sets. Considering first the prior group, we indeed observe in Figure 4.6 a level stretch consisting only of censored observations between $t^*_{(n)}$ and $t_{(n)}$, but is it long enough? We calculate the observed value of q_n as

$$q_n = \frac{N_n}{n} = \frac{71}{121} = 0.5868.$$

(The interval of length $t_{(n)} - t^*_{(n)}$ to the left of $t^*_{(n)}$ contains all 71 uncensored observations in the data.) The estimated value of p for this data is around 0.70. We thus look at Table B.6, for $p = 0.7$. It is clear that a value of q_n equal to 0.5868 substantially exceeds the 95% point, 0.48, corresponding to $n = 100$, $p = 0.7$ and $B = 4$ and is close to the 90% point, 0.59, with $B = 6$. An interpolation suggests that 0.5868 exceeds the 95% point, 0.55, for $n = 100$, $p = 0.7$ and $B = 5$. Thus there is only a 5% chance, approximately, of observing a value of 0.5868 or larger of q_n in an exponential survival population with censoring distribution $U[0, 5]$. As $F_0(\tau_G) = 1 - e^{-5} = 0.9933$ with $B = 5$ is quite close to 1, we are confident that follow-up has been sufficient for the prior conviction group, as would accord with our intuition from Figure 4.6.

Next we look at the data for those released with no prior convictions (Figure 4.6), where p is estimated to be 0.45. When we come to apply the q_n-test, we find that there is only one uncensored observation (i.e. $t^*_{(n)}$ itself) in the interval $(2t^*_{(n)} - t_{(n)}, t^*_{(n)}]$. Thus the observed value of q_n is

$$q_n = \frac{N_n}{n} = \frac{1}{296} = 0.0034.$$

This is far less than the 90% point with $n = 300$ and $B = 2$, either for $p = 0.4$ or for $p = 0.5$. So it appears that we cannot reject the possibility that $\tau_{F_0} > \tau_G$, that follow-up is insufficient. In fact, 0.0034 is quite close to the 10% points in Tables B.3 and B.4, suggesting considerable evidence that the follow-up has *not* been sufficient. However, this conclusion is based on information from one data point only, the failure at around 10 years, and we notice that this point, corresponding to $t^*_{(n)}$, is quite remote from the second largest uncensored failure time. This is one way in which 'outliers' can be manifested in this kind of data: the individual failing at time $t^*_{(n)}$ far remote from the others may come from a different population to the other individuals. There are various ways to test for this, and we will develop one of them in Chapter 5, a simple method based on comparing the largest uncensored observation to the second largest, when a parametric model is fitted to the data. Omission of the individual failing at around 10 years reduces substantially the evidence against sufficient follow-up for the remainder of the data, as we will see in Section 5.5.

3. Other examples We briefly set out here the results of testing H_{01} and H_{02} for some of the small data sets plotted in Chapter 1. In each case we will find, as seems reasonable from visual assessment of the figures in Chapter 1, that H_{01} is rejected but the test of follow-up is inconclusive so that we have insufficient evidence to be confident of the presence of immunes. Given the small sizes of the data sets, this seems unavoidable. The calculations for these data sets are very easy and the reader may like to check them as exercises.

(i) For the data in Figure 1.1 (those receiving the drug treatment in Table 1.1), we have $n = 21$, and

  (a) $\hat{p}_n = 0.5578$ lies below any tabulated 5% point with $n = 20$ (and below any 1% point except in one case) in Tables A.1 and A.2, regardless of the value of μ, so H_{01} is confidently rejected.

  (b) $q_n = 4/21 = 0.1905$ lies between the 10% and 90% points for $n = 20$, $p = 0.6$ and $B = 2, 4$ or 6 in Tables B.4 and B.5, so the test for H_{02} is inconclusive: there is no strong evidence either in support of or against H_{02}.

(ii) For the data in Figure 1.2:
Group 1 has $n = 42$; we find

  (a) $\hat{p}_n = 0.1019$, which is well below any tabulated 1% point with $n = 40$ in Tables A.1 and A.2, so H_{01} is strongly rejected.

(b) $q_n = 5/42 = 0.1190$ is between the 10% and 90% points for $n = 40$, $p = 0.2$ for any B in Table B.1, so the data is inconclusive for deciding H_{02}, according to our test statistic.

Group 2 has $n = 36$; we find

(a) $\hat{p}_n = 0.5004$ is below any tabulated 1% point with $n = 40$ in Tables A.1 and A.2 so H_{01} is rejected.

(b) $q_n = 2/36 = 0.0556$; for $p = 0.5$ and $n = 40$, 0.0556 is between the 10% and 90% points for $B = 2$ or 4, but close to the 10% point for $B = 6$ and well below the 10% point for $B = 8$, so there is some evidence against H_{02}, but it is not enough to conclude that follow-up is insufficient.

(iii) For the data in Figure 1.3, we have $n = 120$; we find

(a) $\hat{p}_n = 0.2417$ is well below any 1% point with $n = 100$ in Tables A.1 and A.2, so H_{01} is strongly rejected.

(b) $q_n = 9/120 = 0.0750$ is between the 10% and 90% points for $n = 100$, $p = 0.2$ and $B = 2$, 4 or 6 in Table B.1, so the result is inconclusive; we have no strong evidence either in support of or against H_{02}.

APPENDIX NOTES TO CHAPTER 4

1. Large-sample justification of the q_n-test

In Section 4.3, the use of q_n was motivated in the case $p < 1$. However, it turns out that even when $p = 1$, i.e. F is proper, large values of N_n or of q_n support H_{02}. This is proved in Theorem 4.5 under a technical condition which we state as follows. Define the increasing function

$$h(a) = \int_{(\tau-a,\tau]} [1 - G(t-)]\, \mathrm{d}F(t) \tag{4.33}$$

when $0 < a < \tau = \min\{\tau_{F_0}, \tau_G\} < \infty$. We say that h is of *dominated variation* (cf. Bingham *et al.* 1987, p. 54) as $a \downarrow 0$ if $h(a) > 0$ and for all $\lambda > 0$,

$$\limsup_{a\downarrow 0} \frac{h(\lambda a)}{h(a)} < \infty. \tag{4.34}$$

This is a rather mild regularity assumption on the behaviour of h near 0, as we discuss below. Standard methods as in Bingham *et al.* show that (4.34) is equivalent to the existence of $M > 0$ and $a_0 = a_0(M) > 0$ such that

$$h(2a) \leq Mh(a) \tag{4.35}$$

whenever $0 < a \leq a_0$.

Theorem 4.5 *Assume F is continuous on, and assigns positive mass to, intervals $(\tau - a, \tau)$ for all $a \in (0, a_0]$ for some $a_0 > 0$. Suppose also that $h(a)$ is of dominated variation as $a \downarrow 0$. Then if $N_n \to \infty$ in probability as $n \to \infty$, we must have $\tau_{F_0} < \tau_G$, thus H_{02} is true. Conversely, if $0 < p < 1$ and $\tau_{F_0} < \tau_G$, then $N_n \to \infty$ almost surely as $n \to \infty$.*

Proof Suppose $N_n \xrightarrow{P} \infty$, but by way of contradiction that $\tau_{F_0} \geq \tau_G$. Then we have

$$P\{N_n > 2b\} \to 1 \tag{4.36}$$

for each $b > 0$. Given a sequence $a_n \downarrow 0$, define events A_n by

$$A_n = \{\tau - a_n < t^*_{(n)}\}$$

where $\tau = \tau_{F_0} \wedge \tau_G = \tau_G$. The idea of the proof is to choose a_n so that A_n has relatively high probability; this is possible by (4.33) and (4.34), which will tell us that there is positive probability of observing uncensored observations arbitrarily close to τ. On A_n, however, N_n is bounded with high probability, as we will show, contradicting (4.36).

To carry out this program, fix b so large that

$$\frac{1}{b} + \mathrm{e}^{-b/M} < \frac{1}{2}$$

where M is as in (4.35). Then define $a_n = a_n(b)$ by

$$a_n = \inf\left\{a > 0 : h(2a) \geq \frac{b}{n}\right\},$$

where h is defined in (4.33). The continuity of F near τ implies that h is continuous near 0, so

$$h(2a_n) = \frac{b}{n}.$$

We have by (4.26) that

$$P\{A_n^c\} = P\{t^*_{(n)} \leq \tau - a_n\} = (1 - \mathrm{P}\{\tau - a_n < t_1^* \leq u_1\})^n$$

$$= \left\{1 - \int_{(\tau - a_n, \tau]} [1 - G(t-)]\, \mathrm{d}F(t)\right\}^n = [1 - h(a_n)]^n.$$

Thus if n is large enough we have by (4.35) that

$$P\{A_n^c\} \leq \left[1 - \frac{h(2a_n)}{M}\right]^n = \left[1 - \frac{b}{Mn}\right]^n \leq \mathrm{e}^{-b/M}, \tag{4.37}$$

so A_n has relatively high probability, as claimed. To obtain a bound for N_n on A_n, note that $t^*_{(n)} > \tau - a_n$ on A_n, while $t^*_{(n)} \leq t_{(n)} \leq \tau$ almost surely. Thus

$$(2t^*_{(n)} - t_{(n)}, t^*_{(n)}] \subseteq (\tau - 2a_n, \tau]$$

when A_n occurs. Define random variables

$$Y_{in} = I\{\tau - 2a_n < t_i^* \le \tau, t_i^* \le u_i\}$$

which are independent for $1 \le i \le n$. Since N_n is the number of uncensored t_i in $(2t_{(n)}^* - t_{(n)}, t_{(n)}^*]$, we have $N_n \le \sum_{i=1}^n Y_{in}$ on A_n. Since $\tau_G \le \tau_{F_0}$ we have $u_1 \le \tau$ almost surely and so

$$\mathrm{E}(Y_{in}) = P\{\tau - 2a_n < t_1^* \le u_1\} = h(2a_n) = \frac{b}{n}$$

and

$$\mathrm{Var}(Y_{in}) \le \mathrm{E}(Y_{in}^2) = \mathrm{E}(Y_{in}) = \frac{b}{n}.$$

It follows by Chebychev's inequality and (4.37) that

$$\begin{aligned} P\{N_n > 2b\} &\le P\{N_n > 2b, A_n\} + P\{A_n^c\} \le P\left\{\sum_{i=1}^n Y_{in} > 2b, A_n\right\} + \mathrm{e}^{-b/M} \\ &\le P\left\{\sum_{i=1}^n Y_{in} - b > b\right\} + \mathrm{e}^{-b/M} \le \frac{n\mathrm{Var}(Y_{in})}{b^2} + \mathrm{e}^{-b/M} \\ &\le \frac{1}{b} + \mathrm{e}^{-b/M} < \frac{1}{2}. \end{aligned}$$

This contradicts (4.36), as asserted.

Conversely suppose $\tau_{F_0} < \tau_G$, $0 < p < 1$, and F_0 attributes positive mass to intervals $(\tau_{F_0} - a, \tau_{F_0})$, for all $a > 0$, a small enough. Given $\varepsilon \in (0, \tau_G - \tau_{F_0})$, we can choose n so large that, for all but finitely many n, almost surely,

$$\tau_{F_0} - \varepsilon < t_{(n)}^* \le \tau_{F_0} < \tau_G - \varepsilon < t_{(n)} \le \tau_G. \tag{4.38}$$

This is possible because $t_{(n)} \to \tau_G$ almost surely even if $\tau_{F_0} < \tau_G$ when there are immunes in the population, while $t_{(n)}^* \to \tau_{F_0}$ almost surely by (4.28). When (4.38) holds

$$(2t_{(n)}^* - t_{(n)}, t_{(n)}] \supseteq [\tau - \eta + \varepsilon, \tau - \varepsilon]$$

where $\eta = \tau_G - \tau_{F_0} > 0$ and $\tau = \tau_{F_0} \wedge \tau_G = \tau_{F_0}$. Hence

$$N_n \ge \sum_{i=1}^n I\{\tau - \eta + \varepsilon < t_i^* \le \tau - \varepsilon, t_i^* \le u_i\} = \sum_{i=1}^n Z_i, \text{ say.}$$

Now the Z_i are i.i.d with

$$E(Z_1) = \int_{(\tau-\eta+\varepsilon, \tau_G]} [F((\tau - \varepsilon) \wedge u) - F(\tau - \eta + \varepsilon)]\,\mathrm{d}G(u)$$

$$\geq \int_{[(\tau+\varepsilon),\tau_G]} [F(\tau-\varepsilon) - F(\tau-\eta+\varepsilon)]\,\mathrm{d}G(u)$$

$$= [1 - G(\tau+\varepsilon)][F(\tau-\varepsilon) - F(\tau-\eta+\varepsilon)].$$

We can choose ε so that the last quantity is positive, since $\tau < \tau_G$ and F attributes positive mass to all intervals $(\tau - a, \tau)$. Thus $E(Z_1) > 0$. Now by the strong law of large numbers we have

$$\frac{1}{n}N_n \geq \frac{1}{n}\sum_{i=1}^{n} Z_i \to E(Z_1) > 0$$

almost surely as $n \to \infty$, which implies $N_n \to \infty$ almost surely. ■

2. Dominated variation of a function

Conditions related to the dominated variation of a function such as we specify in (4.34) are extensively studied in Bingham *et al.* (1987), for example. We will simply note here that (4.34) holds under very general circumstances, as follows.

(i) When $\tau_{F_0} < \tau_G$, we have $G(\tau) < 1$, hence (4.34) reduces to

$$\limsup_{a\downarrow 0} \frac{F(\tau) - F(\tau-\lambda a)}{F(\tau) - F(\tau - a)} < \infty. \tag{4.39}$$

This is true for example if for some $k > 0$, $F^{(k)}(\tau-) \neq 0$, as can be proved by Taylor expansion or L'Hôpital's rule.

(ii) The more important case is of course $\tau_{F_0} \geq \tau_G$. If $\tau_{F_0} \geq \tau_G$ but G has nonzero mass at τ_G, i.e. $G(\tau_G-) < 1$, then again (4.34) reduces to (4.39) and hence holds very widely. If $\tau_{F_0} \geq \tau_G$ and G is continuous at $\tau_G = \tau$, we can apply L'Hôpital's rule (assuming $F'(t) > 0$ in $(\tau - a, \tau)$ for some $a > 0$). Then (4.34) reduces to

$$\limsup_{a\downarrow 0} \frac{[1 - G(\tau-\lambda a)]F'(\tau-\lambda a)}{[1 - G(\tau - a)]F'(\tau - a)} < \infty, \tag{4.40}$$

which again holds very widely, e.g. if both F and G have nonzero left derivatives of some order at τ.

Conditions (4.39) and (4.40) mean that F and G have essentially only algebraic rates of change near τ, and thus hold for many commonly used survival distributions.

3. More about the q_n-test

The q_n-test that we suggest for sufficient follow-up in Section 4.3 was preceded, in Maller and Zhou (1994), by a related test which we called an α_n-test. In terms of the notation in Section 4.3, q_n, N_n and α_n are related by $q_n = (1 - N_n/n)^n = (1 - \alpha_n)^n$. From the point of view of hypothesis testing, q_n, N_n and α_n are equivalent since they are monotone functions of each other. We chose to tabulate the percentiles of q_n in Tables B.1–B.8 rather than those of N_n or α_n simply because the normalisation of N_n by n to give q_n produces numbers, between 0 and 1, of a convenient size. Now it was suggested in Maller and Zhou (1994) on the basis of not very good evidence that a reasonable procedure might be to take the level of significance, α, as an approximate critical point for the α_n-test, that is, rejecting H_{02}^c if $\hat{\alpha}_n < \alpha$. But in the light of the simulations reported in Section 4.3 we can easily see that such an approximation is not a good one. In fact, it turns out that the test proposed in Maller and Zhou (1994) is far too conservative. So for the present we propose the q_n-test for sufficient follow-up, referring to Tables B.1–B.8 for its approximate percentiles.

We do not expect q_n to be the last word, however. It is only a rough test which could be improved on. As a simple count of numbers of uncensored observations, it probably shares with some similar nonparametric tests a lack of power in detecting alternatives. The discrete nature of its distribution, clear from Tables B.1–B.8, leads to a lack of discrimination between the percentiles, and its bimodal nature is also unfortunate. Test statistics similar in nature to q_n which use information in the KME relevant to its levelling at its right extreme more effectively are probably preferable.

4. Nonparametric approaches to the two-sample problem

The following papers have specifically studied the two-sample problem for survival data with immunes.

Gray and Tsiatis (1989) assume a mixture model for survival times in two independent samples and derive an 'optimal linear rank test' for the hypothesis that the two immune proportions are the same. When cure rates are less than 50%, they show it to be substantially more efficient than the log-rank test, which is a well-known nonparametric test for differences in survival curves valid whether or not immunes are assumed to be present (Collett (1994, p. 40)). The statistic formed from the difference in the maximum values of the Kaplan–Meier estimators for the two samples is less efficient.

Laska and Meisner (1992) suggest a cure model in which the observations on survival times and censor indicators are augmented by the possibility of observing that an individual is in fact cured. This is taken to be so if the observed survival time exceeds some prespecified value. They assume a nonparametric mixture model for the survival times, show that the maximum likelihood estimate of the susceptible proportion is the value of the KME at its right extreme, and obtain the likelihood ratio statistic for testing nonparametrically a difference in susceptible proportions between two (or more) independent samples. They compare the behaviour of this statistic with that of the log-rank statistic and with the statistic proposed by Gray and Tsiatis (1989).

CHAPTER 5

Parametric Models for Single Samples

5.1 PARAMETRISING THE SURVIVAL DISTRIBUTION

In this chapter we will assume that a parametric model has been specified for the distribution F_0 of the lifetimes of the susceptible individuals. By this we mean that $F_0(t)$ is given as a function of t which also depends on the value of one or more *parameters*. This function may simply be 'guessed' on the basis of visual examination of the data, or suggested by a model for some other aspect of the data (such as a Poisson process model for the time-to-event process), or we may try a standard survival model just to see if it fits. There is a large variety of such models commonly used in survival analysis (e.g. Cox and Oakes 1984, Ch. 2), of which perhaps the simplest is the exponential distribution. We will couch much of our development in this chapter in terms of the family of Weibull distributions, of which the exponential is a special case. The Weibull formula is easy to write down, and it often turns out to describe reasonably well survival time distributions which are observed in practice. (We give some examples later; see also Collett (1994) for the Weibull in 'ordinary' survival analysis, and the discussion in Section 1.4).

There are theoretical reasons for the suitability of the Weibull as a distribution for survival times — see the appendix notes to this chapter — which we do not wish to emphasise too much, because there are also examples where the Weibull distribution does *not* provide a good description of the survival times (again, examples will be given later). In practice, we often begin by fitting a Weibull or exponential distribution to data (unless it is obviously contraindicated) as a reference distribution, or as a baseline against which other models can be compared. We could equally well use the gamma distribution, which is another generalisation of the exponential distribution, but the Weibull turns out to serve us fairly well.

The Weibull and exponential mixture models

Suppose then that the survival times of the susceptibles have the Weibull cumulative distribution function (c.d.f.)

$$F_0(t) = 1 - e^{-(\lambda t)^\alpha}, \qquad t \geq 0, \tag{5.1}$$

where λ and α are the associated *parameters* of the distribution, that is, constants involved in the distribution which are almost always unknown and to be estimated. We always have (and keep) $\alpha > 0$ and $\lambda > 0$. Under the independent censoring model introduced in Section 2.1, the c.d.f. F of the true survival times then has the parametric representation

$$F(t) = pF_0(t) = p(1 - e^{-(\lambda t)^\alpha}), \qquad t \geq 0. \tag{5.2}$$

This brings in another parameter, p, the susceptible proportion, multiplying $F_0(t)$.

What about the censoring distribution G? In some applications we may wish to assume a parametric form for G also, but for the most part we would like to estimate the parameters describing F *free from any assumptions whatsoever about* G. To a large extent we can do this, but we must also realise that the properties of the estimators obtained will depend in some ways on the form of G; in particular, on how 'heavy' are the tails of G compared with those of F. This corresponds to the 'heaviness' of the censoring, and thereby to how much information regarding the survival distribution is lost by the censoring.

We will discuss this further in Section 5.2, when we give some large-sample properties of maximum likelihood estimators of the parameters in (5.2). But the next step is to look at the likelihood method itself.

The likelihood method for survival data

When a parametric model such as (5.1) has been specified (or postulated) as the distribution from which a set of data has been drawn, we can estimate the parameters involved by the method of *maximum likelihood*.

The *likelihood function* of the sample is the joint probability density function (or the probability mass function, if the postulated distribution is discrete) of the random variables on which the data points constitute observations, evaluated at the data points themselves. The likelihood function, considered as a function of the unknown parameters, can be maximised over the parameter space of admissible parameter values to produce *maximum likelihood estimates* (MLEs) of the parameters. The rationale for this procedure is, briefly, that we wish to choose those parameter values for which the probability of observing the sample among the postulated family of distributions is the greatest. This produces estimates which are 'good' according to certain theoretical desiderata, and just as importantly, the procedure does seem to work well in many practical cases.

To write down the likelihood function for survival data of the type we are interested in, we have to work out the joint probability distribution of the *censored* failure times. Consider the case of Type II censoring (Section 2.1) in which n items are put on test at time 0 and observation ceases after a predetermined number $r < n$ of failures have been observed. As usual, let $t_1, t_2, \ldots, t_n$ be independent random variables representing the observed, possibly censored, lifetimes of individuals $1, 2, \ldots n$, and let

$$t_{(1)} \leq t_{(2)} \leq \cdots \leq t_{(n)} \tag{5.3}$$

be their order statistics. Supposing the survival c.d.f. of the t_i to be a continuous distribution (such as the Weibull or exponential) with density $f(t) = F'(t)$, the inequalities in (5.3) are in fact strict (almost surely), so what we observe are failure times $t_1, t_2, \ldots, t_n$ which can be ordered as

$$t_{(1)} < t_{(2)} < \cdots < t_{(r)} = t_{(r+1)} = \cdots = t_{(n)}. \tag{5.4}$$

Using arguments from the theory of order statistics (e.g. Lawless 1982; Reiss 1989) we see that the joint probability density function of the observed random variables $t_{(1)}, t_{(2)}, \ldots, t_{(r)}$ is

$$f(t_1, \ldots, t_r) = \frac{n!}{(r-1)!}\left(\prod_{i=1}^{r} f(t_i)\right)(1 - F(t_r))^{n-r}. \tag{5.5}$$

Here $t_1, t_2, \ldots t_r$ are real numbers satisfying $0 < t_1 < t_2 < \cdots < t_r$.

We can write equation (5.5) in terms of all the original (unordered) survival times $t_1, t_2, \ldots, t_n$ by making use of the observed censor indicators $c_1, c_2, \ldots c_n$. Recall that $c_i = 1$ if t_i is uncensored and $c_i = 0$ otherwise. For the simple Type II censoring we are currently considering, we have $c_i = 1$ if $t_i \leq t_{(r)}$, and $c_i = 0$ otherwise. From (5.5) we can then write for the likelihood function

$$L_n(t_1, t_2, \ldots, t_n) = k\left(\prod_{i=1}^{n} (f(t_i))^{c_i}(1 - F(t_i))^{1-c_i}\right) \tag{5.6}$$

where k is a constant factor not depending on t_i, c_i or on any unknown parameters. Note that in writing (5.6) we are indulging in the common abuse of using the same notation t_i for the random variable and its observed value.

A very similar expression results from Type I censoring, in which n items are followed from time 0 to a predetermined time $\Delta > 0$, by which time a random number, R say, have failed. In this case we observe uncensored failure times $t_{(1)} < \cdots < t_{(R)}$, with the remaining $n - R$ times censored at Δ. It is not hard to show that the joint density of $t_{(1)}, \ldots, t_{(R)}$, and R, is the same as is given by (5.5) but with the factor $(1 - F(t_r))^{n-r}$ replaced by $(1 - F(\Delta))^{n-r}$ and r replaced by R. Thus the likelihood again has the form (5.6) if we define $t_i = \Delta$ when $c_i = 0$, for the censored observations.

Equation (5.6) suggests what is so, that the general principle for writing down the likelihood of a sample, some of whose observations are censored (right-censored), is to multiply in a factor $f(t_i)$ for any uncensored observation t_i, where f is the density or mass function of the survival time, and to multiply in a factor of $1 - F(t_i)$ for any censored observation t_i. Here $1 - F(t_i)$ can be thought of as substituting the probability of observing a value larger than t_i, for the density of t_i, when t_i is censored.

Likelihoods for the 'ordinary' Weibull and exponential survival models

As an example, it is easy to write down the likelihood function for a sample of censored survival times $t_1, t_2, \ldots, t_n$ (with censor indicators $c_1, c_2, \ldots, c_n$) from a Weibull distribution *in which no immunes are present*. By differentiating (5.1) we get for the density of $F(t)$ the function

$$f(t) = \alpha\lambda^\alpha t^{\alpha-1} e^{-(\lambda t)^\alpha}, \qquad t > 0. \tag{5.7}$$

Thus we obtain from (5.6) that

$$\begin{aligned} L_n(t_1, t_2, \ldots, t_n, \lambda, \alpha) &= k \prod_{i=1}^{n} (\alpha\lambda^\alpha t_i^{\alpha-1} e^{-(\lambda t_i)^\alpha})^{c_i} (e^{-(\lambda t_i)^\alpha})^{1-c_i} \\ &= k \prod_{i=1}^{n} (\alpha\lambda^\alpha t_i^{\alpha-1})^{c_i} e^{-(\lambda t_i)^\alpha} \end{aligned} \tag{5.8}$$

on recalling, from (5.1), that $1 - F(t) = e^{-(\lambda t)^\alpha}$. We call (5.8) *the likelihood for the (ordinary) Weibull model for survival times*. The utility of the Weibull model for analysing survival data is well exemplified by Aitkin *et al.* (1989, pp. 280–287) and by Collett (1994, Ch. 4).

If we set $\alpha = 1$ in (5.8) we obtain the likelihood for the 'ordinary' exponential model in the form

$$L_n(t_1, t_2, \ldots t_n, \lambda) = k \prod_{i=1}^{n} \lambda^{c_i} e^{-\lambda t_i}. \tag{5.9}$$

Now define

$$n_u = c_1 + c_2 + \cdots + c_n, \tag{5.10}$$

which is the number of uncensored observations, and $\bar{t} = (t_1 + t_2 + \cdots + t_n)/n$, which is the sample mean of the t_i. Then (5.9) takes the simple form

$$L_n(t_1, t_2, \ldots t_n, \lambda) = k\lambda^{n_u} e^{-\lambda n \bar{t}}. \tag{5.11}$$

We will always assume that at least one observation is uncensored, so $n_u > 0$. Equation (5.11) is a relatively simple function of λ, and at this stage it is the dependence of L_n on λ which we wish to emphasise, so we will write (5.11) briefly as

$$L_n(\lambda) = k\lambda^{n_u} e^{-\lambda n \bar{t}}. \tag{5.12}$$

Likelihoods for the Weibull and exponential mixture models

For a second example, much closer to our current interests, consider the *Weibull survival model with immunes*. Since this can be obtained as a 'mixture' of the distributions of the survival times of susceptible and immune individuals, we will call it the *Weibull mixture model*.

In this case the survival time distribution is the three-parameter distribution given by (5.2), that is,

$$F(t) = p(1 - e^{-(\lambda t)^{\alpha}}), \qquad t \geq 0. \tag{5.13}$$

The parameter space, the range of possible values, is $(0, 1]$ for p, and $(0, \infty)$ for each of λ and α. Notice that we allow, as usual, the possibility that $p = 1$, corresponding to an absence of immunes in the population; in that case the mixture model reduces to the 'ordinary' Weibull model for survival used in the previous section. The density of $F(t)$ is given by differentiating (5.13) as

$$f(t) = p\alpha\lambda^{\alpha}t^{\alpha-1}e^{-(\lambda t)^{\alpha}},$$

so from (5.6) we have for the likelihood of $t_1, \ldots, t_n$ and $c_1, \ldots c_n$ the expression

$$L_n(t_1, \ldots, t_n, \lambda, p, \alpha) = k\prod_{i=1}^{n}(p\alpha\lambda^{\alpha}t_i^{\alpha-1}e^{-(\lambda t_i)^{\alpha}})^{c_i}(1 - p + pe^{-(\lambda t_i)^{\alpha}})^{1-c_i} \tag{5.14}$$

(where k is a constant).

Setting $\alpha = 1$ in (5.14) thus gives us the likelihood for the *exponential mixture model* in the form

$$L_n(t_1, t_2, \ldots, t_n, \lambda, p) = k\prod_{i=1}^{n}(p\lambda e^{-\lambda t_i})^{c_i}(1 - p + pe^{-\lambda t_i})^{1-c_i}. \tag{5.15}$$

This corresponds to the c.d.f.

$$F(t) = p(1 - e^{-\lambda t}), \qquad t \geq 0,$$

obtained by setting $\alpha = 1$ in (5.13). When, in addition, $p = 1$, (5.15) reduces to the likelihood for the 'ordinary' exponential model (5.9), but when $p < 1$ no further simplification is possible.

Example: rearrest data

A large data set collated by the Crime Research Centre of the University of Western Australia contains the entire arrest records up till June 1993 of all West Australians who were arrested once or more after April 1975; see Broadhurst and Loh (1995) for a further description of the data. (Follow-up has since been extended to June 1995.) As for reimprisonment, the probability of an individual's rearrest following the first arrest strongly depends, in the Western Australian jurisdiction at least, on various characteristics of the individuals, and particularly on their sex and race (Aboriginal or non-Aboriginal).

To illustrate a Weibull mixture model fit to a subset of these data, we show in Figure 5.1 the Kaplan–Meier estimator of the times to first rearrest for the male non-Aborigines in this data set, a total of 97 537 men.

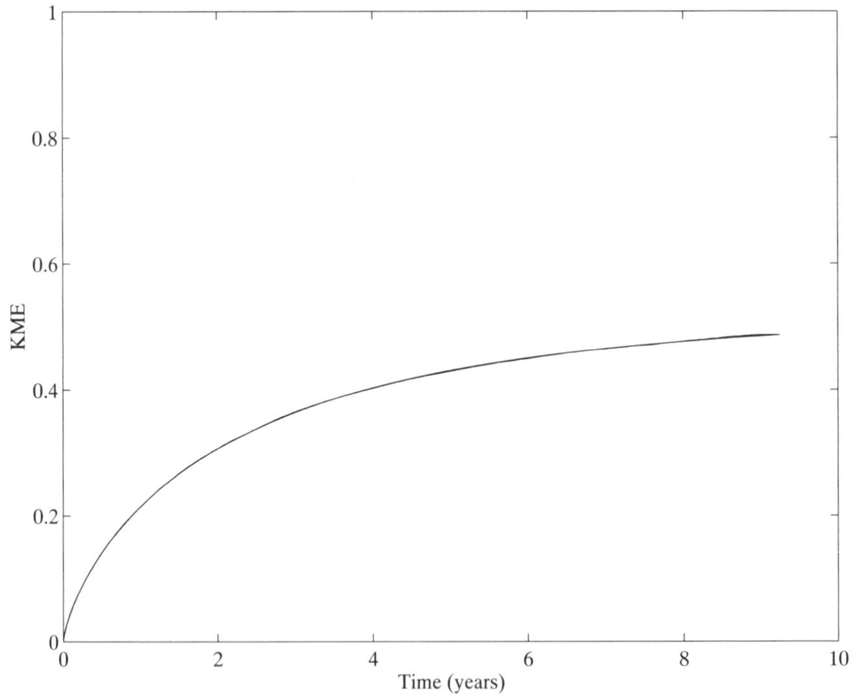

Figure 5.1 Rearrest data, with fitted Weibull mixture model.

A Weibull mixture model fitted to the data is also plotted in Figure 5.1, but it fits so well that it cannot be visually distinguished from the KME, which for such a large data set is practically continuous; see Broadhurst and Loh (1995) for other similar graphs.

The Weibull parameters for the fitted model are

$$\tilde{p} = 0.5180, \qquad \tilde{\lambda} = 0.4319, \qquad \tilde{\alpha} = 0.7470.$$

Since the shape parameter α is close to 1 (and less than 1) the data is exponential-like in appearance, as can be seen from Figure 5.1.

5.2 ESTIMATING PARAMETERS AND TESTING HYPOTHESES

The likelihood function L_n can be used to estimate parameters, as in the above example, and also to test hypotheses about them. Suppose that a general likelihood contains parameters which we will denote generically by the symbol $\boldsymbol{\theta}$. $\boldsymbol{\theta}$ may be a single parameter or more generally a vector of parameters. The range of permissible values of $\boldsymbol{\theta}$ is the *parameter space* of interest. For example, in the ordinary Weibull model with no immunes (5.1), the parameter vector is $\boldsymbol{\theta} = (\lambda, \alpha)$, with parameter space $(0, \infty) \times (0, \infty)$. Allowing for a possible proportion

$1 - p$ of immunes produces the three-parameter Weibull mixture model given in (5.2), with parameter vector $\boldsymbol{\theta} = (p, \lambda, \alpha)$, a three-dimensional parameter ranging over $(0, 1] \times (0, \infty) \times (0, \infty)$.

Maximum likelihood estimates

Having decided on, hypothesised or postulated a parametric form for the survival distribution, and written down the likelihood function of the sample as above, we estimate the parameters by maximising the likelihood over the permissible range of parameter values (usually the whole parameter space). We can, at least in principle, always maximise the likelihood $L_n(\theta)$ over the parameter space and obtain an MLE $\tilde{\theta}_n$, based on a sample size n. (For the time being we will not emphasis the dependence of $\tilde{\theta}_n$ on n, and simply write it as $\tilde{\theta}$.) In some cases there are *unique* values of the parameters which maximise the likelihood, and we then speak of 'the' *maximum likelihood estimates* (MLEs) of the parameters. The likelihood may even be concave over the parameter space, certainly ensuring the existence of unique MLEs.

For computational reasons it is easiest to deal with the *log-likelihood*, since this turns the product, which is always present in a likelihood formed from independent observations, into a sum. We will use the notation $l_n(\cdot)$ for log-likelihoods. As a simple illustrative example, consider the likelihood for censored exponential observations (without an immune component present) written in (5.12). This gives for the corresponding log-likelihood the formula (omitting the constant $\log k$)

$$l_n(\lambda) = \log L_n(\lambda) = n_u \log \lambda - \lambda n \bar{t} \qquad (\lambda > 0).$$

We can maximise this by some elementary calculus. We obtain by differentiation

$$\frac{\mathrm{d} l_n(\lambda)}{\mathrm{d}\lambda} = \frac{n_u}{\lambda} - n\bar{t} \tag{5.16}$$

and

$$\frac{\mathrm{d}^2 l_n(\lambda)}{\mathrm{d}\lambda^2} = -\frac{n_u}{\lambda^2}. \tag{5.17}$$

For this example the parameter space is $(0, \infty)$. (5.17) shows that $l_n(\lambda)$ (hence $L_n(\lambda)$) is a concave function of λ on $(0, \infty)$ (second derivative negative) and (5.16) then shows that a *unique, explicitly obtained* maximum $\tilde{\lambda}_n$ is given by

$$\tilde{\lambda}_n = \frac{n_u}{n\bar{t}} = \frac{c_1 + \cdots + c_n}{t_1 + \cdots + t_n}. \tag{5.18}$$

We call $\tilde{\lambda}_n$ the (unique) maximum likelihood estimator (MLE) of λ in the ordinary exponential survival model. Note that $\tilde{\lambda}_n > 0$ since we assumed $n_u > 0$, i.e. not all observations are censored.

The above situation with a unique, explicitly obtained MLE is the simplest that can be hoped for. Next best is that *unique* parameter estimate(s) exist and can be obtained, not explicitly as in (5.18), but numerically, by some numerical technique such as the Newton–Raphson procedure. In the Newton–Raphson procedure we attempt to find a zero of the first derivative of the log-likelihood; other techniques (such as the simplex method) try to maximise the likelihood or log-likelihood more directly. The EM technique, which is particularly well adapted to mixture distributions, can also be used to find the maximum of the log-likelihood.

There is always *some* maximum of the likelihood, so an MLE always exists in a broad sense, but it may be infinite in one or more components, and there are many situations in which it is not even known whether unique MLEs exist (or it is known that unique estimates do *not*), or where (unique) MLEs can only be shown to exist with high probability in large samples. This is the case with the exponential mixture model with immunes (5.15), which we discuss further in Section 7.3. The simple exponential example above produces an MLE *interior* to the parameter space ($(0, \infty)$ in that case) but the likelihood more generally may be maximised on a boundary, and indeed this can occur for the exponential or Weibull mixture models with immunes, since a 'boundary' maximum value of the log-likelihood corresponding to a value of $\tilde{p}_n = 1$ may well be found. This situation is not pathological—in fact it is of interest to us since we will take it as indicative of an absence of immunes in the data if it occurs, and we will wish to assess the amount of confidence we can attach to such a conclusion.

In order to answer such specific questions, we need also to consider the testing of hypotheses related to the parameters, as we now do.

Likelihood ratio tests and the deviance

Suppose we wish to test a hypothesis H_0 which specifies that some or all components of the parameter vector $\boldsymbol{\theta}$ take on particular fixed values. For example, if $\alpha = 1$ the Weibull mixture model reduces to the exponential, or if $p = 1$ the Weibull mixture model reduces to the ordinary Weibull model for survival. This corresponds to restricting the parameters to a subset of the whole parameter space. The likelihood method tells us to maximise the likelihood under the hypothesis H_0, that is, over the subspace of the parameters corresponding to H_0, and produce another estimate $\tilde{\boldsymbol{\theta}}_{H_0}$, say. Then form the ratio

$$\frac{L_n(\tilde{\boldsymbol{\theta}}_{H_0})}{L_n(\tilde{\boldsymbol{\theta}}_n)} \tag{5.19}$$

so as to compare the value of the likelihood at the hypothesised parameter value with the maximum value the likelihood can take; of course this ratio does not exceed 1. The philosophy behind the likelihood ratio (LR) test is that if H_0 is plausible, the ratio in (5.19) should not be too far below 1.

Conversely, values of the ratio much smaller than 1 suggest that H_0 is implausible. Actually it turns out to be convenient to take logs again. We denote by $-2l_n(\tilde{\boldsymbol{\theta}}_{H_0})$ the *deviance* of the model fitted under H_0, and let

$$d_n = 2(l_n(\tilde{\theta}_n) - l_n(\tilde{\theta}_{H_0})) \qquad (5.20)$$

be the *deviance difference* statistic; of course d_n is nonnegative. Values of d_n close to 0 suggest that H_0 is plausible, whereas large values of d_n suggest that H_0 is implausible. But how large is large enough? In order to assess this, and thus whether or not the hypothesised parameter values are plausible in view of the data, we need to know the distribution of d_n, or at least, the higher percentiles of this distribution. We will reject H_0 at the 5% significance level, for example, if the observed value of d_n is larger than the 95th percentile of its distribution under H_0. Unfortunately it is seldom the case that we can write down the distribution of d_n explicitly, or even tabulate its percentiles, and so we resort to large sample approximations and/or simulations. These, along with a discussion of the large-sample properties of the MLEs themselves, are given in Chapters 7 and 8. But next we describe a numerical method for finding the MLEs of the parameters in a Weibull mixture model, and assessing their precision.

Finding the MLEs with the Newton–Raphson procedure

As described in Section 5.1, the Weibull mixture model is fitted to a sample of survival times $t_1, t_2, \ldots, t_n$, with associated censor indicators $c_1, c_2, \ldots, c_n$, by maximising the likelihood function (5.14) for variations in the three parameters p, λ and α. The resulting MLEs cannot be displayed in explicit formulae. Instead, we have to use some iterative or searching procedure to find them numerically. There are various methods of doing this; we will not describe them in detail (some notes are given in Appendix Note 4) but a program written in Fortran for our own use employs the Newton–Raphson technique, and it works extremely well to fit the Weibull or exponential mixture model to the data sets described in this book as well as to many other data sets we have analysed.

We will merely briefly note here that Newton–Raphson works as follows. Suppose we 'guess' an initial value $\boldsymbol{\theta}_1$ of the parameter vector $\boldsymbol{\theta}$ (= (p, λ, α) here) which maximises the likelihood, or equivalently, the log-likelihood, $l_n(\boldsymbol{\theta})$. We then calculate an 'updated' version of $\boldsymbol{\theta}$, which we will call $\boldsymbol{\theta}_2$, from the equation

$$\boldsymbol{\theta}_2 = \boldsymbol{\theta}_1 + F_n^{-1}(\boldsymbol{\theta}_1) S_n(\boldsymbol{\theta}_1). \qquad (5.21)$$

Here $S_n(\boldsymbol{\theta}_1)$ is the vector of first derivatives of $l_n(\boldsymbol{\theta})$ with respect to the parameters p, λ, and α:

$$S_n(\boldsymbol{\theta}) = \frac{\partial l_n(\boldsymbol{\theta})}{\partial \boldsymbol{\theta}} = \begin{bmatrix} \dfrac{\partial\, l_n(\theta)}{\partial p} \\ \dfrac{\partial l_n(\theta)}{\partial \lambda} \\ \dfrac{\partial l_n(\theta)}{\partial\, \alpha} \end{bmatrix} \qquad (5.22)$$

and $F_n(\theta)$ is minus the second derivative matrix of $l_n(\theta)$ with respect to p, λ and α:

$$F_n(\boldsymbol{\theta}) = -\frac{\partial^2 l_n(\theta)}{\partial \boldsymbol{\theta}^2} = -\begin{bmatrix} \frac{\partial^2 l_n(\theta)}{\partial p^2} & \frac{\partial^2 l_n(\theta)}{\partial p \partial \lambda} & \frac{\partial^2 l_n(\theta)}{\partial p \partial \alpha} \\ \frac{\partial^2 l_n(\theta)}{\partial \lambda \partial p} & \frac{\partial^2 l_n(\theta)}{\partial \lambda^2} & \frac{\partial^2 l_n(\theta)}{\partial \lambda \partial \alpha} \\ \frac{\partial^2 l_n(\theta)}{\partial \alpha \partial p} & \frac{\partial^2 l_n(\theta)}{\partial \alpha \partial \lambda} & \frac{\partial^2 l_n(\theta)}{\partial \alpha^2} \end{bmatrix}. \tag{5.23}$$

In (5.21), $F_n^{-1}(\theta_1)$ denotes the inverse matrix of $F_n(\theta_1)$, assumed to exist.

For computational reasons (and also for theoretical reasons; see the large-sample analyses of Chapters 7 and 8) it is better to reparametrise p, λ and α as

$$\beta = \log\left(\frac{p}{1-p}\right), \qquad \delta = \log(\lambda) \text{ and } \gamma = \log(\alpha). \tag{5.24}$$

Here β, δ and γ are new parameters that are equivalent to the original parameters in the sense that the old parameters can be obtained from the new by the one-to-one inverse transformations

$$p = \frac{e^\beta}{1+e^\beta}, \qquad \lambda = e^\delta \text{ and } \alpha = e^\gamma. \tag{5.25}$$

We then take as $\boldsymbol{\theta}$ the 3-vector (β, δ, γ) and write (5.22) and (5.23) in terms of (β, δ, γ) rather than (p, λ, α). The transformation (5.24) allows β, δ and γ to vary unrestrictedly in $(-\infty, \infty)$, and is more convenient, computationally, for this reason.

The Newton–Raphson technique proceeds, then, by calculating a new value $\theta_2 = (\beta_2, \delta_2, \gamma_2)$ from the old, $\boldsymbol{\theta}_1 = (\beta_1, \delta_1, \gamma_1)$, by means of (5.21). We then substitute $\boldsymbol{\theta}_2$ for $\boldsymbol{\theta}_1$ in the right-hand side of (5.21) and repeat the procedure to get still another value $\boldsymbol{\theta}_3$. Repeating the procedure N times, we find that the $(N+1)$th iterate $\boldsymbol{\theta}_{N+1}$ is given by

$$\boldsymbol{\theta}_{N+1} = \boldsymbol{\theta}_N + F_n^{-1}(\boldsymbol{\theta}_N) S_n(\boldsymbol{\theta}_N). \tag{5.26}$$

If the sequence of iterates $\boldsymbol{\theta}_N$ converges to a value, $\tilde{\boldsymbol{\theta}}$ say, then $\tilde{\boldsymbol{\theta}}$ satisfies

$$\tilde{\boldsymbol{\theta}} = \tilde{\boldsymbol{\theta}} + F_n^{-1}(\tilde{\boldsymbol{\theta}}) S_n(\tilde{\boldsymbol{\theta}}) \tag{5.27}$$

(assuming continuity of $F_n^{-1}(\boldsymbol{\theta})$ in $\boldsymbol{\theta}$, which holds in our case). Equation (5.27) implies that

$$S_n(\tilde{\boldsymbol{\theta}}) = 0, \tag{5.28}$$

and since $S_n(\boldsymbol{\theta})$ is the first derivative vector of the log-likelihood $l_n(\boldsymbol{\theta})$, (5.28) shows that $\tilde{\boldsymbol{\theta}}$ is a stationary point of $l_n(\boldsymbol{\theta})$ and, with any luck, a maximum.

A number of things can go wrong with the above procedure, in general. Note that we assume the matrix of second derivatives $F_n(\boldsymbol{\theta})$ has an inverse at each of the values $\boldsymbol{\theta}_N$ and at the stationary point $\tilde{\boldsymbol{\theta}}$. If $F_n(\boldsymbol{\theta})$ has no inverse at a point $\boldsymbol{\theta}$, i.e. it is singular at that point, the likelihood surface is not concave near the stationary point, and we expect to have difficulties. Remember also that the parameter p is restricted to vary in the interval $(0, 1]$. It may be the case that the likelihood is maximised on the boundary $p = 1$. This is indicative of data which has a small, or zero, proportion of immunes, and is not an anomalous situation. In terms of the reparametrisation (5.24) by which p is transformed to $\beta = \log(p/(1-p))$, a 'boundary maximum' of the likelihood is signalled by the Newton–Raphson method moving the estimate of β closer and closer to $+\infty$ with each iteration. This is easily detected in a program such as the Fortran program mentioned above.

The Newton–Raphson procedure requires for its implementation the analytical calculation of the derivatives of $l_n(\boldsymbol{\theta})$ up to the second order, so as to be able to find $S_n(\boldsymbol{\theta})$ and $F_n(\boldsymbol{\theta})$ in a computer program. These derivatives may be tedious to calculate, but a benefit is that the matrix $F_n^{-1}(\tilde{\boldsymbol{\theta}})$, that is, the inverse of minus the second derivative matrix of $l_n(\theta)$, evaluated at the MLE $\tilde{\boldsymbol{\theta}}$, is a consistent estimate of the variance–covariance matrix of $\tilde{\boldsymbol{\theta}}$, under some conditions which we will specify later; see for example Theorems 8.2 and 8.3. If the Newton–Raphson technique is not used to maximise the likelihood, or at least if some estimate of the information matrix is not made, some other method of assessing the precision of estimation of $\tilde{\boldsymbol{\theta}}$ is needed—we will not consider this problem.

In a similar way to that described above, we can fit the exponential mixture model by maximising the likelihood (5.15), or its logarithm, with respect to the parameters p and λ. This time the derivative vector $S_n(\boldsymbol{\theta})$ of the log-likelihood is only two-dimensional, and minus the second derivative matrix, $F_n(\boldsymbol{\theta})$, of the log-likelihood is a 2×2 matrix—we simply omit the terms involving α in (5.22) and (5.23)—but otherwise the iterative procedure specified by (5.26) is exactly the same. Once again, all going well, the iterative procedure produces an MLE $\tilde{\boldsymbol{\theta}} = (\tilde{p}, \tilde{\lambda})$ which maximises the likelihood. This may again be a boundary estimate, in that we may have $\tilde{p} = 1$. The precision of the parameter estimates $\tilde{p}$ and $\tilde{\lambda}$ can be obtained from their estimated variance–covariance matrix, which is $F_n^{-1}(\tilde{\boldsymbol{\theta}})$, just as before.

Example: Kersey et al. leukaemia data—group 1 (allogeneic transplants)

It may be helpful to illustrate some of the above ideas on a data set. We will demonstrate the fitting of Weibull and exponential mixture models to Group 1 (allogeneic transplants) of the leukaemia data previously described in Section 4.3. Then to illustrate a hypothesis test of interest on this data, we will compare the Weibull and exponential mixture model fits by testing the hypothesis that the Weibull shape parameter α equals 1.

Using the program described above, we find that the best Weibull fit to the allogeneic transplant data, according to the likelihood method, has parameter estimates

$$\tilde{p} = 0.7311, \quad \tilde{\lambda} = 1.4518 \text{ and } \tilde{\alpha} = 0.9452. \tag{5.29}$$

The Weibull curve with these parameters is shown, together with the Kaplan–Meier estimator calculated from the data, in Figure 5.2. Note that in

(5.29) and elsewhere, unless otherwise stated, survival times will be measured in years. For transformation to other time-scales, the parameter λ should be rescaled accordingly. The log-likelihood changes by an additive constant if the data is rescaled, which is inessential as far as deviance differences are concerned. Constant multiples of the likelihood are ignored, so the factor k in (5.14) and (5.15) is taken as 1.

Visually, the fit in Figure 5.2 seems good. (Tests for the goodness of fit of the exponential mixture model to data are considered in Section 5.4.) We note from Figure 5.2 that the fitted Weibull is quite exponential in appearance; this is also suggested by the closeness of $\tilde{\alpha}$ to 1 in (5.29), because when its shape parameter α equals 1, the Weibull distribution reduces to the exponential distribution. So a reasonable simplification will occur if we can demonstrate that the data is just as well fitted, to within its intrinsic accuracy, by an exponential distribution as by a Weibull. In other words, we wish to test the hypothesis $H_0 : \alpha = 1$ that the Weibull reduces to its submodel, the exponential.

To do this we use a likelihood ratio test, or more conveniently, the deviance test. The value of the deviance for the Weibull fit with parameters given by (5.29) is obtained from the computer program as $-2 \log \tilde{L} = 92.30$. Fitting an

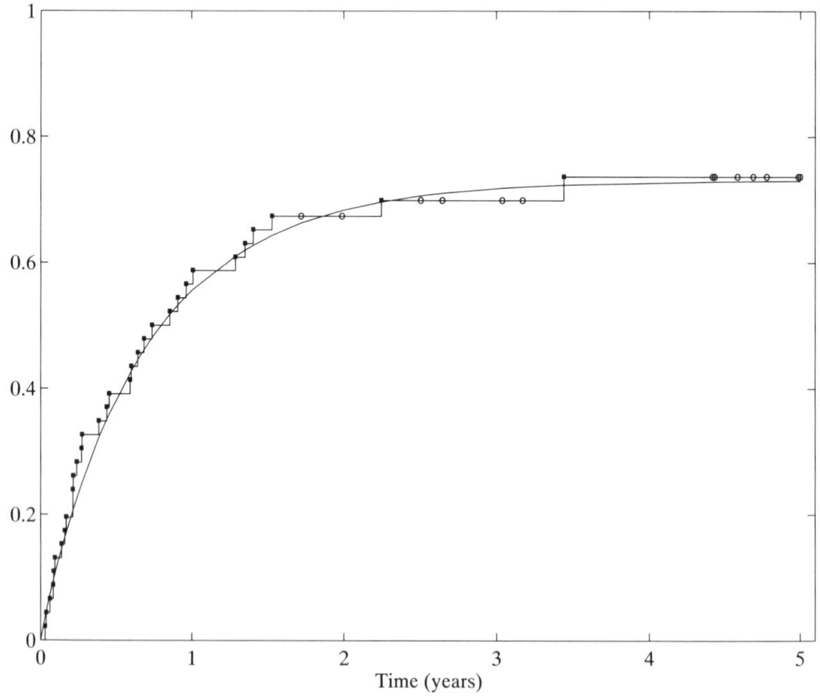

Figure 5.2 Leukaemia data, Group 1 (allogeneic transplants) with fitted Weibull mixture model; the fit appears good. (Compare with Figure 5.3.)

exponential mixture model to the data by the Newton–Raphson method described previously (in our case this is done by the same computer program) produces the following estimates:

$$\breve{p} = 0.7289 \text{ and } \breve{\lambda} = 1.4332. \tag{5.30}$$

These are little different to the corresponding estimates in (5.29). The deviance associated with this exponential mixture model fit is $-2\log\breve{L} = 92.46$. To test H_0 we calculate the deviance difference

$$d_n = (-2\log\breve{L}) - (-2\log\tilde{L}) = 92.46 - 92.30 = 0.16. \tag{5.31}$$

The shape parameter, α, may take all positive values, $\alpha > 0$, so $\alpha = 1$ is interior to the permissible parameter space for α. We will suggest in Section 5.3 that deviance differences for hypothesis tests such as this, where parameters are interior to the parameter space, have in large samples approximately a chi-square distribution with degrees of freedom equal to the difference in the number of parameters fitted under the two models. In this case, we are comparing the three-parameter Weibull mixture model with the two-parameter exponential mixture model, so d_n in (5.31) is taken as, approximately, an observation on a chi-square random variable with one degree of freedom (χ_1^2). The 95th percentile of χ_1^2, obtained from standard tables of the chi-square distribution is 3.84, and the observed deviance difference, 0.16, is substantially smaller than this. So we accept H_0, and conclude that the exponential mixture model produces a fit not significantly worse than the Weibull mixture model. Or to put it another way, the extra α shape parameter of the Weibull is not required for a description of the data which is adequate at the 95% confidence level; we might as well take it as being equal to 1, to within the order of accuracy of the data.

It would be sensible at this stage to plot the fitted exponential distribution along with the Weibull on Figure 5.2, to compare them visually, but there is little point because the fitted curves are not distinguishable on the graph to the order of accuracy of the plot. Note that, although we have so far decided that the exponential mixture model is *not significantly worse* than the Weibull, we have not yet assessed the *adequacy of either* as a descriptor of the distribution of the data. This involves a goodness of fit test for the Weibull or exponential which we shall discuss in Section 5.4.

5.3 TESTING PARAMETRICALLY FOR THE PRESENCE OF IMMUNES

In Section 4.2 we discussed a *nonparametric* test for the hypothesis H_{01}, that there are no immunes present in the population. As we pointed out there, standard tests may not behave as we expect, at least in respect to their distributional properties, since we are testing a hypothesis which specifies that a parameter is located at a boundary of the parameter space. Certainly 'standard' asymptotic theory does not apply in this situation.

The same is true under the *parametric* setups of Section 5.2, but in some parametric cases, in particular for the exponential distribution, we are able to work out the large-sample distribution of the likelihood ratio test for the presence or absence of immunes. This discussion of the large-sample properties of the test we will use is deferred to Chapter 7. For now we will just formulate the test, and illustrate its use on some data sets, anticipating the large-sample properties of Chapter 7.

Problems associated with testing at the boundary of the parameter space only appear when we have to calculate the (large-sample) *distribution* of the test statistic. The procedure itself is quite straightforward. Assume the independent censoring model and, as usual, write the distribution of survival times $F(t)$ as

$$F(t) = pF_0(t), \tag{5.32}$$

where $F_0(t)$ is the distribution of the survival times of the susceptibles. The proportion of immunes in the population is $1 - p$, so a test of

$$H_0 : p = 1 \tag{5.33}$$

is a test for the presence of immunes. When a parametric model such as the Weibull or exponential is assumed for $F_0(t)$, fitted to the data, and used as the basis for a likelihood ratio or deviance test, we will call this a *parametric test for the presence of immunes*. We will exemplify it in this section for survival data which is well described by an exponential distribution, but of course the method can be applied in general—provided we are confident that a fitted distribution is a good model for the data, and that the (large-sample) theory applies.

Suppose now that

$$F_0(t) = 1 - e^{-\lambda_0 t}, \qquad t \geq 0,$$

and the model (5.32) has been fitted to the data by the method of maximum likelihood as described above. This produces MLEs $\tilde{p}$ and $\tilde{\lambda}$, say, and a value $-2\log\tilde{L}_n$ of the deviance for the maximised likelihood. There is, as always, a possibility that the likelihood is maximised in p for $\tilde{p} = 1$, i.e. a boundary estimate of p may occur. As we mentioned in Section 5.1, the program used to fit the model must allow for this to happen. If it does indeed occur, so be it: our MLE in this case is taken as $\tilde{\boldsymbol{\theta}} = (\tilde{\lambda}, 1)$, or in general, we write it as $\tilde{\boldsymbol{\theta}} = (\tilde{\lambda}, \tilde{p})$. If a boundary estimate does indeed occur, we will show (in Section 7.3) that the corresponding MLE of λ is then precisely the estimate we would obtain by fitting the ordinary exponential model directly to the data.

Under the hypothesis H_0 in (5.33), we take $p = 1$, in which case the exponential mixture model simply reduces to the ordinary exponential model. Maximising the likelihood this time results in the explicit estimate given by (5.18), which we write in the present context as

$$\tilde{\lambda}_{H_0} = \frac{n_u}{n\bar{t}}. \tag{5.34}$$

So we write $\tilde{\theta}_{H_0} = (\tilde{\lambda}_{H_0}, 1)$ and form the deviance statistic

$$d_n = 2(l_n(\tilde{\theta}_n) - l_n(\tilde{\theta}_{H_0})). \tag{5.35}$$

Testing for 'large' values of d_n is equivalent to the likelihood ratio test of H_0. But how large is large? As we discussed above, since we are testing a boundary hypothesis, the large-sample distribution of d_n under H_0 is not a chi-square distribution, as it usually is when testing an interior hypothesis.

Note that, if a boundary estimate of p occurs when calculating $\tilde{\theta}_n$, we have $d_n = 0$ identically because in this case $\tilde{\lambda}$ is equal to $\tilde{\lambda}_{H_0}$ as given by (5.34). It turns out that, asymptotically, $d_n = 0$ will occur in about 50% of samples when H_0 is true. In approximately the other 50% of the samples, an interior maximum $\tilde{p}_n < 1$ of the likelihood will occur. More precisely, the large-sample distribution of d_n is, under reasonable conditions on the censoring, a 50–50 mixture of a chi-square random variable with one degree of freedom and a point mass at 0. This result is discussed in more detail in Section 7.3. For now, let us assume it to be so and demonstrate the test on some data.

Example: Kersey et al. leukaemia data — group 1 (allogeneic transplants)

The KMEs of both groups in this data set are shown in Figure 4.3, and in Section 5.2 we demonstrated that the allogeneic group appears to be well fitted by an exponential mixture model (Figure 5.2). The parameter estimates for the exponential mixture model fitted to this group are

$$\tilde{p} = 0.7289 \text{ and } \tilde{\lambda} = 1.4332, \tag{5.36}$$

as in (5.30), and the deviance for the maximal likelihood is 92.46. To fit the ordinary exponential model we take $\tilde{p}_{H_0} = 1$ and calculate $\tilde{\lambda}_{H_0}$ from (5.34), resulting in

$$\tilde{\lambda}_{H_0} = 0.4728. \tag{5.37}$$

The pair $(\tilde{\lambda}, \tilde{p})$ certainly appears different from the pair $(\tilde{\lambda}_{H_0}, 1)$, but let us test this by calculating the value of d_n, as specified by (5.35). For this data d_n has value

$$d_n = 115.43 - 92.46 = 22.97.$$

Here 115.43 is the value of $-2\log \tilde{L}_{H_0}$, the maximised likelihood for the fitted ordinary exponential model. This may be obtained directly from formula (5.9) (with $k = 1$). Now we ask: is d_n too large plausibly to be an observation on a random variable which is a 50–50 mixture of a chi-square random variable with one degree of freedom, and a point mass at 0? The 95th percentile $c_{0.95}$ of the distribution of such a random variable is given by

$$\tfrac{1}{2} + \tfrac{1}{2}P(\chi_1^2 \le c_{0.95}) = 0.95. \tag{5.38}$$

So $c_{0.95}$ satisfies $P(\chi_1^2 \le c_{0.95}) = 0.9$. From standard tables of χ_1^2 we then obtain

$$c_{0.95} = 2.71. \tag{5.39}$$

Since $22.97 > 2.71$ we reject H_0 at the 5% significance level and consider that there is strong evidence of $p < 1$ for this data.

In Figure 5.3 we show the ordinary exponential fit plotted along with the KME for this data. The figure leaves little doubt as to the poor quality of the fit of the model without immunes. So we will certainly include the parameter p in our model of this data. Recall that in Section 4.2 a nonparametric test for the presence of immunes in this data set also rejected the null hypothesis of no immunes with high confidence. The q_n test for sufficient follow-up in Section 4.3 did not support the hypothesis that the KME for the data has levelled sufficiently to conclude the presence of immunes, but that nonparametric test was strongly influenced by the single uncensored observation at around 3.2 years. If we are prepared to assume an exponential mixture model for the data, as appears reasonable from Figure 5.2, then the more powerful *parametric* test of $H_{01} : p = 1$ strongly suggests the presence of immunes.

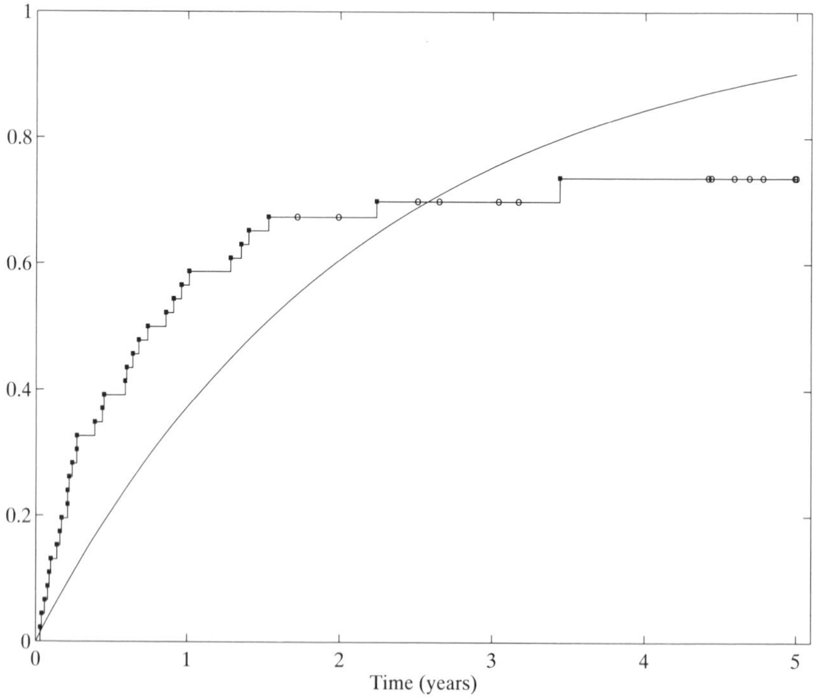

Figure 5.3 Leukaemia data, Group 1 (allogeneic transplants) with fitted ordinary exponential model; a poor fit. (Compare with Figure 5.2.)

Example: recidivism data

Now we illustrate our parametric test for the presence or absence of immunes on some recidivist data. This set consists of the recidivism times (time to first return to prison after first release) of 62 Western Australian prisoners. The data forms part of the Western Australian Crime Research Centre database which was discussed in Section 4.3. The current small subgroup consists of male non-Aboriginal prisoners who were released without parole and were unmarried at the commencement of their prison term; it is also a subset of a larger subgroup of 216 prisoners which will be analysed in Section 6.4 for the effects of parole and marital status on recidivism.

For the present group of 62, we fit both the Weibull mixture model and the ordinary Weibull model by the method of Section 5.2. The following maximum likelihood estimates were obtained for the mixture model:

$$\tilde{p} = 0.84,\ \tilde{\lambda} = 0.036 \text{ and } \tilde{\alpha} = 0.79.$$

The minimised value of $-2\log L$ was 310.65, and a 95% confidence interval for p, calculated by the method described in Section 5.2, was (0.12, 0.91). (See also Table 6.11 and the note in Section 6.4 to the effect that, for criminology data such as we are currently considering, it is common to measure time in 'months' (30-day periods), a scale change which only affects the values of λ and the value of the likelihood of the fitted model in simple ways.)

We see from the above estimates that the value of p, the probability of ultimate failure, or equivalently, that of $1 - p$, the immune proportion, is very poorly determined for this data. Furthermore, it seems that p is quite likely to take a value as large as 0.91, very close to its boundary value of 1. This suggests that we should not be too confident of the presence of immunes, although the fact that the confidence interval does not quite include 1 tends to suggest that they may be present.

In any event, let us test for this formally by the method outlined above. The MLEs for an ordinary Weibull fit to the data were

$$\hat{\lambda} = 0.0237 \text{ and } \hat{\alpha} = 0.74,$$

and the minimised value of $-2\log L$, the deviance, was 310.86. (Of course there is no estimated value of p for this model!) Comparing the deviance value with the corresponding value of 310.65 for the Weibull mixture model gives a deviance difference of $310.86 - 310.65 = 0.32$, which is far smaller than the value of 2.71 needed for significance of this statistic; see (5.39) and recall the boundary nature of the hypothesis test. In other words, the formal test provides no evidence of immunes.

Figure 5.4 shows the KME of the failure times, the fitted ordinary Weibull distribution and the fitted Weibull mixture distribution. (The Weibull mixture distribution is also shown in Figure 6.5—the present data set is Group (2,1) of that figure.) In keeping with the nonsignificance of the likelihood ratio statistic,

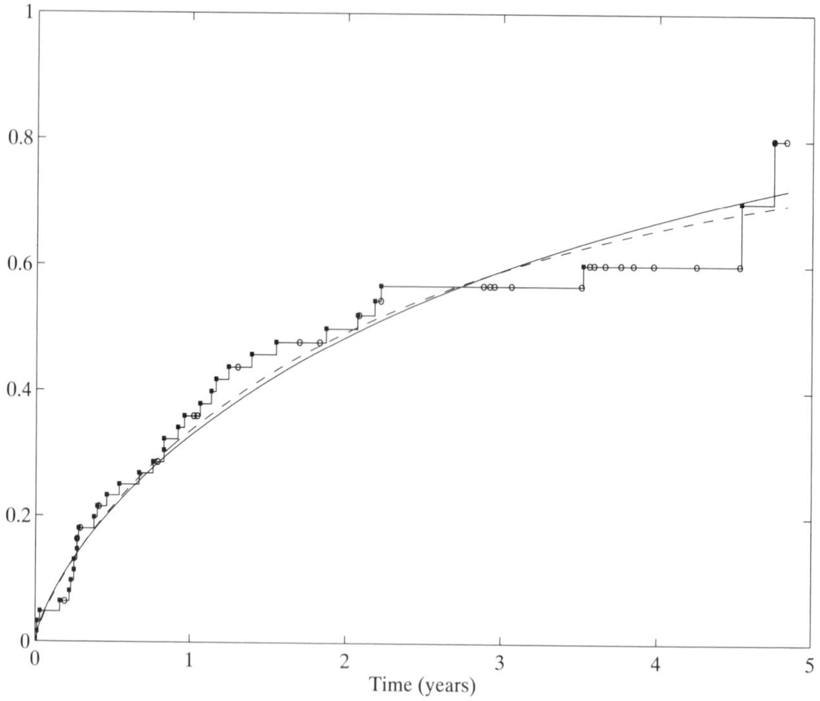

Figure 5.4 Time to first recidivism data, No parole/unmarried group, with fitted Weibull models; the fits are not significantly different to each other, according to a likelihood ratio test.

the Weibull mixture curve in Figure 5.4 describes the data little better than the ordinary Weibull curve. The ordinary Weibull distribution has compensated for the slight indication of levelling off in the KME, evident in Figure 5.4, by decreasing the estimated values of λ and α, thus making the curve rise less steeply (via the smaller λ) but changing its shape by increasing the curvature (via the smaller α) to maintain a similar value of the deviance. Neither of the Weibull models produces a very good fit, and we hesitate to draw conclusions from them. Insofar as the ordinary Weibull model produces almost as high a likelihood value as the Weibull mixture model, we might have to revise our initial idea that there may be immunes present.

We can reconcile the interpretations to some extent by considering follow-up. There are three late failures in the data (out of a total of 24) and we should suspect from this that follow-up may not be sufficient to determine whether or not immunes are present.

The q_n-test we introduced in Section 4.3 to test for sufficient follow-up has a value of q_n given by

$$q_n = \frac{N_n}{n} = \frac{1}{62} = 0.0161,$$

which is smaller than the 5% point in Table B.7 with $n = 60$, $p = 0.8$ and $B = 2$; the latter has value 0.0167. So there is strong evidence that follow-up for this set of data is far from sufficient to allow us to conclude the presence of immunes. (See also the further discussion concerning this data set in Section 6.4.)

5.4 DIAGNOSTICS: TESTING FOR GOODNESS OF FIT

In this section we propose a method of testing the goodness of fit of a parametric distribution to censored survival data, which we have found to be very convenient to apply. It is similar to a test devised by Filliben (1975) for the normality of uncensored data, which is especially easy to implement on a computer and has the extra advantage of providing a clear visual assessment of the goodness of fit.

Filliben suggested calculating the correlation coefficient between the order statistics of a sample and the expected values (or median values) of the order statistics for a sample of the same size from the standard normal distribution. The value of the correlation coefficient forms the basis of the hypothesis test — values close to 1 indicate a close fit, those close to 0 indicate a poor fit — and a plot of the order statistics against their expected values (or medians), calculated under the null hypothesis of a normal distribution, gives the visual display.

Our suggested test for censored survival times is a variant of this. Consider the hypothesis

$$H_0 : F = \tilde{F}$$

where $\tilde{F}$ is a specified cumulative distribution function such as the exponential. Let

$$t_{(1)} \le t_{(2)} \le \cdots \le t_{(n)}$$

be the ordered sample of survival times. Suppose for the moment that there is no censoring. A P-P (probability) plot of the data is a plot of $\tilde{F}(t_{(i)})$ against $i/(n+1)$, $1 \le i \le n$. This should be close to a straight line if H_0 is true, the rationale being that when $F = \tilde{F}$, $\tilde{F}(t_i)$ has the uniform distribution, $U[0, 1]$, over [0,1], so $\tilde{F}(t_{(i)})$ has expectation the same as the expectation of the ith order statistic in a sample of n from $U[0, 1]$, which is $i/(n+1)$. In other words, $\tilde{F}(t_i)$ should be approximately proportional to i if H_0 is true, giving close to a straight line for the P-P plot.

Now let $\hat{F}_n$ be the empirical distribution function of the sample $t_1, \ldots, t_n$. Since $\hat{F}_n(t_{(i)}) = i/n$, the P-P plot is equivalent to plotting $\tilde{F}(t_{(i)})$ against $\hat{F}_n(t_{(i)})$. When the data are subject to censoring, the same argument leads us to expect to obtain, when H_0 is true, a near-straight line with slope close to 1, by plotting $\tilde{F}(t_{(i)})$ against $\hat{F}_n(t_{(i)})$, provided we take $\hat{F}_n$ as the KME. When $\tilde{F}(t_{(n)})$ is close to $\hat{F}_n(t_{(n)})$, as it is in our case, the correlation coefficient r between $\tilde{F}(t_{(i)})$ and $\hat{F}_n(t_{(i)})$, $1 \le i \le n$, should provide an appropriate measure of the goodness of fit. Thus we propose to use this correlation coefficient to test the goodness of fit. Note that we include the censored observations in calculating the correlation.

Example: Kersey et al. leukaemia data—groups 1 and 2

In Section 5.2 we showed how to fit the exponential mixture distribution to Group 1 (allogeneic transplants) of the data of Kersey *et al* (see Figure 5.2). Now we test for the goodness of fit of the model for both groups. First consider Group 1. For a fit of the exponential mixture distribution given by

$$\tilde{F}(t) = \tilde{p}\left(1 - \mathrm{e}^{-\tilde{\lambda}t}\right) \tag{5.40}$$

we obtain maximum likelihood estimates $\tilde{p}$ and $\tilde{\lambda}$ as given by (5.36). The values of $\tilde{F}(t_{(i)})$ and $\hat{F}(t_{(i)})$ are plotted against each other in Figure 5.5.

Visually, the fit appears reasonable, in that the line is fairly near to the 45° line. For the correlation coefficient between $\tilde{F}(t_{(i)})$ and $\hat{F}(t_{(i)})$ we find a value of $r = 0.9959$. On the other hand, for Group 2 (autologous transplants), $\tilde{F}(t)$ is given by (5.40) with MLEs

$$\tilde{p} = 0.83 \text{ and } \tilde{\lambda} = 2.55. \tag{5.41}$$

An exponential mixture model fit to the Group 2 data appears poor, as can be judged from Figure 5.6 (and a Weibull mixture model fit is, visually, little better).

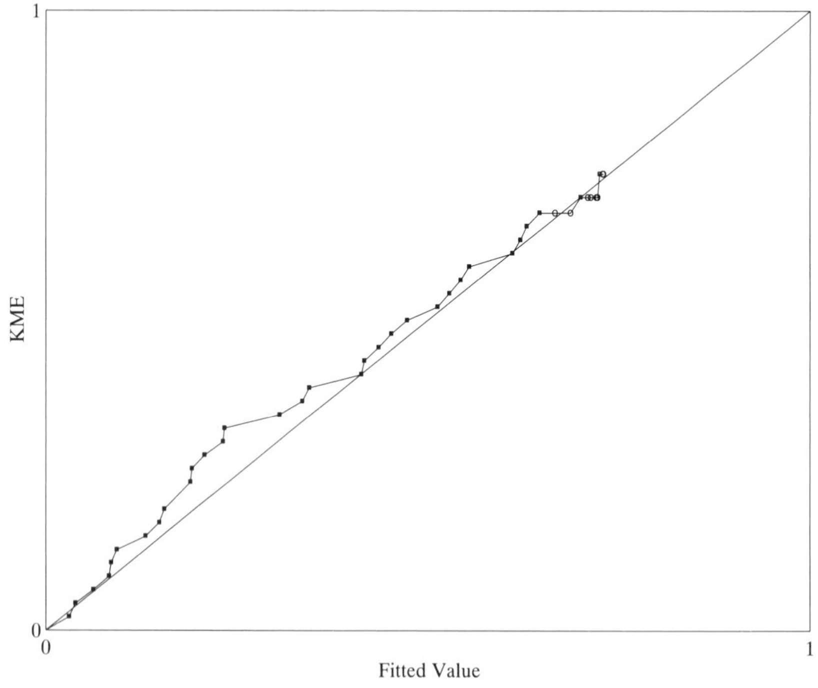

Figure 5.5 P-P plot of KME versus the fitted value of an exponential mixture model for Group 1 (allogeneic transplants) leukaemia data; judged to be a good fit by the correlation coefficient test. The straight line is the line $y = x$.

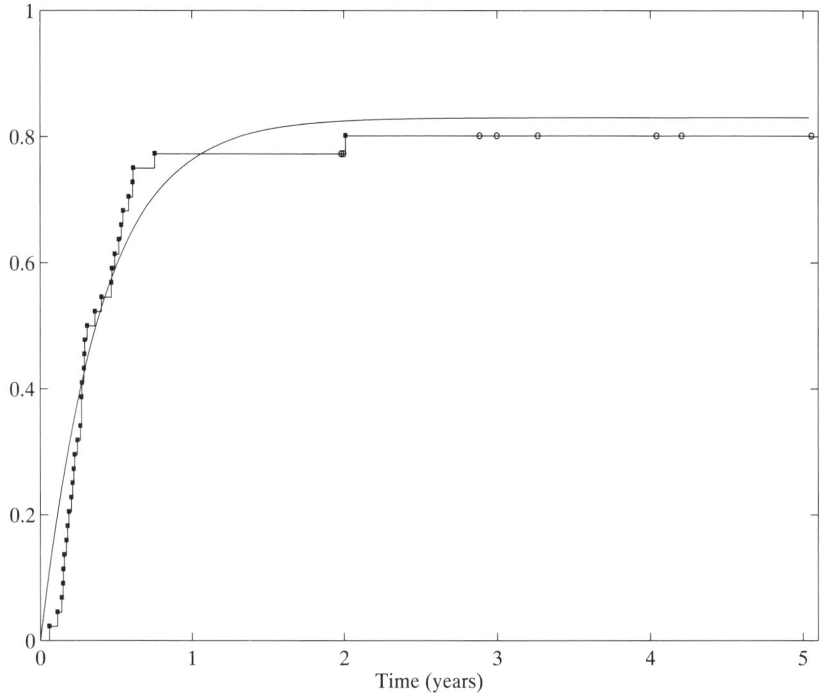

Figure 5.6 Leukaemia data, Group 2 (autologous transplants) with fitted exponential mixture model.

In fact, it appears that none of the usual parametric survival distributions would fit this data well, with its almost uniform (and fast!) initial failure rate. Figure 5.7 shows $\tilde{F}(t_{(i)})$ plotted against $\hat{F}_n(t_{(i)})$, for Group 2. We find that the correlation coefficient r for the comparison of $\hat{F}_n(t)$ with $\tilde{F}(t)$ has the value 0.9709.

The above values of r suggest a good fit for Group 1 and a poor fit for Group 2, and now we wish to test rigorously for these effects. In the case of normally distributed, uncensored data, it was relatively straightforward for Filliben (1975) to simulate and tabulate the percentage points of the distribution of the test statistic r. When survival times are censored, however, we must consider the effect of the unknown censoring distribution on the null distribution of r. The problem is exacerbated by the fact that we (currently) have no theoretical knowledge of the distribution of r.

Simulating the percentiles of r

In the absence of some theoretical guidance concerning the distribution of r, we proceed by doing some simulations, just as we did in Chapter 4 when trying to assess the null distributions of the nonparametric tests considered there. In Chapter 4 we used both the exponential and the uniform as censoring

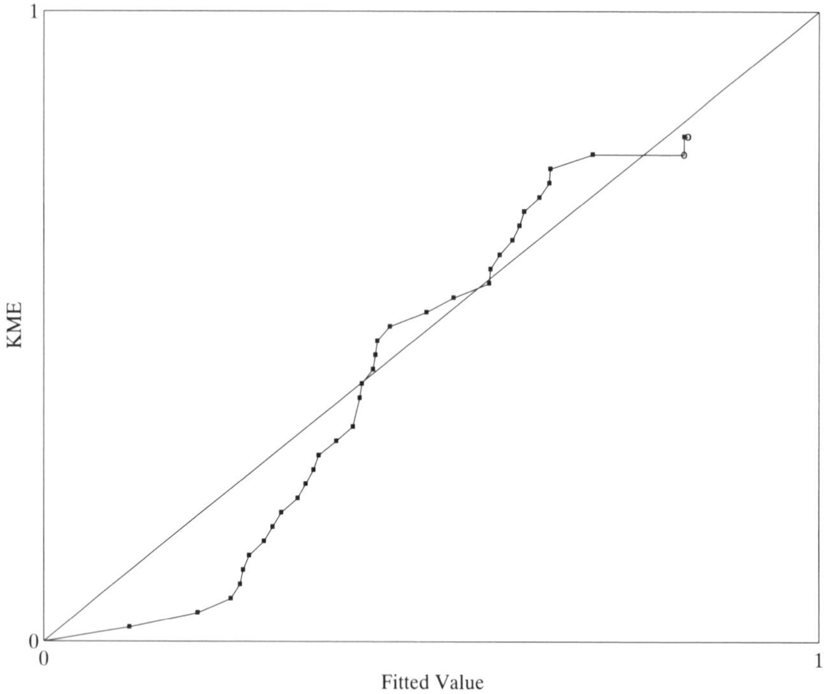

Figure 5.7 P-P plot of KME versus the fitted value of an exponential mixture model for Group 2 leukaemia data; judged to be a poor fit by the correlation coefficient test.

distributions, and we will begin by considering them both again. For Table 5.1 we generated $n = 50$ survival times from an exponential distribution with mean 1 (we can always transform the exponential to this situation) and values of the susceptible proportion, p, of 0.4, 0.6, 0.8 and 1. The censoring distributions were exponential with mean $\mu = 1$ or 2, corresponding to heavy and light censoring, and uniform with mean of 1 or 2. We took 1000 repetitions of samples of such censored survival times, and for each repetition calculated the Kaplan–Meier estimator $\hat{F}_n(t)$, then the correlation coefficient r of $\hat{F}_n(t_i)$ with the corresponding values of the underlying survival distribution

$$\tilde{F}(t_i) = p\left(1 - \mathrm{e}^{-t_i}\right), \qquad t \geq 0.$$

Table 5.1 shows the 5th and 80th percentiles of the distribution of r as estimated from the 1000 repetitions. The table suggests that light censoring or large values of p tend to produce larger correlation coefficients than heavy censoring or small values of p, as we might intuitively expect. Moreover, for a given immune proportion, the form of the censoring distribution does not seem to have much impact on the r values; the percentiles do not differ too much for censoring distributions as different as the exponential and the uniform, though

Table 5.1 Percentiles of the correlation coefficient.

		Censoring distribution			
		$\exp(1/\mu)$	with	$U[0, B]$	with
	p	$\mu = 1$	$\mu = 2$	$B = 2$	$B = 4$
	0.4	0.973	0.984	0.980	0.987
80%	0.6	0.982	0.989	0.986	0.991
	0.8	0.986	0.992	0.989	0.993
	1	0.990	0.994	0.991	0.994
	0.4	0.848	0.914	0.874	0.926
5%	0.6	0.904	0.942	0.925	0.956
	0.8	0.933	0.966	0.948	0.969
	1	0.957	0.974	0.964	0.975

the percentiles of the exponential distributions are generally slightly smaller than those of uniform distributions with the same means. This is due to the skewness of the exponential distribution towards 0, which tends to make the censoring slightly heavier than for the uniform distribution.

For practical purposes, however, the kind of censoring distribution seems to have little effect on the percentiles of r, at least based on the evidence of Table 5.1. Consequently, for a more extensive set of simulations, we restricted ourselves to a uniform censoring distribution. These more extensive simulations are reported in Tables C.1 to C.9 at the end of the book. For those, we used 5000 repetitions, and a more extensive set of values of B, the right extreme of the uniform censoring distribution. To illustrate their use, we consider again the leukaemia data shown in Figures 5.2 and 5.5. (Further illustrations are given in Chapter 6.)

Example: Kersey et al. leukaemia data (continued)

Let's continue our analysis of the Kersey *et al.* data, from the previous section. Table 5.1 suggests that for sample sizes of order 50 from an exponential mixture distribution with $p = 0.8$, values of r exceeding the 80th percentile of its null distribution under a light $U[0, 4]$ censoring, which is 0.993, indicate a good fit. (Indeed a correlation coefficient greater than the 10th percentile would be considered to be indicative of a reasonable fit.) The observed value of $r = 0.9959$, by exceeding the 80th percentile, thus lends support to our visual deduction of a good fit of the exponential mixture model for the allogeneic group.

The value of $r = 0.9709$ for the autologous group just exceeds the tabulated 5th percentile of 0.969 for $p = 0.8$ and a light uniform, $U[0, 4]$, censoring (see Table 5.1), though it is close to this value. The correlation test does not reject the exponential fit for this data set, just, so we infer that a P-P plot like that of Figure 5.7 is just on the borderline of what is considered to be a good fit to a data set of size about 50, at the 5% level, as far as Table 5.1 can tell us. But a more thorough investigation (in Appendix Note 3) lends support to our visual observation that the fit is poor.

5.5 DIAGNOSTICS: TESTING FOR AN OUTLIER

The good fit of a parametric model, such as the exponential mixture model, to a set of data may be spoiled by a single observation which is aberrant from the others, or atypical in terms of the distribution of the remaining data. Outliers may occur in any statistical analysis, and a large literature is devoted to discussing their detection, and how to deal with them once detected.

In the context of a parametric mixture distribution fitted to censored survival data, outliers may manifest themselves as large uncensored observations. We may thus spot them as occurring near the right-hand end of the Kaplan–Meier plot of the data, where the KME has essentially levelled off, and all but the possible outlier(s) are censored observations. They may have the effect of decreasing the value of the q_n-statistic, which we use to test for sufficient follow-up in the data, to the extent that it becomes nonsignificant, whereas to all appearances, apart from the possible outlier, the KME has levelled off. In other words, we may be forced to conclude that we have insufficient follow-up ($\tau_{F_0} > \tau_G$ in the notation of Section 2.2) due to the occurrence of one or a few late failures among a number of large, censored observations.

Example: recidivism data

As an example, consider the data set plotted in Figure 5.8, which are the data for the no-prior group we discussed in the second example of Section 4.3. The data used in this example comes from a subgroup of prisoners who had no convictions prior to the serious sexual offence, and the event of interest is a return to prison for any offence following a prisoner's release from prison for a serious sexual offence. The survival time, which we call a failure time in this context, is the time taken for that return to prison, if it occurs. Releasees not returning to prison within the limit of follow-up time represent censored observations.

Note that the KME of the observed failure times appears to have levelled off after about 7.5 years, but that one late failure occurs at about 10 years. Also shown in Figure 5.8 is a fitted Weibull mixture model; it has estimated parameters

$$\tilde{p} = 0.4583, \tilde{\lambda} = 0.3042 \text{ and } \tilde{\alpha} = 0.8485. \tag{5.42}$$

In general statistical practice, when outliers occur, we try to account for their presence. Barnett and Lewis (1984) provide an extensive discussion of various methods of doing so. Perhaps the first thing to check is the accuracy of the data — is the aberrant observation merely a recording or keying error? This often turns out to be the case, in some situations, but for the prison or arrest data analysed in this book it is rarely the answer, since the official records are kept with scrupulous accuracy and the computer files are cross-checked in various ways. In the present case the recidivism at around 10 years in Figure 5.8 did occur, but still we feel that it is aberrant from the other points. How can we formalise this intuition?

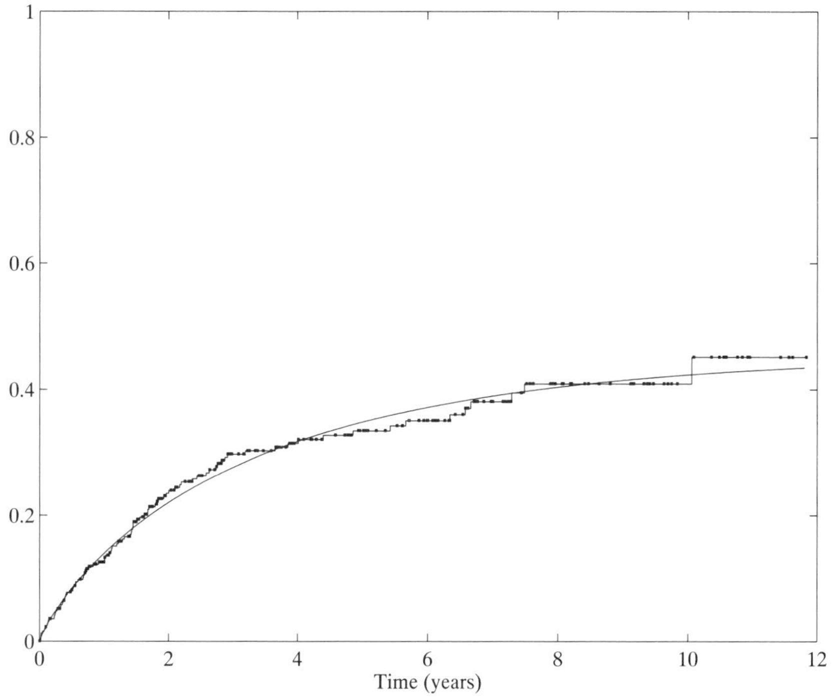

Figure 5.8 Recidivism data: no-prior group with fitted Weibull mixture model.

'Outliers' can really only be defined in terms of some model which accounts reasonably well for the main body of the data, that is, for all the observations except the putative outlier, and with respect to this model the outlier must be extremal in some way. In the case of the no-prior data, we notice that the failure at around 10 years in Figure 5.8 is quite distant from the next largest failure time, which occurs at about 7.5 years. At earlier times than this, the failures are fairly evenly spread. This observation suggests that we test for the extremism of the largest (uncensored) failure time by comparing it with the second largest *uncensored* failure time. We need to assume some distribution for the main body of the data, and we will use the fitted Weibull distribution with parameters given in (5.42), even though, as Figure 5.8 indicates and as we will see below, in some respects it is not completely suitable as a description of the data.

Leaving aside this issue for the moment, let us give a general method for comparing the largest uncensored observation with the second largest uncensored observation in a set of survival data. We need some notation for these observations. Assume our usual i.i.d. censoring model. Then for the survival times $t_1, t_2, \ldots, t_n$ with censor indicators $c_1, c_2, \ldots, c_n$, define random variables

$$S_i = t_i c_i, \qquad 1 \leq i \leq n.$$

The S_i are i.i.d., since t_i^*, u_i and thus the random variables $t_i c_i = \min(t_i^*, u_i) I(t_i^* \leq u_i)$ are i.i.d. The nonzero values of S_i are the uncensored observations, and all censored observations are collected at the zero value of the S_i. Order the S_i so that

$$S_n^{(n)} \leq \cdots \leq S_n^{(1)}. \tag{5.43}$$

Then the largest and second largest uncensored observations can be written as

$$t_{(n)}^* = S_n^{(1)} \text{ and } t_{(n)}^{**} = S_n^{(2)}.$$

Under the independent censoring model, we can calculate the probability that $t_{(n)}^*$ and $t_{(n)}^{**}$ differ by more than a certain amount, as follows. Let $0 < x < y$ and write

$$\begin{aligned} P\{t_{(n)}^{**} \leq x < y < t_{(n)}^*\} &= P\{S_n^{(2)} \leq x < y < S_n^{(1)}\} \\ &= P\left\{\max_{j \neq i} S_j \leq x < y < S_i, \text{ for some } i \leq n\right\} \\ &= nP\left\{\max_{2 \leq j \leq n} S_j \leq x < y < S_1\right\} \\ &= nP\{S_1 > y\}[P\{S_1 \leq x\}]^{n-1}. \end{aligned} \tag{5.44}$$

We further calculate

$$P\{S_1 > y\} = P\{u_1 \geq t_1^* > y\} = \int_{(y,\infty)} (1 - G(t-))\,\mathrm{d}F(t) \tag{5.45}$$

where, as usual, $F(t)$ is the distribution of the t_i^* and G is the censoring distribution, that is, the distribution of the u_i. As usual, let

$$H(t) = 1 - (1 - F(t))(1 - G(t))$$

be the distribution of the observed survival times $t_i = \min(t_i^*, u_i)$. Then from (5.45) we have the formula

$$P\{S_1 > y\} = \int_{(y,\infty)} (1 - H(t-)) \frac{\mathrm{d}F(t)}{1 - F(t-)}. \tag{5.46}$$

We wish to estimate this function from the data, so we have to estimate the functions F and H. The function F will be estimated by the distribution we have fitted to the survival times; in the case of the no-prior data, this is the Weibull mixture distribution with parameters $\tilde{p}$, $\tilde{\lambda}$ and $\tilde{\alpha}$ specified by (5.42). Call this distribution $\tilde{F}_n$, so we have

$$\tilde{F}_n(t) = \tilde{p}(1 - \mathrm{e}^{-(\tilde{\lambda} t)^{\tilde{\alpha}}}), \qquad t \geq 0. \tag{5.47}$$

To estimate H we simply use the empirical distribution function estimator

$$\hat{H}_n(t) = \frac{1}{n} \sum_{i=1}^{n} I(t_i \leq t), \qquad t \geq 0. \tag{5.48}$$

By standard theory (e.g. Shorack and Wellner 1986, p. 106), or simply from the strong law of large numbers, $\hat{H}_n(t)$ is an unbiased and consistent estimator of $H(t)$, for each $t > 0$.

Substitute

$$1 - \hat{H}_n(t) = \frac{1}{n}\sum_{i=1}^{n} I(t_i > t)$$

and $F(t) = \tilde{F}_n(t)$ into (5.46) to get an estimated value of $P\{S_1 > y\}$ given by

$$\tilde{P}\{S_1 > y\} = \frac{1}{n}\sum_{i=1}^{n} I(t_i > y)\int_{(y,t_i)} \frac{\mathrm{d}\tilde{F}_n(t)}{1 - \tilde{F}_n(t-)}. \tag{5.49}$$

Since $\tilde{F}_n(t)$ is continuous we can evaluate the integral to obtain the relatively simple expression

$$\tilde{P}\{S_1 > y\} = \frac{1}{n}\sum_{i=1}^{n} I(t_i > y)\log\left(\frac{1 - \tilde{F}_n(y)}{1 - \tilde{F}_n(t_i)}\right). \tag{5.50}$$

Into this we can further substitute the explicit expression (5.47) for $\tilde{F}_n$, in order to calculate values of $\tilde{P}\{S_1 > y\}$.

In the present circumstances we need only evaluate this probability at the values of the largest and second largest uncensored observations, which for the no-prior data are

$$t^*_{(n)} = 10.0575 \text{ years and } t^{**}_{(n)} = 7.4904 \text{ years}. \tag{5.51}$$

Using (5.44) we then obtain

$$\begin{aligned} &\tilde{P}\{t^{**}_{(n)} \le 7.4904 < 10.0575 \le t^*_{(n)}\} \\ &\quad = n\tilde{P}\{S_1 > 10.0575\}(1 - \tilde{P}\{S_1 > 7.4904\})^{n-1} \end{aligned} \tag{5.52}$$

in which the sample size $n = 296$. We calculate the probabilities from (5.50) to find

$$\tilde{P}\{S_1 > 10.0575\} = 0.000\,455 \text{ and } \tilde{P}\{S_1 > 7.4904\} = 0.015$$

and then (5.52) gives

$$P\{t^{**}_n \le 7.4904 < 10.0575 \le t^*_n\} = 296(0.000\,455)(1 - 0.015)^{295} = 0.001\,56.$$

This is a small p-value — much smaller than the conventional 0.01 level, in fact — so we decide at the 99% level of confidence that the observed $t^*_{(n)}$ and $t^{**}_{(n)}$ do not represent the largest and second largest uncensored observations from the distribution given by (5.47). Since $t^*_{(n)}$ is aberrant, we are entitled to regard it as an 'outlier'.

Given that we have no doubt of the correctness of the data on the failure at 10.0575 years, what does it mean to classify it as an 'outlier'? One interpretation is that it may be drawn from a different distribution to the rest of the data. This may be due to dependence of the data point on an underlying factor or combination of factors which has not been allowed for in the analysis — or has not even been measured.

All the prisoners representing the data set plotted in Figure 5.8 are homogeneous with respect to gender (they are male), number of prior imprisonments (none) and, substantially, with respect to their offence types (serious sexual); but there are plenty of other factors, on which we have information, which vary for them. The age of a prisoner at release from prison is an important one; others are a prisoner's employment status or marital status at release or at rearrest. Many other factors are also known to affect the recividism of prisoners to a greater or lesser degree. Examination of the covariates associated with the aberrant data point in this example, however, suggests no likely explanation.

An unobserved factor or factors, such as psychological state during the release, propensity to associate with other recividists, etc., would perhaps provide an alternative explanation in this case, but we have no way of assessing this from the available information. In the end, we simply accept that the data point is aberrant, and ask, if this is so, how much it may have affected the conclusions we have drawn from the data following our analysis of it.

One of the important aspects of an analysis of censored survival data that we stressed in Section 2.3 is to test for sufficient follow-up. When we applied the q_n-test in Section 4.3 to the no-prior data in Figure 5.8, we found considerable evidence that $\tau_{F_0} > \tau_G$, that is, of insufficient follow-up. Of course this test is highly sensitive to a late failure such as occurs in this data, since we are required to count the number N_n of uncensored observations in the interval $(2t^*_{(n)} - t_{(n)}, t^*_{(n)}]$ (where $t_{(n)}$ is the largest observation). If, however, we omit the largest uncensored observation — deemed to be outlying — we obtain for the remaining 295 points a value for $q_n = N_n/n$ of

$$q_n = \frac{13}{295} = 0.0441.$$

This is well above the 10% point in Table B.3, corresponding to $p = 0.4$, $B = 8$ and $n = 300$, which is 0.0133, so the evidence against $\tau_{F_0} \leq \tau_G$ is much weaker now. It is not unreasonable to assume that follow-up is sufficient, if we are prepared to disregard the one late failure, although the evidence for $\tau_{F_0} \leq \tau_G$ is not strong either.

If we now refit the Weibull mixture model with this outlier omitted we obtain parameter estimates

$$\breve{p} = 0.4309, \quad \breve{\lambda} = 0.3541, \quad \breve{\alpha} = 0.8846 \tag{5.53}$$

which are somewhat different to those given in (5.42), especially in respect to λ. So the outlier, if that is what it is, makes some difference to the inference we wish to draw from the parametric analysis of this data set.

Readers can decide for themselves how satisfying they find the above procedure. Interpreting the no-prior data as being modelled by a failure distribution such as the Weibull mixture model with parameters estimated by (5.53), provided one outlying observation is omitted, seems to us a reasonable summary. Others may prefer a more formal approach, such as fitting a more sophisticated mixture model in which an extra component to the distribution is allowed. Such models have been used with great success by Aitkin (1996).

In the case of the no-prior data, we would expect that an extra component in the survival data would indeed be found by the method of Aitkin, and turn out to be represented by the observation that we have deemed to be outlying—since the mixture method really amounts to formalising our informal method of testing for an outlier, when there is only one such observation.

5.6 OTHER SURVIVAL MODELS

In Section 5.1 we suggested that the exponential distribution, or perhaps its more general relative, the Weibull distribution, be used as a 'baseline' or touchstone model for the description of survival data. The range of survival data that can be modelled by the exponential distribution itself is very limited, because the 'shape' of the exponential distribution—clearly shown in Figure 5.2—cannot be altered. In particular, it is not altered by changing the value of λ in the representation

$$1 - e^{-\lambda t}, \qquad t \geq 0$$

for the c.d.f. of the exponential distribution. The exponential *mixture* model, of form

$$p(1 - e^{-\lambda t})$$

has an extra parameter, but likewise its *shape* cannot be changed by varying p and/or λ. Changing the value of p simply changes the value at which the distribution levels off.

However, the Weibull family, as parametrised in (5.1), is much more flexible in shape. The parameter α in (5.1) is called the *shape parameter* of the Weibull, because the appearance of the Weibull c.d.f. depends crucially on its value. When $\alpha < 1$, the Weibull has a somewhat exponential appearance: its density is monotone, decreasing from a value of infinity at $t = 0$ to a value of 0 at $t = \infty$. When $\alpha = 1$ the Weibull reduces to the exponential, with its characteristic shape, but when $\alpha > 1$, the Weibull density has a single mode at a positive value of t, and its c.d.f. is sigmoid in shape. For schematic plots which show the shapes of the Weibull p.d.f. and c.d.f. for various values of α, see for example Nelson (1982, p. 37).

Consequently, the Weibull can be used to model survival data for which the KME has a sigmoidal shape, in which case the estimate of α obtained after fitting the Weibull by maximum likelihood, or by some other method, should exceed 1. Likewise, the Weibull can approximate to a quite differently shaped exponential-like KME when $\alpha \leq 1$. Similar comments apply to the Weibull mixture model with c.d.f. as given by (5.2); the extra parameter p merely changes the value at which the Weibull levels off, and does not affect its shape.

The parameter λ in the Weibull approximately measures the rate of failure of the data. It is approximately inversely proportional to the median of the survival times, since

$$\text{median of Weibull} = \frac{(\ln 2)^{1/\alpha}}{\lambda}.$$

This quantity depends to some extent on α, also, but the main determinant is λ, especially for most of the data sets we consider, which have α close to 1. For $\alpha = 1$, the above approximate statements are exact, since the Weibull reduces to the exponential distribution in that case.

We give, throughout the book, various examples of data sets which are well fitted by an exponential or Weibull mixture distribution, and the reader can see from the accompanying figures how the shape of the Weibull changes so as best to model the data; see for example Figures 6.2 and 6.3.

But what if, after fitting a Weibull (mixture) distribution to data, one finds by visual inspection or by a goodness of fit test, such as we outlined in Section 5.4, that the Weibull is a poor description of the data? There are various other parametric models in common usage for survival data, and we will list some of them here, without discussing their properties in detail, since it turns out that, for most of the data sets we discuss in the book, the Weibull is quite an adequate model. This is not to say that survival data which is not well fitted by the Weibull does not exist — it certainly does — but we can illustrate well enough for our purposes the main points concerning the analysis of survival data with immunes from data sets which are approximately exponential or Weibull distributed. Nevertheless, we will briefly discuss now some other commonly used models for survival data.

Any distribution concentrated on $t \geq 0$ can potentially be used as a survival distribution, furthermore, each one can be made into a mixture model — so as to allow for the presence of immunes — exactly as we did for the exponential and the Weibull distributions. We just allow the original 'proper' distribution to level off at some value p below 1. Commonly used distributions in 'ordinary' survival analysis are the lognormal, the gamma (including the chi-square distribution and various generalised gamma distributions), the logistic and various 'extreme value distributions'. These are listed, for example, in Nelson (1982), Lawless (1982) and Kalbfleisch and Prentice (1980). Kalbfleisch and Prentice also formulate what they call a 'generalised F distribution' which includes most commonly used (continuous) distributions as special cases (or as limiting cases). The family of distributions can be further extended by the introduction of an extra parameter to shift the time origin of the data, but this has not proved to be necessary for the data we analyse.

In Section 1.4 we discussed a number of areas in which data has been analysed or modelled with an allowance for immunes. Apart from the exponential mixture model, which was used in a number of those analyses, we know of Schmidt and Witte's (1988) application of the lognormal mixture model to recidivist data (it fitted a little better than a Weibull mixture model), and a few applications, some quite recent, using the Gompertz distribution as a model for survival data with immunes. This is quite an interesting model on which we expand briefly here.

The Gompertz distribution is a version of one of the class of 'extreme value' distributions. In the parametrisation used by Cantor and Shuster (1992), it takes the form

$$F(t) = 1 - \exp\left\{\frac{\alpha}{\beta}(1 - e^{\beta t})\right\}, \qquad t \geq 0. \tag{5.54}$$

Here $\alpha > 0$, but β may take either sign. When $\beta > 0$, the c.d.f. is proper, with $F(\infty) = 1$, but when $\beta < 0$, F is improper with mass at infinity equal to $1 - e^{\alpha/\beta}$. This suggests its use as a model for data containing an immune component, with the possibly attractive feature that one need not make any a priori assumption as to the existence or otherwise of immunes; one may fit (5.54), by maximum likelihood, for example, and 'let the data tell us' the value of β. A value of $\beta > 0$ corresponds to 'ordinary' survival data with immunes, whereas a value of $\beta < 0$ points to a possible immune proportion in the data.

We have only limited experience to report in this direction, but it seems to hint that the function in (5.54) is not flexible enough as a general model for survival data. Our evidence comes from the paper of Fallen (1996), who fitted the Weibull mixture model and the improper Gompertz model (i.e. with $\beta < 0$) to five sets of criminological data. He considered a set of data reproduced by Maltz (1984); the two sets of data given in Maller and Zhou (1994) (taking those prisoners with prior convictions and those without priors as separate sets of data); and two data sets available from Schmidt and Witte (1988): their 1978 set of data and an updated 1980 set. Fallen found that the improper Gompertz provided a significantly better fit than the Weibull to none of these sets of data, and for a couple of data sets it was significantly worse. Interestingly, he also fitted a mixture model, based on the *proper* Gompertz (i.e. with $\beta > 0$) and a mass at infinity to represent immunes. This model was significantly better than the *improper* Gompertz for a couple of the data sets mentioned above, and was also better (though not significantly better) than the Weibull mixture model for a couple of sets. Given this fairly scant evidence, preliminary indications do not support the view that the Gompertz (either in its proper or improper form) is likely to fit as wide a variety of data sets as the Weibull, or that the idea of allowing the data to tell us whether the data contains immunes or not, via the sign of β, is likely to be useful.

Yukovlev (1994) suggests the use of an improper failure time distribution of the form $1 - e^{-\theta F(t)}$, where F is a proper c.d.f., which he derives from a simple stochastic model for tumour latency time. This model also is not a mixture model, as we use the term. It is a generalisation of the Gompertz distribution (5.54) (see also Cantor and Shuster (1994)) and may be used in a similar way.

APPENDIX NOTES TO CHAPTER 5

1. The Weibull distribution

The Weibull distribution is named for Wallodi Weibull, who popularised its use, especially in engineering contexts, as a model for the strengths and lifetimes of materials such as metals and fibres. Prior to Weibull's tremendously influential papers (Weibull 1939, 1951), it seems that most such data had been forced into the mould of the normal or 'error' distribution, possibly after some transformation, with consequent calculation of means and standard deviations, primarily. Lifetime data such as was used in actuarial applications seems to have been assumed exponentially distributed, in the main, to the extent that distributional properties were considered at all. Nowadays the use of the Weibull distribution as a model in the fields of reliability and actuarial statistics alone accounts for much more literature than we can survey here, and that is additional to its applications in the area of medical statistics.

It seems that Weibull's seminal (1951) paper was rejected on its original submission to a major statistical journal (it eventually appeared in the *Journal of Applied Mechanics*). Graduate students and others may or may not draw some comfort from this.

2. Parametric or nonparametric?

Should we use a parametric or nonparametric approach to the analysis of survival data? It is fairly well established in statistical consciousness, we think, that a parametric model is superior in that it usually gives greater power (in the technical sense of more often rejecting an incorrect null hypothesis) than a nonparametric approach, but that this is at the expense of restricting the analysis to a model which may (and should!) fit the data well but can never be a perfect representation of it.

As we have seen, there are important aspects of both approaches which we should consider in the analysis of survival data; in particular, the KME provides a very valuable nonparametric descriptive representation of the data, in any circumstance, whereas models such as the Weibull or exponential mixture model are especially amenable to dealing with covariate data, as we shall see in Chapter 6, and to theoretical analysis, as we show in Chapters 7 and 8.

If a data set is not well fitted by any of the parametric models that we try, we are left with no choice but to use the nonparametric approach. To highlight the comparison in our context, we can consider the virtues of using the nonparametric estimate $\hat{p}_n$, the maximum of the KME, as an estimator of p, the proportion of susceptibles, rather than the MLE, $\tilde{p}_n$, obtained from a parametric model such as the exponential. How much are we likely to 'lose' if we use $\hat{p}_n$ rather than $\tilde{p}_n$ in a situation in which the latter is in fact justified?

The answer to this question will depend on sample size as well as on the types of survival and censoring distributions from which the data is generated, but we can throw light on some special cases. Somewhat surprisingly, $\hat{p}_n$ performed quite well by comparison with $\tilde{p}_n$ in a small simulation experiment carried out by Ghitany *et al.* (1995), in which the underlying survival and censoring distributions were exponential. They considered the accuracy and precision of the two estimators over a range of values of p_0, the true susceptible proportion, and of λ_0, the failure rate of the susceptibles, finding that the estimators performed similarly for values of p_0 between 0.5 and 0.9, but that $\hat{p}_n$ was the more precise estimator for small values of p_0 and relatively small sample sizes. Both estimators performed reasonably well near the boundary $p_0 = 1$.

We remark that an earlier paper by Ghitany (1993) gives some information on the extent to which follow-up or the lack of it affects the precision of estimation of p_0 and λ_0 in an exponential mixture model; see also Goldman (1984, 1991) in this context.

For testing the difference in cured proportions between two groups, Sposto *et al.* (1992) assumed the parametric exponential and Weibull mixture models of our next chapter, and compared by simulations the properties of the likelihood ratio test, the Wald statistic (Gill (1993 p. 403)) and the score test (the standardised first derivative of the log-likelihood) with the statistic formed from the difference in the maximum values of the Kaplan–Meier estimators for the two samples. They found that the difference statistic often performed well despite being biased in some circumstances, through the Wald statistic was as good or better than the others, at least when the survival functions of the two samples did not cross.

3. Further notes on the goodness of fit tests for the Kersey et al. data

For both sets of the leukaemia data discussed in Section 5.4, the goodness of fit statistic for the exponential distribution, r, was quite close to significance — in the case of the allogeneic group, we accept the goodness of fit, whereas for the autologous group, we just about reject it. For data sets as small as this, we realise that a definitive answer is unlikely; nevertheless, readers will probably feel that some further analysis is indicated. We can do this by considering more closely the censoring patterns in the data, and tailoring our simulations of the percentiles of r to follow them a little better.

Notice that the censoring in the data for both groups is in fact very light — there are no censored observations before 18 months, whereas only a small proportion of individuals survived beyond that time (see Table 4.1). So we are led to simulate r values for samples with lighter censoring, which more closely resembles that in the data. We did two more simulations of 1000 repetitions, drawing survival times from improper exponential distributions with parameters given by (5.36) and with $n = 46$ for Group 1 (allogeneic), and by (5.41) and with $n = 44$ for Group 2 (autologous). The censoring distribution was chosen as $U[1.5, 5]$ (in years) for both groups. For Group 1, the simulated 80th percentile of the null distribution of r was 0.9952, which, compared with an observed value of 0.9959, confirms the good fit of the exponential distribution to this group. For Group 2, the simulated 5th percentile of the null distribution of r was 0.9766, above the observed value of 0.9709 for the group (in fact, the value of 0.9709 is slightly less than the 2nd percentile of the simulated r values). On this basis we reject the exponential fit to the autologous data at a 5% (or even 2%) level of significance.

These more detailed simulations confirm our intuitions concerning the goodness of fit of the exponential to the allogeneic data, and the lack of fit of that distribution to the autologous data.

Finally, we mention that there are various other ways of testing the goodness of fit in survival data which we will not survey here. These are commonly based on extensions of empirical distribution function methods (see, for example, Guilbaud 1988).

4. *Computational issues*

We mentioned in Section 5.2 that a Fortran program was developed for our own use in fitting the Weibull mixture model to single samples. It also fits the more complex factorial models described in the next chapter, handling up to two-factor analyses with possible interaction between the factors. A comparable commercial package is not available to our knowledge, but readers with access to modern powerful software packages such as the computer package GLIM (Baker and Nelder 1978, Baker 1985), or to modern computational packages, should have little trouble in writing Newton–Raphson, EM or simplex routines to accomplish the task. Fallen (1996) has a program written in Visual Basic.

5. *Testing goodness of fit for other than exponential distributions*

In Tables C.1–C.9 the simulated percentiles of the goodness of fit statistic are based on the correlation coefficient r between fitted values calculated using an exp(1) null distribution and the KME. But we can also use the tables for other distributions than exp(1) because the value of r is invariant under monotone transformations of the data. To see why, write $F_e(t) = 1 - e^{-t}$ as the c.d.f. of the exp(1) distribution and let the null distribution whose goodness of fit we wish to assess be specified under H_0 by $\tilde{F}(t)$. Consider the transformation $t_i \to h(t_i)$ where $h(t) = F_e^{-1}(\tilde{F}(t))$. Let $\hat{F}_n$ denote the KME of the original data t_i and $\hat{F}_n^*$ the KME of the transformed data $h(t_i)$. The KME depends only on the order of the observations and on the censoring indicators c_i, and these are not changed by the transformation $t_i \to h(t_i)$ if $h(t)$ is a nondecreasing function. As a result, we have $\hat{F}_n^*(h(t_i)) = \hat{F}_n(t_i)$ and

$$F_e(h(t_i)) = F_e(F_e^{-1}(\tilde{F}(t_i))) = \tilde{F}(t_i)$$

for each i. Thus the r defined for the original data t_i to test $H_0 : F = \tilde{F}$, i.e. the correlation coefficient between $\tilde{F}(t_i)$ and $\hat{F}_n(t_i)$, is the same as that defined for the transformed data to test $H_0 : F = F_e$, i.e. the correlation coefficient between $F_e(h(t_i))$ and $\hat{F}_n^*(h(t_i))$.

We will use this property in the next chapter to test goodness of fit of the Weibull distribution.

CHAPTER 6

The Use of Concomitant Information

6.1 SURVIVAL DATA WITH COVARIATES

In many practical situations, survival data (in the form of survival or failure times of individuals or components) comes with associated or *concomitant* information on which the survival times are thought to depend. In medical data, for example, this information may be by design, in that the effect of one or more treatments is being compared with the effect on survival times of a 'control' (a null or standard treatment); or it may occur incidentally, via some natural classification of patients, perhaps into male or female, young or old, etc. These classifications may be crossed with or nested in each other in various ways. Similarly, in criminology we may measure the time to return to prison of an individual after his/her release from prison for the first time, and the distribution of this recidivism time may (and usually does) depend on the sex and race of the prisoner, the type of offence, the release type (parole or not) and/or any of a host of other classification variables. For this sort of data, too, there may be 'designed' treatment groups, in which prisoners are allocated to groups, e.g. therapeutic counselling groups, whose success is to be measured by the extent to which they increase the average recidivism times of the groups, and/or decrease the overall proportion of recidivists, compared with those not receiving the counselling.

Likewise, in most other situations in which survival time data occurs (reliability theory, etc.; see the examples in Section 1.3), there will be some, often a great deal of, such auxiliary information. Furthermore, this information may come in the form of an observation on a continuous or almost continuous variate—the blood pressure of a patient before a treatment is applied, or the amount of money a prisoner has on release—rather than in the form of a discrete grouping variate such as those outlined above.

In this chapter we give a method for the analysis of survival data with immunes and covariates, followed by numerous examples with the aim of illustrating its practical utility and also of drawing attention to some pitfalls. We consider only relatively simple 'factorial' models which describe the effect of single factors, or factors taken two at a time, on survival, but the methods work in general. Before giving examples, let us specify more precisely what we consider to be the purposes of fitting a model with covariates. In practice, the following motivations are important:

(a) To obtain a convenient summary of the data in its simplest form, but with due regard to significant differences between groups, or to the effect of a continuous covariate; and this in turn depends on items (b) and (c).
(b) To distinguish between or to test for differences between groups, or to test for the effect of continuous covariates on survival.
(c) To allow for (or 'to remove the effect of', or 'to control for') the effects of certain explanatory variables, in assessing the effects of other explanatory variables, on survival.

We may take items (a) to (c) as ends in themselves, or as initial steps in the further modelling or prediction of the survival/failure mechanism.

The above considerations apply to any data analysis in which a dependent variable is to be related to explanatory variables, perhaps linearly by multiple regression, or nonlinearly by some nonlinear regression; or by a more specific nonlinear setup such as a generalised linear model or a model with immunes and covariates, such as the one which is introduced in the next section. For the case of the model with immunes, the explanatory variables may effect the survival distributions in one of two ways:

(a) Via the probability that an individual is immune.
(b) Via the rate of failure, or the mean or median survival time, or similar, for susceptible individuals.

Our analysis must attempt to disentangle these effects as best it can. Many readers will already be well versed in the fitting of multiple regressions, or in the analysis of variance and covariance of normally distributed data, and will be aware of the difficulties and dangers inherent in fitting and interpreting even such relatively well-understood models. Mosteller and Tukey's (1977) book on regression, especially Chapter 13, 'Woes of Regression Coefficients', is indispensable and salutary reading in this regard. (See also Tukey (1977).)

When we move to the more complicated generalised linear models or to survival models with immunes, these difficulties remain (in rather similar form, due to the 'generalised linear' way in which the covariables will be assumed to affect the survival distribution of susceptibles, and the probabilities of being immune), and indeed are magnified by the extra nonlinearity inherent in the 'link functions' (in generalised linear model terminology) used for the nonnormally distributed data. We will try to elucidate some of the issues, after first presenting some examples of fitting parametric models to data in which immunes may or may not be present.

Before introducing the models, we will give some notation and terminology. We allow the (possibly censored) survival times to depend on *covariates* (also called explanatory variables or regressor variables), by associating with each individual i in the data set a vector containing this auxiliary information, which occurs in the form of observations on, or measurements of, discrete and/or continuous variables. So the structure of the data will be as in Table 6.1, below.

We collect the covariate information on individual i into a *covariate vector*

$$\mathbf{x}_i = (x_{i1}, x_{i2}, \dots, x_{ip})$$

so that our data consists of n observations of the form $(t_i, c_i, \mathbf{x}_i)$. The question of how the covariates affect survival times may then be formalised parametrically by letting the distribution of each t_i depend on the covariate vector $\mathbf{x}_i$ in some way. Then, as usual, we estimate the parameters involved (by maximum likelihood, typically) and test hypotheses about them (by likelihood ratio tests) with, as usual, the minimum possible assumptions on the censoring mechanism which resulted in the censored observations.

As an example, the Gehan–Freireich data given in Table 1.1 (see also Figure 1.1) can now be recognised as being of the form of Table 6.1. The 'survival' times t_i are in this case the times to recurrence of leukaemia (following initial diagnosis and treatment) of the patients, censored ($c_i = 0$) if in fact the leukaemia did not recur within the period of observation. The 'covariate' vector $\mathbf{x}_i$ here reduces to a (one-dimensional) discrete variable x_i taking the value 1 if individual i is in the control group, or the value 2 if i has received a drug treatment. This is the simplest possible (nontrivial) version of a 'covariate'. In Section 4.3 we decided that there is reasonable evidence of a proportion of immunes among those receiving the drug treatment, although follow-up is not sufficient to be really decisive. From Table 1.1 we can also see that there is an obvious difference between the control and drug groups: all observations in the control group are uncensored, so there is no evidence of an immune proportion for this group. Given this difference, we do not discuss further detailed covariate analysis of this data.

As a more complex example, consider the ovarian cancer data listed in Table 6.2, in which the survival times are measured (in years) from the time of treatment of the 26 women in the trial, censored if the patient was still alive at the end of the trial.

Each survival time in Table 6.2 is accompanied by a censor indicator and four covariates. The covariates are (i) the chemotherapy treatment of the patient; (ii) the patient's age at the beginning of the trial; and (iii) and (iv) the values of two other binary indicators (extent of residual disease and a performance

Table 6.1

Survival time	Censor indicator	Covariate information
t_1	c_1	$x_{11}, x_{12}, \dots, x_{1p}$
t_2	c_2	$x_{21}, x_{22}, \dots, x_{2p}$
$\vdots$	$\vdots$	$\vdots$
t_n	c_n	$x_{n1}, x_{n2}, \dots, x_{np}$

Table 6.2 Ovarian cancer data with covariates representing Treatment group 1 (x_1), age (x_2), extent of residual disease (x_3) and performance indicator (x_4).

i	t_i	c_i	x_{i1}	x_{i2}	x_{i3}	x_{i4}
1	0.1616	1	1	72	2	1
2	0.3151	1	1	74	2	1
3	0.4274	1	1	66	2	2
4	0.7342	1	1	74	2	2
5	0.9014	1	1	43	2	1
6	0.9671	1	2	63	1	2
7	1.0000	1	2	64	2	1
8	1.0329	0	2	58	1	1
9	1.1534	0	2	53	2	1
10	1.1808	1	1	50	2	1
11	1.2274	0	1	56	1	2
12	1.2712	1	2	56	2	2
13	1.3014	1	2	59	2	2
14	1.3068	0	1	64	2	1
15	1.5425	1	2	55	1	2
16	1.7479	1	1	56	1	2
17	2.0384	0	2	50	1	1
18	2.1068	0	2	59	2	2
19	2.1096	0	2	57	2	1
20	2.2000	0	1	39	1	1
21	2.3425	0	1	43	1	2
22	2.8493	0	1	38	2	2
23	3.0301	0	1	44	1	1
24	3.0932	0	2	53	1	1
25	3.3041	0	2	44	2	1
26	3.3616	0	2	59	1	2

indicator). Three of these covariates are what we will call 'factors': they take on integer values $1, 2, 3, \ldots$, up to some fixed number, and are merely indicators of which of a number of (cross-classified) groups an individual may belong to. The treatment, residual disease and performance indicators in the ovarian cancer data are of this type; indeed they are all factors consisting of two *levels*, which describe membership of one or another of two groups. More generally, a factor may consist of more than two levels, indicating membership of one of a number of groups. By contrast, the age of a patient in the ovarian cancer data is an example of a *continuous* variate, which potentially may take on any value (positive, in this case).

In reality, limitations of measurement restrict the possible values of a continuous variate, and indeed the ages in Table 6.2 are listed accurate only to within 1 year, so the age variate in that example can be thought of as a (discrete) factor with a relatively large number of levels. We can *discretise* a continuous variate simply by dividing its range into a small number of intervals — for example, the age variate defines a factor of two levels according to whether a patient is under or over 60 years of age — and it is in fact convenient to do this for purposes of data display. The discretising of a continuous variate into a discrete variate necessarily sacrifices some of the information in it, and the resulting analysis is dependent on the particular intervals we choose (usually subjectively) for discretisation. We need to keep this in mind when interpreting the results even of 'looking' at the data in this way.

We have demonstrated often enough in the previous chapters that we place great store in some form of display of the data, and for survival data with covariates this is just as important or even more important than for single-group data. We can reveal differences, if any, in survival distributions between the levels of a factor (i.e. between the subgroups of the data which make up the factors) simply by plotting the KMEs for each of the sets of survival times separately on the same graph. Of course, the significance, or otherwise, of these differences must be assessed by some form of hypothesis test such as we will propose.

6.2 COMPARING TWO GROUPS USING THE WEIBULL MIXTURE MODEL

Many of the ideas of the analysis of data with covariates can be illustrated with data in which there are only two groups, which are often designated generically as a 'treatment' group and a 'control' group. As an example, the reader may like to keep in mind the Gehan–Freireich data mentioned in the last section, and consistent with the aims set out there, think in terms of comparing the two groups, generally, with respect to (a) the rates of failure of susceptibles and (b) the immune proportions, if any, in the groups.

For the moment we will assume as a parametric model for survival times the Weibull mixture distribution discussed in detail in Section 5.1:

$$F(t) = p\left(1 - e^{-(\lambda t)^{\alpha}}\right), \qquad t \geq 0. \tag{6.1}$$

Rates of failure of the susceptibles are measured mainly by the magnitude of the parameter λ in the Weibull model, whereas the magnitude of the immune proportion is measured by $1 - p$. There is a further comparison that can be made between the groups: (c) the shape parameter α.

The interpretation of a difference in shape parameters, if it exists, will be discussed and illustrated later.

The distribution assumed for survival times from each group, then, is of the form (6.1), and we can phrase the above comparisons (a), (b) and (c) in terms of the three-parameters of interest, p, λ and α, by letting them vary between the two groups. Thus we assume the observations in Treatment group 1 come from the distribution

$$p_1\left(1 - e^{-(\lambda_1 t)^{\alpha_1}}\right), \qquad t \geq 0,$$

and those in Treatment group 2 come from the distribution

$$p_2\left(1 - e^{-(\lambda_2 t)^{\alpha_2}}\right), \qquad t \geq 0,$$

where, in general, the parameters describing the two groups may be different: we allow $p_1 \neq p_2, \lambda_1 \neq \lambda_2$ and $\alpha_1 \neq \alpha_2$. We will assume that $0 < p_1 < 1$ and $0 < p_2 < 1$ for the time being. Suppose, by renumbering the observations if necessary, that the first n_1 observations out of the total of n come from Treatment group 1, and the remaining $n_2 = n - n_1$ come from Treatment group 2.

Models 000, 111, etc.

The hypotheses of interest are of the form $p_1 = p_2$, $\lambda_1 = \lambda_2$, $\alpha_1 = \alpha_2$, and/or $p_1 = 1$, $p_2 = 1$, or $p_1 = p_2 = 1$, and these may be taken in conjunction with each other. These encapsulate the questions we want the data to answer for us. Because of the large number of models to be considered, we will adopt a certain coding system which we hope will be evocative. First, we may consider that parameters p_1 and p_2 (the immune proportions for the two groups) do not differ between the groups, regardless of whether λ_1 and λ_2, or α_1 and α_2, are equal. This hypothesis can be formulated as

$$H_{011} : p_1 = p_2,$$

in which case we write the common value of p_1 and p_2 as p, assumed to be in (0, 1). We will describe the distribution of survival times under H_{011} as follows.

Model 011 The p's of the two groups are the same, but the λ's and α's may differ.

Alternatively, we may wish to consider the possibility that $\lambda_1 = \lambda_2$, regardless of the values of the other parameters. This is Model 101.

Model 101 The λ's of the two groups are the same, but the p's and α's may differ.

Associated with Model 101 we also have a null hypothesis H_{101}, and similarly for the other models.

Next, the shape parameters of the Weibull may be the same, regardless of the other parameters.

Model 110 The α's of the two groups are the same, but the λ's and p's may differ.

There are also three models in which two pairs of parameters are equal.

Model 001 $p_1 = p_2$ and $\lambda_1 = \lambda_2$, but α_1 may differ from α_2.

Model 010 $p_1 = p_2$ and $\alpha_1 = \alpha_2$, but λ_1 may differ from λ_2.

Model 100 $\lambda_1 = \lambda_2$ and $\alpha_1 = \alpha_2$, but p_1 may differ from p_2.
There are two models left.

Model 111 p_1 and p_2, λ_1 and λ_2, and α_1 and α_2, may differ.
This model simply allows each group to have its own description in terms of the p, λ and α parameters.

Model 000 $p_1 = p_2$, $\lambda_1 = \lambda_2$ and $\alpha_1 = \alpha_2$.
Model 000 requires no differences in parameters between groups; in other words, it specifies that the distribution of survival times is identical for each group, i.e. there is simply no distinction between treatment groups in their survival time distributions.

Fitting the models by maximum likelihood

The above plethora of models is required if we are to investigate all possible forms of Weibull distributions which may fit our two-group data. As examples will show, we do need to investigate them carefully. We also need to be systematic in fitting them, and in testing the hypotheses they represent. A way of doing this is as follows.

Let us first consider the fitting of Model 111, which allows the most general situation: possibly different p's, λ's and α's. The likelihood has the form (cf. equation (5.14))

$$\begin{aligned} L_{111} &= L_n(p_1, p_2, \lambda_1, \lambda_2, \alpha_1, \alpha_2, t_1, \ldots, t_n) \\ &= k\prod_{i=1}^{n_1}\left(p_1\alpha_1\lambda_1^{\alpha_1}t_i^{\alpha_1-1}\mathrm{e}^{-(\lambda_1 t_i)^{\alpha_1}}\right)^{c_i}\left(1-p_1+p_1\mathrm{e}^{-(\lambda_1 t_i)^{\alpha_1}}\right)^{1-c_i} \\ &\quad\times\prod_{i=n_1+1}^{n}\left(p_2\alpha_2\lambda_2^{\alpha_2}t_i^{\alpha_2-1}\mathrm{e}^{-(\lambda_2 t_i)^{\alpha_2}}\right)^{c_i}\left(1-p_2+p_2\mathrm{e}^{-(\lambda_2 t_i)^{\alpha_2}}\right)^{1-c_i} \end{aligned} \qquad (6.2)$$

(where k is a constant not depending on the parameters).

Here we applied the rule for writing down a likelihood of censored data established in Section 5.1: multiply together the contributions from each individual, as the postulated density for the group the individual belongs to, if the survival time is uncensored, or as the tail of the probability distribution for the group the individual belongs to, if the survival time is censored. We have to maximise L_{111}, or more conveniently its logarithm l_{111}, for variations in all six parameters.

Note that, for Model 111, the likelihood factorises into two completely separate components, one depending only on p_1, λ_1 and α_1, the other only on p_2, λ_2 and α_2. What's more, those components are, apart from a constant factor, just the likelihoods of each of the two groups considered separately. It follows that the six parameter estimates $\tilde{p}_1$, $\tilde{p}_2$, $\tilde{\lambda}_1$, $\tilde{\lambda}_2$, $\tilde{\alpha}_1$ and $\tilde{\alpha}_2$, say, obtained under Model 111, are just those that would be obtained by fitting the Weibull mixture model to each group separately, according to the method discussed in Section 5.1. Furthermore the value of the maximised log-likelihood will be (up to a nonessential constant) the sum of the maximised log-likelihoods from each of the two groups. We will write the likelihood as

$$\tilde{L}_{111} = L_{111}(\tilde{p}_1, \tilde{p}_2, \tilde{\lambda}_1, \tilde{\lambda}_2, \tilde{\alpha}_1, \tilde{\alpha}_2). \tag{6.3}$$

This factorisation into separate components does not occur, however, for the other models, since the groups have one or more parameters in common, and this must be taken into account when maximising the likelihood. Suppose for example that we wish to test H_{011}, as specified above, so we fit Model 011, in which $p_1 = p_2$ but the λ's and α's are allowed to differ. The likelihood now takes the form

$$\begin{aligned} L_{011} = k \prod_{i=1}^{n_1} \left(p\alpha_1 \lambda_1^{\alpha_1} t_i^{\alpha_1 - 1} \mathrm{e}^{-(\lambda_1 t_i)^{\alpha_1}} \right)^{c_i} \left(1 - p + p\mathrm{e}^{-(\lambda_1 t_i)^{\alpha_1}} \right)^{1-c_i} \\ \times \prod_{i=n_1+1}^{n} \left(p\alpha_2 \lambda_2^{\alpha_2} t_i^{\alpha_2 - 1} \mathrm{e}^{-(\lambda_2 t_i)^{\alpha_2}} \right)^{c_i} \left(1 - p + p\mathrm{e}^{-(\lambda_2 t_i)^{\alpha_2}} \right)^{1-c_i} \end{aligned} \tag{6.4}$$

where p is the common value of p_1 and p_2, and k is a factor not depending on the parameters.

If we fit Model 011 by the methods described in Section 5.1, for example, we will obtain only five parameter estimates: $\breve{p}$, $\breve{\lambda}_1$, $\breve{\lambda}_2$, $\breve{\alpha}_1$ and $\breve{\alpha}_2$, say, from which we can calculate a corresponding maximised value of the likelihood

$$\breve{L}_{011} = L_{011}(\breve{p}, \breve{\lambda}_1, \breve{\lambda}_2, \breve{\alpha}_1, \breve{\alpha}_2). \tag{6.5}$$

In this case both groups contribute information or 'add strength' to the estimation of p, and this also modifies the estimates of λ_1, λ_2, α_1 and α_2. (Examples of this will be given below.)

Comparing models by analysis of deviance

In order to test H_{011} we simply compare the values of the likelihoods $\tilde{L}_{111}$ and $\breve{L}_{011}$ via their ratio, or more conveniently, via the deviance statistic

$$d_n = 2\left(\log \tilde{L}_{111} - \log \breve{L}_{011} \right).$$

According to the asymptotic theory to be outlined in Section 8.4, if the sample is large enough, d_n will have as approximate distribution a χ^2 distribution with one degree of freedom, where

$$\begin{aligned} 1 = \text{no. of parameters estimated under } H_{111} \text{ (i.e., 6)} \\ - \text{ no. of parameters estimated under } H_{011} \text{ (i.e., 5)}. \end{aligned}$$

(Note that the asymptotic theory of d_n falls into the class of 'interior' problems, in the terminology of Section 5.2, since the hypotheses H_{111} and H_{011} specify values of p_1, p_2 and p strictly between 0 and 1.)

Consequently, a test for H_{011}, against the alternative H_{111}, is simply to compare d_n with the appropriate percentile of the χ^2 distribution with one degree of freedom. If the test is to be performed at the 95% significance level, we simply look and see whether d_n exceeds 3.84 or not. If it does, we reject H_{011} and decide that a model which specifies equal values of p_1 and p_2, the susceptible proportions of the treatment and control groups, does not provide an adequate description of the data; in other words, the immune proportions differ significantly between the two groups. Alternatively, if we accept H_{011}, separate estimates of p_1 and p_2 are superfluous to a reasonable description of the data — we can condense them into a single estimate $\breve{p}$, with no significant loss of information.

All of this analysis, and the others to be described below, depends of course on the Weibull mixture model providing a good description of the data. We should test the goodness of fit of the Weibull distribution to each group separately — i.e. effectively test the goodness of fit of Model 111 — by some method such as that described in Section 5.4. Note further that we have assumed so far that the estimates of p_1 and p_2 under Model 111, or of p under Model 011, are strictly less than 1. However, part of our analysis should be to see whether this is actually the case. This can be tested for each group separately, either nonparametrically as in Section 4.2, or parametrically as in Section 5.2, after fitting an exponential, Weibull or some other mixture model to each group. If we find at this stage that one group possesses an immune proportion (at some level of confidence) whereas the other does not, we can take this as evidence of a difference between the groups in respect to their immune proportions. There may still be an interest in fitting Model 011, however, because 'strength' is contributed towards estimation of the parameters concerned by the pooling of groups together.

Regardless of whether there is a difference in p's or not, we may wish to compare λ_1 and λ_2 and/or α_1 and α_2 between the two groups, and we can do this by the method outlined earlier. If we wish to examine the plausibility of hypothesis H_{101}, we first fit Model 101 by maximising the likelihood L_{101}, which is obtained by equating λ_1 and λ_2 to λ, say, in (6.2), but leaving p_1, p_2, α_1 and α_2 free to vary. This results in estimates $\breve{p}_1$, $\breve{p}_2$, $\breve{\lambda}$, $\breve{\alpha}_1$ and $\breve{\alpha}_2$. We than evaluate (6.2) with $p_1 = \breve{p}_1$, $p_2 = \breve{p}_2$, $\lambda_1 = \lambda_2 = \breve{\lambda}$, $\alpha_1 = \breve{\alpha}_1$ and $\alpha_2 = \breve{\alpha}_2$, obtaining a value of the maximised likelihood $\breve{L}_{101}$, say. Following this we form the deviance statistic

$$d_n = 2\left(\log \tilde{L}_{111} - \log \breve{L}_{101}\right) \tag{6.6}$$

and compare it with the 95th percentile of a χ_1^2 random variable, 3.84. This provides the likelihood ratio test for H_{101} versus H_{111}. In an exactly analogous way, we can test H_{110}, H_{100}, H_{010}, H_{001} and H_{000}, against H_{111}. Note that fitting Model 000 simply means fitting the Weibull mixture model to the whole data set, ignoring the separation into groups, by the single-group method of Section 5.2.

We can also test various of the restricted hypotheses against each other. For example, a test of H_{011} versus H_{001} compares values of λ between the two groups, having already restricted the values of p_1 and p_2 to be equal, but allowing the values of α to differ. This comparison would be tested by the deviance statistic

$$d_n = 2(\log \tilde{L}_{011} - \log \breve{L}_{001}), \tag{6.7}$$

in a notation analagous to that used in (6.6), and similarly for other comparisons. Another logical variant is to begin by fitting the model with the fewest parameters, 000, then to add parameters one by one until no further improvement in fit (i.e. significant decrease in deviance) is obtained. We will illustrate various versions of these methods in the examples which follow.

6.3 EXAMPLES OF TWO-GROUP ANALYSES

A two-group analysis: breast cancer data

The data in Figure 6.1 represent the survival times, following surgery, of 45 breast cancer patients. This data was analysed by Leathem and Brooks (1987).

The data, which is also tabulated in Table 6.3, is classified into two treatment groups: Group 1 = − (negative staining) and Group 2 = + (positive staining). Figure 6.1 shows the KMEs together with Weibull mixture models fitted separately to each treatment group.

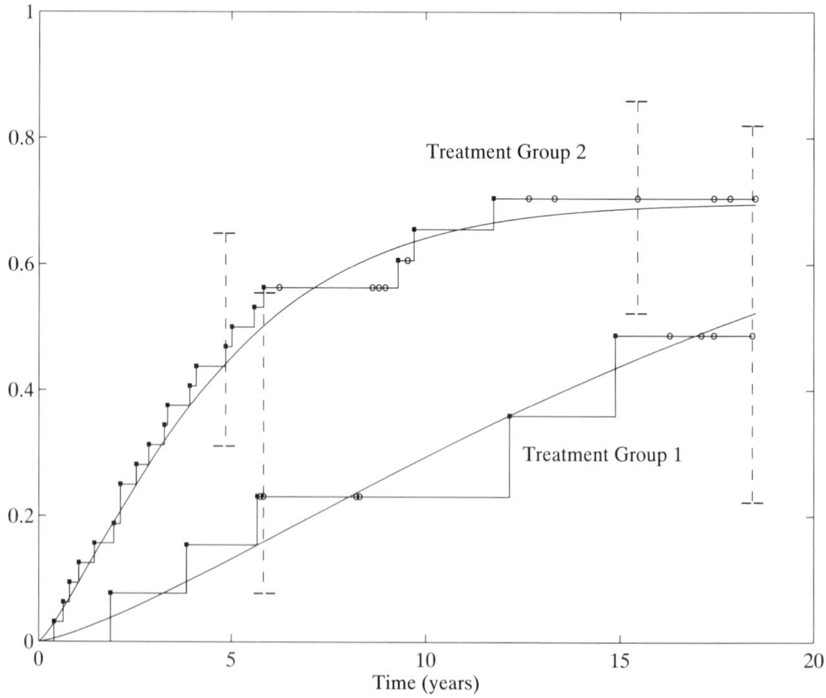

Figure 6.1 Breast cancer data: Group 1 = −(negative staining), Group 2 = +(positive staining); dotted lines indicate 95% confidence intervals on the KMEs.

Table 6.3 Breast cancer data: Group 1 = negative staining, Group 2 = positive staining; t_i is the survival time in years of patient i.

Group 1			Group 2					
No.	t_i	c_i	No.	t_i	c_i	No.	t_i	c_i
1	1.8904	1	1	0.4110	1	17	5.5890	1
2	3.8630	1	2	0.6575	1	18	5.8356	1
3	5.6712	1	3	0.8219	1	19	6.2466	0
4	5.7534	0	4	1.0685	1	20	8.6301	0
5	5.8356	0	5	1.4795	1	21	8.7945	0
6	8.2192	0	6	1.9726	1	22	8.9589	0
7	8.3014	0	7	2.1370	1	23	9.2877	1
8	12.1644	1	8	2.1370	1	24	9.5342	0
9	14.8767	1	9	2.5479	1	25	9.6986	1
10	16.2740	0	10	2.8767	1	26	11.7534	1
11	17.0959	0	11	3.2877	1	27	12.6575	0
12	17.4247	0	12	3.3699	1	28	13.3151	0
13	18.4110	0	13	3.9452	1	29	15.4521	0
			14	4.1096	1	30	17.4247	0
			15	4.8493	1	31	17.8356	0
			16	5.0137	1	32	18.4932	0

A number of points seem visually apparent in Figure 6.1:

1. The Weibull fit to each group is good (and this will be verified below by the method of Section 5.4).
2. The two groups differ greatly in their survival patterns: individuals in Group 2 (+) fail much more quickly than those in Group 1 (−) (although the 95% confidence intervals for the KME plotted in a couple of places in Figure 6.1 overlap substantially).
3. There seems reasonable visual evidence of an immune proportion for Group 2 (+), but less so for Group 1 (−), where deaths continue at a low rate for up to 15 years after surgery.

We shall see that the above impressions are fairly well borne out by a formal two-group analysis, in which we fit the three-parameter Weibull mixture model and compare parameters as described in Section 6.2. Table 6.4 lists the parameter estimates and values of the deviance, $-2 \log L$, i.e. minus twice the log-likelihood evaluated at the fitted parameters, for various Weibull mixture models, using the nomenclature defined in Section 6.2.

These models were fitted by the Fortran program described in Section 5.2, beginning in this case with the simplest model, Model 000, then progressing to the more complex models.

Model 000 (row 1 of Table 6.4) is a single Weibull mixture model fitted to the whole data set, thus in effect ignoring the separation into the + and − treatment groups. At the other extreme, Model 111 (row 5 of Table 6.4) allows different

Table 6.4 Two-group analysis of deviance for breast cancer data (Leathem and Brooks 1987).

Row	Model	No. of parameters	Treatment group	$\tilde{p}$	$\tilde{\lambda}$	$\tilde{\alpha}$	$-2\log\tilde{L}$
1	000	3	(both)	0.67	0.1595	1.19	184.96
2	010	4	−	0.70	0.0642	1.30	178.63
			+	0.70	0.2087	1.30	
3	011	5	−	0.70	0.0662	1.41	178.57
			+	0.70	0.2088	1.27	
4	110	5	−	0.93	0.0473	1.28	178.57
			+	0.70	0.2090	1.28	
5	111	6	−	0.83	0.0544	1.34	178.56
			+	0.70	0.2092	1.27	
6	101	5	−	0.48	0.1315	1.42	182.48
			+	0.81	0.1315	1.05	
7	001	4	−	0.66	0.1683	1.07	184.90
			+	0.66	0.1683	1.24	
8	100	4	−	0.46	0.1583	1.18	182.59
			+	0.75	0.1583	1.18	

values of p, λ and α, for each of the groups, effectively fitting a Weibull mixture model separately to each group. The in-between models have one or more of the parameters p, λ and α common to the two groups.

The final four columns of Table 6.4 list the MLEs of the parameters for each model fitted, and the corresponding values of the deviance. Differences in deviance are equivalent to likelihood ratio tests for the hypotheses corresponding to the fitted models, and constitute approximately observations on chi-square random variables whose degrees of freedom equal the difference in the numbers of parameters fitted in each model.

To see how this works, consider the first two rows in Table 6.4. They contain the information we need for Model 000 and for Model 010, in which only the λ parameters are permitted to differ between the two treatment groups. We chose to examine this model first because a difference in survival rates is the most striking feature in Figure 6.1.

Model 010 allows different λ's between the two groups, but restricts the p's and α's to be the same. The estimated values of the four-parameters can be read from row 2 of Table 6.4 as

$$\tilde{p} = 0.70, \qquad \tilde{\lambda}_1 = 0.0642, \qquad \tilde{\lambda}_2 = 0.2087, \qquad \tilde{\alpha} = 1.30.$$

The value of $-2\log L$ for these parameters is 178.63. By comparison, Model 000 produced estimates

$$\breve{p} = 0.67, \qquad \breve{\lambda} = 0.1595, \qquad \breve{\alpha} = 1.19,$$

with a value of $-2 \log L$ of 184.96. The deviance difference is $184.96 - 178.63 = 6.33$, and we take it as approximately an observation on a chi-square random variable with $4 - 3 = 1$ degree of freedom. Since 6.33 considerably exceeds 3.84, the 95th percentile of χ_1^2, we reject the hypothesis tested by this procedure: that the λ values of the two treatment groups are the same (whilst the p and α values are also kept the same), and decide that we must allow different λ values for an adequate description of the data.

Next we ask if different values of α are also required, so we now allow the extra flexibility of the 011 model. Fitting it results in the parameter estimates in row 3 of Table 6.4. We see that the estimates of the α parameters for each group, 1.41 and 1.27, are little different, and indeed the deviance has only decreased from 178.63 to 178.57, a small and insignificant change of 0.06 (by comparison with 3.84) for the addition of the extra α parameter. Thus we decide that two α parameters are not necessary for adequate description of the data, or equivalently, the two treatment groups do not differ in respect to the shape parameters of their survival distributions. Likewise, comparing the 010 model with the 110 model gives a deviance change of only $178.63 - 178.57 = 0.06$, again nonsignificant, showing that different values of p are not required either.

We also include the estimates for the other models in Table 6.4. Notice that, by comparison with the 010 model, the other four-parameter models, 001 and 100, have deviances which are larger by amounts $184.90 - 178.63 = 6.27$ and $182.59 - 178.63 = 3.96$, both of which are significant at the 5% level. Thus we are not inclined to accept either the 001 or 100 model as being better modellers of the data than the 010 model.

This procedure has produced a model, 010, which gives as good a description of the data in terms of maximising the likelihood, as we can get with the six possible Weibull mixture parameters, three for each treatment group—a model which is also *necessary* for adequate description of the data, in the sense that reducing it to the next simplest model, the 000 model, results in a significantly smaller likelihood or, equivalently, a significantly larger deviance.

How does this help our interpretation of the effect of the treatments on survival? Referring back to Figure 6.1, a difference in survival rates λ between the treatment groups is certainly apparent. We also decided from the analysis of deviance that there is not a significant difference in immune proportions between the two groups, and this is not inconsistent with Figure 6.1 either. Note that, although the KME for Group 2 appears to have levelled off, the KME for Group 1 appears not to have done so, and the estimate of p for Group 1 is of low precision. The fitted Weibull distributions follow the KMEs fairly faithfully (goodness of fit will be verified more formally below), and if we are prepared to extrapolate the Weibull fit for the Group 2 data, we see that it may well climb to a value similar to Group 1, for large survival times.

In order to throw further light on this aspect, let us now test the hypotheses $H_0 : p_1 = 1$ and $H_0 : p_2 = 1$ separately for each of the two treatment groups. These are boundary hypothesis tests as discussed in Section 5.3. To make this test for

Group 1, we fit the Weibull mixture model to the data for Group 1, obtaining the parameter estimates listed for Group 1 (+) under the 111 model in Table 6.4. The corresponding deviance, not shown in Table 6.4, is 42.63. We compare this with an ordinary Weibull model fitted to the Group 1 data with the statistical package GLIM as described in Aitkin *et al.* (1989). The parameter estimates for this fit are

$$\breve{\lambda} = 0.0437 \text{ and } \breve{\alpha} = 1.28$$

with a corresponding deviance of 42.78. We compare the deviance difference

$$42.78 - 42.63 = 0.15$$

with 2.71, the 95th percentile of the distribution in (5.38) (see (5.39)), obtaining a nonsignificant result. Thus the likelihood ratio test gives no evidence of immunes in Group 1.

For Group 2, the estimates for the ordinary Weibull fit are

$$\breve{\lambda} = 0.0936 \text{ and } \breve{\alpha} = 0.89$$

with a corresponding deviance of 140.83. The deviance for the Weibull mixture model is 135.93, giving a difference of

$$140.83 - 135.93 = 4.90$$

which is significant by comparison with 2.71. Thus there is quite good evidence that $p_2 < 1$, i.e. that there are immunes in Group 2.

These tests suggest a difference between immune proportions for the two groups, in that we accepted that $p_1 = 1$ for Group 1, but decided that $p_2 < 1$. But Figure 6.1 shows that a large extrapolation is needed to draw this conclusion from the Group 1 data. On the other hand, the analysis of deviance based on the pooled data given in Table 6.4 found no difference between the groups in respect to their p parameters. We accept this latter conclusion, in that the pooled test is more powerful than those based on the individual groups. Given this, and the scarcity of data in Group 1, we might then be prepared to combine the two groups for the purposes of the boundary test, in which case we compare the sums of the two deviances for each group. This gives a difference of

$$(42.78 + 140.83) - (42.63 + 135.93) = 183.61 - 178.56 = 5.05$$

which is again significant by comparison with 2.71. Thus there appears to be evidence of immunes in this data, considered overall.

The above analysis clearly requires some caveats. First we should keep in mind that we have applied asymptotic results to a finite sample of data, and the data in Group 1 are indeed sparse—there are only 13 observations, 8 of them censored. This concerns us most, perhaps, when we treat as observations on chi-square random variables the deviance differences between the fitted models. It is not presently known just how accurate this is, but it is no more or less a problem for the Weibull mixture model than it is for the analyses of deviance of generalised linear models, or for the ordinary survival analyses of censored lifetime data which are used so frequently in practice. We are therefore no worse off with the mixture model in this respect, and we must make the most of the data we have.

It is interesting to compare the results of our analysis of deviance with those of Collett (1994, pp. 126, 131), who performed an 'ordinary' Weibull survival analysis on this data. We reach the same conclusion as Collett, concerning the difference in rates of survival, namely that there are significantly different λ parameters between the two groups (the significance was not quite at the 0.05% level in Collett's case, whereas it is high in our analysis) — but of course the ordinary analysis gives no clues as to the existence of an immune proportion for Group 2 (+) (and whether follow-up is sufficient to be confident of this or not). Collett did not compare α values — the shapes of the Weibull distribution — between the two groups.

In statistics we must discuss the precision of estimation of any procedure. To do this here, we report approximate 95% confidence intervals for the unknown parameter values in the fitted models. These were calculated by the method discussed in Section 5.2, using the inverse of the observed information matrix of the data. For the 010 model the estimates with the corresponding 95% confidence intervals are

$$\tilde{p} = 0.70\ (0.50, 0.85), \qquad \tilde{\lambda}_1 = 0.0642\ (0.0272, 0.1518),$$

$$\tilde{\lambda}_2 = 0.2087\ (0.1385, 0.3144), \qquad \tilde{\alpha} = 1.30\ (0.92, 1.82).$$

These show that precision, as would be expected from such a small sample, is not high.

For comparison with the parametric analysis, we now make the *nonparametric* tests for the presence of immunes in each of the two groups, using the method of Section 4.2. For Group 1 the maximum value of the KME is $\hat{p}_1 = 0.4872$, from a sample of size $n = 13$ of which 8, or about 60%, are censored. We should thus refer the value 0.4872 to Table A.1 with $n = 20$ and μ a little smaller than 1, so that $1/(1 + \mu)$ is about 0.6. Since 0.4872 is less than 0.5699, the 5th percentile corresponding to $n = 20$ and $\mu = 1$, there is some evidence of the presence of immunes in this group. Table A.2 with $B = 2$ gives the same conclusion. Likewise, for Group 2, $\hat{p}_2$ is 0.7047, $n = 32$ and there are 11, or about 1/3, censored observations. Table A.1 with $n = 30$ and $\mu = 2$ gives the first percentile as 0.7431, leading us to reject the hypothesis that $p_2 = 1$. Table A.2 with $B = 4$ gives the same conclusion.

However, follow-up is not sufficient for a definite conclusion from this data. The nonparametric statistic $q_n = N_n/n$ for testing sufficient follow-up takes values of

$$2/13 = 0.15 \text{ for Group 1}$$

$$5/32 = 0.16 \text{ for Group 2.}$$

Neither of these values provide strong evidence for or against the hypothesis that follow-up is sufficient, on comparing them with the percentiles listed in

Tables B.1 to B.7. For Group 1, we have $\hat{p}_1 = 0.4872$ and $n = 13$, and the 10th and 90th percentiles for $p = 0.5$ and $n = 20$ are 0.05 and 0.35, which bracket the observed q_n value of 0.15. Similarly, for Group 2, we have $\hat{p}_1 = 0.7047$ and $n = 32$, and the 10th and 90th percentiles for $p = 0.7$ and $n = 20$ are 0.05 and 0.40, which again bracket the observed q_n value of 0.16.

To complete this analysis we will consider the goodness of fit of the Weibull distributions to this data. Visual assessment of Figure 6.1 would lead us to accept the Weibull with no qualms, but let us illustrate the more formal procedure for goodness of fit discussed in detail in Chapter 5. Tables C.1 to C.9 contain the simulated percentage points of the correlation coefficient statistic r, between observed (KME) and fitted (Weibull) distributions, when the true survival distribution is exponential and a range of uniform censoring distributions are allowed. As we discussed in Section 5.4, the percentage points of r do not seem to vary much with the form of the censoring distribution, and seem to be largely determined by the sample size n, the susceptible proportion p and the extent of the censoring. The present fitted Weibull distributions have shape parameters close to 1 (Table 6.4), hence are close to exponential, but in any case we can use the values in Tables C.1 to C.9 for a goodness of fit test, as we showed in Appendix Note 5 to Chapter 5 that the correlation coefficient statistic is invariant to the choice of distributions after transformation.

To carry out the test, we calculate the correlation coefficients r between the KME and the fitted values from Model 111 for each treatment group separately, obtaining

$$r = 0.9812 \text{ for Group 1}$$

$$r = 0.9912 \text{ for Group 2.}$$

Comparing the observed value of 0.9812 with the percentage points in Table C.7 (with $p = 0.8$, since the estimated p is 0.83 for Group 1), we see that 0.9812 is well above any 20% point for $n = 20$ (note that the percentage points tend to be smaller with smaller n). Hence we decide that the Weibull mixture model provides quite a good fit to the Group 1 data. For Group 2, a value of 0.9912 substantially exceeds any 20% point for $n = 30$ in Table C.6 (we look under $p = 0.7$, as $\tilde{p} = 0.70$ for Group 2), indicating a good fit of the Weibull mixture model to the Group 2 data as well.

A two-group analysis with a number of factors: ovarian cancer data

The data in Table 6.2 are given by Therneau (1986) and Edmunson *et al.* (1979), and are also analysed by Collett (1994, Table 4.6, pp. 141–144). They consist of 26 survival times (in years) of ovarian cancer patients from a study of different chemotherapy treatments. The covariates are as follows:

1. Treatments (x_1)
   - Treatment group 1 = cyclophosphamide alone

   - Treatment group 2 = cyclophosphamide combined with adriamycin
2. Age of patient at beginning of trial, in years (x_2)
3. Residual disease (x_3)
   - Residual group 1 = residual disease incompletely excised
   - Residual group 2 = residual disease completely excised
4. Performance status (x_4)
   - Performance group 1 = good
   - Performance group 2 = poor

The main interest in this experiment centres on the effect of the chemotherapy treatments, possibly after taking into account the different ages, residual disease scores and performance status of the patients.

Let us begin by looking (literally) at the data in Table 6.2. Our first impression is that it is a small set: 26 individuals with 4 associated covariates on each is, generally speaking, not a lot of information per covariate. Just how much can we expect this data to tell us? Collett's (1994, p. 143) analysis proceeds by fitting the 'ordinary' Weibull model with no immunes. He decides that there is an effect on survival of the age of a patient, but only a slight (nonsignificant) effect of the treatment. But, simply by looking down the first two columns of the data, we notice that large survival times tend to be censored, so there is some evidence of the existence of an immune component. We will see that treating this data as a Weibull mixture model will lead to somewhat different conclusions from Collett's, and will also reveal some interesting aspects of the data that do not show up in the ordinary Weibull analysis.

Let us continue 'looking' at the data by examining the Kaplan–Meier empirical distributions of the survival times. In Figure 6.2 these are plotted separately for each treatment group, together with fitted Weibull mixture models.

We are immediately struck by the following observations:

(i) A possible levelling of the distributions, indicative of immune or cured proportions in each group.

(ii) An apparent difference between the two groups in rates of failure, for deaths occurring up to about 2 years after the beginning of the trial, but then little apparent difference beyond this. At a couple of points on the KME plots, 95% pointwise confidence intervals have been drawn, to give an idea of the precision of estimation (not great, with such a small set!). The confidence intervals for the two treatment groups overlap, so a treatment difference, if it exists, is not obvious from this. But the fitting of a covariate model with deviance tests for differences between groups carried out according to the procedure outlined in Section 6.2 may well detect significant differences between the groups, whereas a crude comparison of confidence intervals may not.

We must also remember at this stage that differences apparently due to treatment may in part, or fully, be explained by differences in the other factors; and to a

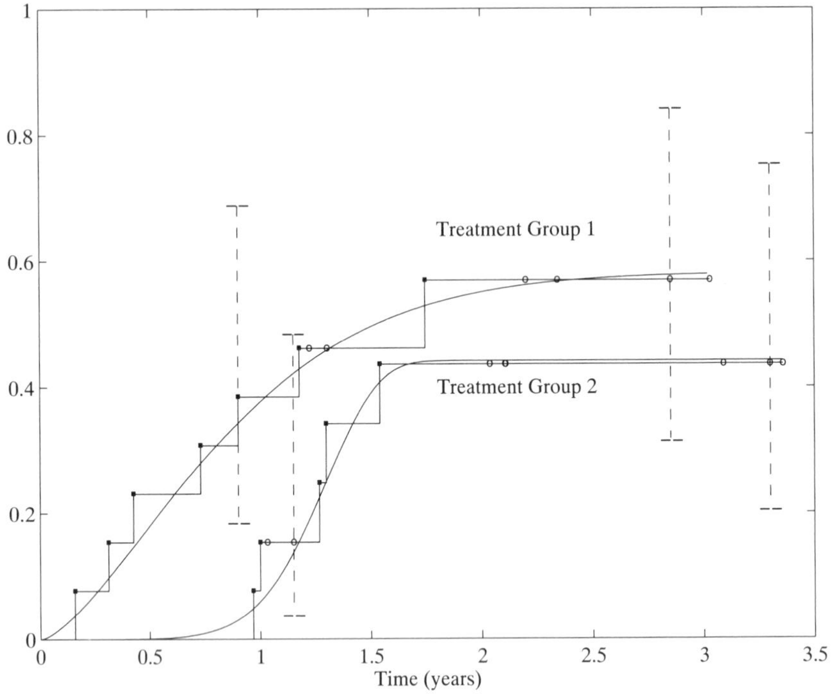

Figure 6.2 Ovarian cancer data: Treatment group 1 = single chemotherapy, Treatment group 2 = combined chemotherapies; dashed lines indicate 95% confidence intervals.

certain extent this will turn out to be so in these data. So before we plunge into an analysis of deviance to test the treatment effects, let us also look at the effects of the other covariates.

In Figure 6.3 we show the KME for each of two age-groups — individuals no more than 60 years old at the beginning of the trial (denoted by '$\leq$60 patients') and individuals older than 60 (denoted by '60+ patients').

From Figure 6.3 we see a distinct difference in survival: the death rate of those over 60 is much higher than for those under 60; furthermore, there appears to be little or no immune proportion (long-term survival) for those over 60, whereas the KME for those under 60 levels off at a value around 0.36. Note that this is precisely the same data as was graphed in Figure 6.2, where there is a suggestion that both groups appear to have immune proportions; there is no contradiction here, since we have reclassified the data according to a different criterion.

Closer inspection of the values in Table 6.2 also reveals that, among the 60+ age-group, all but two patients are from Treatment group 1, and these fail faster, at least in the first year of the trial, as Figure 6.2 shows. The uneven allocation of patients of varying ages to treatment groups may have been a deliberate decision on the part of the experimenter, in that the more rigorous Treatment type 2

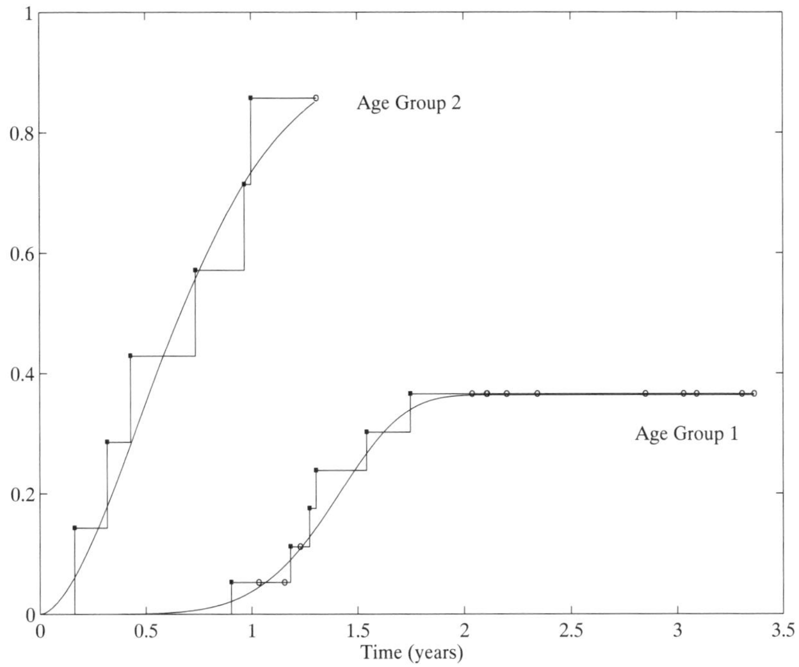

Figure 6.3 Ovarian cancer data: Age-group 1 = under 60 years, Age-group 2 = over 60 years.

(two chemotherapies combined) may be more dangerous to older patients. This considerable confounding between the treatment effects and the age effects warns us to be careful as to whether we attribute significant effects, if any, to treatments or to ages, at least in the first year or so after the commencement of the trial.

Note also that the division of ages into groups representing those aged under or over 60 years is arbitrary — we could have chosen some other age, or divided the data into more than two age-groups, for the purpose of Kaplan–Meier display. But 60 is convenient here since it allocates equal numbers of deaths (six) to each group (although the total numbers of individuals in the age-groups differ, being only 7 for 60+ and 19 for the rest). We discuss this point further below.

This data set is also classified into two groups according to excision of the residual of the disease. Here again there is some confounding with treatments; the first five observations (failures) in Residual group 2 (completely excised) are from Treatment group 1, and it is again possible that the residual condition of these patients influenced the experimenter's decision as to which treatment type was to be applied. Be that as it may, there is some evidence of differences between the two levels of the residual effect, as may be seen by plotting the KMEs for each Residual group (not given here).

Finally, a plot of the data divided into the two performance groups (also not shown here) reveals little evidence of a difference between these two groups, and the treatment groups are more evenly divided between the performance groups, as a perusal of Table 6.1 shows.

To sum up, visual inspection of the data in Table 6.2 and the Kaplan–Meier plots in Figures 6.2 and 6.3 has suggested the following:

(a) There is some evidence that patients in Treatment group 1 have a higher death rate, which commences earlier than that of Treatment group 2. The latter is low until a year from the beginning of the study but then rises quickly over a short period of less than a year. Some 2 years after commencement, failure appears to be completed for both groups, and remaining individuals provide some evidence of an immune component. We noticed, and will keep in mind, that all but two of the highly failure-prone patients over 60 years old received Treatment type 1.

(b) There appears to be a distinct effect due to the age of the patient, those under 60 having a much lower death rate, which does not even begin to take effect until about a year after the commencement of the trial. There is strong visual evidence of an immune proportion for this group. By contrast, those over 60 have a high death rate which begins early in the trial and is virtually complete after a year. There is little evidence of immunes in this group.

(c) There appears to be some difference between residual groups, with a lower proportion of immunes for Residual group 2, the latter group also consisting mainly of Treatment group 1 patients, at least up to 1 year after the commencement of the trial.

(d) For performance groups, we discern a slightly lower survival curve for performance group 1.

Let us now see how the above impressions are borne out by an analysis of the data. We begin with the most important effect—the treatment. Table 6.5 lists the

Table 6.5 Treatment groups analysis: ovarian cancer data.

Row	Model	No. of parameters	Treatment group	$\tilde{p}$	$\tilde{\lambda}$	$\tilde{\alpha}$	$-2\log\tilde{L}$
1	000	3	(both)	0.51	0.8673	2.11	49.31
2	100	4	1	0.59	0.8671	2.11	48.76
			2	0.43	0.8671	2.11	
3	110	5	1	0.57	1.0026	2.09	47.84
			2	0.44	0.7365	2.09	
4	111	6	1	0.58	1.0435	1.50	40.23
			2	0.44	0.7577	6.61	
5	101	5	1	0.65	0.7625	1.29	40.84
			2	0.44	0.7625	6.61	
6	001	4	1	0.53	0.7648	1.35	41.69
			2	0.53	0.7648	6.56	

parameter estimates and values of the deviance, $-2\log L$, for various Weibull mixture models, using the nomenclature defined in Section 6.2. These models were fitted by the Fortran program described in Section 5.2, beginning with the simplest, Model 000, and progressing to the most complicated, Model 111. Recall that Model 000 is in fact a single Weibull mixture model fitted to the whole data set, ignoring the separation into groups. If in fact there are significant differences between groups it will be a poor description of the data, but this is what we want to determine. So the 000 model forms the basis for comparison. In this case, it produces parameter estimates $\tilde{p} = 0.51$, $\tilde{\lambda} = 0.8673$ and $\tilde{\alpha} = 2.11$, with a value of $-2\log \tilde{L}$, evaluated at these parameters, of 49.31. Notice that the estimated Weibull shape parameter $\tilde{\alpha}$ is substantially greater than 1, reflecting the sigmoidal shapes of the curves in Figure 6.2.

Now we add parameters judiciously, watching the change in deviance. Suppose we next fit the 100 model, allowing for different immune proportions but keeping the λ and α parameters the same. From the table we see that the susceptible proportions are estimated to be 0.59 and 0.43, whereas the deviance for this model has decreased to 48.76. As usual we take the difference $49.31 - 48.76 = 0.55$ as approximately an observation on a chi-square random variable whose degrees of freedom equal the difference in the number of parameters fitted under the two models; namely, $4 - 3 = 1$ (see column 3 of Table 6.5). Since $0.55 < 3.84 = \chi^2_{1,0.95}$, the 95th percentile of the χ^2_1 distribution, we see no significant improvement in allowing different p's to describe the two treatment groups.

Next we allow different λ's as well as different p's. This gives the third row in Table 6.5, corresponding to the 110 model. For this, $-2\log L$ decreases from 48.76 to 47.84, a nonsignificant change of only 0.92 for one extra parameter, so now we move to the 111 model, which allows different α's in addition to the different p's and λ's. Here, at last, we get a significant change in $-2\log L$: from 47.84 to 40.23, a difference of 7.61, which is well above the 3.84 needed for significance at the 5% level.

The above shows that no improvement in fit occurs until we allow different α's in the model, suggesting that the treatments differ in that the shapes of their survival distributions differ; but that immune proportions and rates of failure do not differ significantly between the treatment groups.

In order to check this let us now remove parameters from the 111 model and watch the change in $-2\log L$ again. In row 5 of Table 6.5 we keep in the different α and p parameters but restrict the λ's to be the same. The deviance increases only by $40.84 - 40.23 = 0.61$, an insignificant amount. Finally, dropping even the different p's increases the deviance further only from 40.84 to 41.69, another insignificant amount. This demonstrates clearly that a model with different α's is *sufficient* to describe the treatment effect, and also is *necessary*—the latter following now from the comparison of $49.31 - 41.69 = 7.62$, a significant decrease of $-2\log L$ from the 000 model to the 001 model, for the addition of one extra α parameter.

The fitted Model 111 Weibull curves are those shown in Figure 6.2. We will check goodness of fit by the correlation method of Section 5.4 below, and find that the Weibull is, indeed, a reasonable descriptor of the data, according to that test. Next we test, formally, for the presence of immunes in each treatment group by the method of Section 5.3.

The tests for $p_1 = 1$ and $p_2 = 1$ are carried out just as for the breast cancer data analysed in the previous section, so we will only briefly report the results. The parameter estimates for the Weibull mixture model fit to the Treatment group 1 data are shown under the 111 model in Table 6.5. The corresponding deviance, not shown in Table 6.5, is 25.69. An ordinary Weibull model fitted to the Treatment group 1 data produces parameter estimates of

$$\breve{\lambda} = 0.3719 \text{ and } \breve{\alpha} = 0.95$$

with a corresponding deviance of 27.52. We compare the deviance difference $27.52 - 25.69 = 1.83$ with 2.71, obtaining a nonsignificant result. Thus there is little evidence of immunes in Treatment group 1 from this test.

For Treatment group 2, the estimates for the ordinary Weibull fit are

$$\breve{\lambda} = 0.2739 \text{ and } \breve{\alpha} = 1.55$$

with a corresponding deviance of 24.70. The deviance for the Weibull mixture model is 14.54, giving a difference of $24.70 - 14.54 = 10.16$, which is significant by comparison with 2.71. Thus there is strong evidence that $p_2 < 1$, i.e. that there are immunes in Treatment group 2.

These tests suggest, as for the breast cancer data, a difference between immune proportions for the two treatment groups, whereas none was found in the analysis of deviance based on the pooled data. It also seems that the ordinary Weibull distribution is an adequate fit to the Treatment group 1 data. Again, however, although the ordinary Weibull fit may be equally as good as the Weibull mixture model according to the likelihood ratio test, it nevertheless represents a large extrapolation from the range of the data, as Figure 6.2 shows. Consequently, in testing for the presence of immunes, we might wish to consider also the results of nonparametric tests carried out according to the procedure described in Section 5.3. We find that the nonparametric statistic $\hat{p}_n$, the right extreme or maximum value of the KME, has a value of 0.5692 for Treatment group 1 and 0.4359 for Treatment group 2. These are well below any of the 1% points in Table A.2, so there is evidence from this that immunes are present in both groups. However, to be decisive we must also be convinced that follow-up is sufficient. In fact, as we will see below, follow-up appears to be reasonable but not high for either group (better for Treatment group 2 than for Treatment group 1), so we should leave our conclusion regarding the existence of immunes open for Treatment group 1, at least. Of course, we are again attempting to draw inferences from a very sparse set of data.

Next we consider the age analysis. How does belonging to age-group ≤ 60, or 60+, affect survival? The most striking aspects of Figure 6.3 are the lower survival rates for those over 60, and the much higher ultimate proportion dying. Table 6.6 gives the parameter estimates and deviance statistics for the various models.

Row 1 of Table 6.6 simply repeats the 000 model information from Table 6.5; of course it applies here for the age-group analysis, too, as the basis for comparison when no parameters differ between the age-groups. We begin the analysis by looking at the 100 model, which allows for different p parameters between the age-groups, in an attempt to test for the most prominent difference apparent in Figure 6.3. A comparison of deviances gives $49.31 - 41.19 = 8.12$, certainly significant at the 5% level as an observation on χ_1^2, so we have evidence of a difference in values of p between age-groups. Note in fact that the estimate of p for the 60+ age-group is 1.00; a boundary estimate occurred, and this is consistent with what we see in Figure 6.3: no immunes are likely, on visual evidence, to be present.

Is this the only difference between the age-groups? Next we allow λ parameters to differ between the groups, in addition to differing p parameters, but keeping α the same for each group. This is the 110 model, whose deviance is given in row 3 of Table 6.6 as 36.29. This represents a decrease of $41.19 - 36.29 = 4.90$ from the 100 model, is significant at the 5% level and suggests that survival rates also differ between age-groups. The next step is to allow the α parameters too, to differ, reducing the deviance further to 31.05, a change of $36.29 - 31.05 = 5.24$, again significant. It appears that we must allow different p, λ and α parameters for an adequate description of the data, and in fact this is a reasonable model.

We have listed also in Table 6.6 the results of fitting the other models, too, and we note that restricting the λ parameters to be identical does not significantly

Table 6.6 Age-group Analysis: ovarian cancer data.

Row	Model	No. of parameters	Age-group	$\tilde{p}$	$\tilde{\lambda}$	$\tilde{\alpha}$	$-2\log\tilde{L}$
1	000	3	(both)	0.51	0.8673	2.11	49.31
2	100	4	≤ 60	0.35	0.8497	2.12	41.19
			60+	1.00	0.8497	2.12	
3	110	5	≤ 60	0.36	0.6936	2.82	36.29
			60+	0.86	1.3543	2.82	
4	111	6	≤ 60	0.36	0.6902	5.75	31.05
			60+	0.93	1.3060	1.71	
5	101	5	≤ 60	0.36	0.7047	5.58	33.94
			60+	1.00	0.7047	1.25	
6	011	5	≤ 60	0.51	0.6872	5.84	36.66
			60+	0.51	1.4622	1.95	
7	010	4	≤ 60	0.53	0.6501	2.73	41.43
			60+	0.53	1.3765	2.73	
8	001	4	≤ 60	0.53	0.7034	5.67	42.25
			60+	0.53	0.7034	1.24	

worsen the fit: the 101 model, compared with the 111 model, has a deviance difference of only 33.94 − 31.05 = 2.89, which is not significant. So the 101 model is also acceptable as a description of the data. The remaining models, 011, 010 and 001, are clearly inadequate, as the deviances listed in Table 6.6 show.

The results of this analysis are clearly consistent with our visual impression of the data, and we are quite confident of a significant difference between age-groups in ultimate proportion surviving (near 0 for the 60+ group, but around 0.65 for the ≤ 60 group), in survival rates as measured by the λ parameters, and even in the shapes of the survival distributions, as represented by α.

We should complete this analysis of the factors affecting survival in the ovarian cancer data by analysing the effects of 'residual' and 'performance'. We will not go into detail here, but merely mention that for residual, the best fitting model, for the smallest number of parameters, is the 010 model, with a deviance of 44.94. By comparison with the 000 model, deviance 49.31, this suggests a significant difference in survival rates, only, between the residual two groups. For the performance factor, we find no significant differences between the two groups — the deviance for the 111 model is 45.45, a reduction of only 3.86 from the deviance of 49.31 for the 000 model, for the addition of three extra parameters. This is insignificant compared with the upper 5th percentile of a χ^2_3 distribution, which is 7.82.

In order to complete the second aim of the analysis of this data, which is to test for the effects of treatments, after allowing for or controlling for the effects of the other factors, we would have to fit joint models incorporating the effects of treatments and age-groups, say, together. Unfortunately this data set is much too small to allow us to do that. Attempts to maximise the likelihood fail because there are insufficient observations and/or deaths in the Treatment × Age group cells to produce estimates; or if they can be found, then their precision is so small as to be useless. This price has to be paid in fitting the mixture model, with its extra one-third parameters over the ordinary Weibull model, although there are limitations to the numbers of factors that can be fitted jointly in ordinary Weibull models, too. We must use a much bigger data set (see the recidivism data analysed in Section 6.4) to illustrate the fitting of joint factorial models.

At this stage it is worth comparing the results of our mixture model analysis of the ovarian cancer data with those of the ordinary Weibull survival analysis reported by Collett (1994, p. 143). He finds a significant effect of age — with survival rates decreasing with age, as we did — but no other significant effects at all. This is at variance with our findings of a significant difference between treatment groups in the shapes of their survival distribution, and in the possible existence of an immune proportion for Treatment group 2. The difference in shapes of distribution between the two treatment groups, found by the Weibull mixture model, is apparent in Figure 6.2, and can be simply interpreted as a higher death rate for those from Treatment group 1 (single chemotherapy) in the first year or so after the commencement of the trial, compared with a very low rate for those in Treatment group 2 (two chemotherapies combined) over this period.

We remarked earlier that this is in part due to the assignment of most older (60+) patients to the presumably less physically rigorous Treatment group 1. This design quirk, a flaw perhaps from the point of view of the statistician, but possibly an ethical necessity from the point of view of the medical researcher, is in need of highlighting, and our analysis has done that. The mixture model analysis, allowing for the presence of immunes, has produced some quite different results from the ordinary survival analysis.

To complete our investigation of this data set we consider the diagnostic aspects: we test for sufficient follow-up, and for goodness of fit of the Weibull distribution. Since the methods here are simply those outlined in Sections 4.2 and 4.3 and Sections 5.2 and 5.3, for a single group, applied to each of the treatment, age, residual and performance groups considered separately, and since the conclusions can be guessed from Figure 6.2, we merely list the outcomes briefly.

To test for sufficient follow-up, we find that the values of $q_n = N_n/n$ are $4/13 = 0.31$ for Treatment group 1 and $5/13 = 0.38$ for Treatment group 2. On comparing them with the percentage points in Tables B.3 and B.5, we conclude that follow-up is not significantly lacking for either Treatment group. The evidence is stronger for Treatment group 2 than for Treatment group 1, since 0.38 exceeds the 95% point in Table B.3 (with $p = 0.4$) for $B = 2$ and $n = 20$, whereas 0.31 is below the 90% point in Table B.5 (with $p = 0.6$) in similar situations.

For the goodness of fit of the Weibull, the correlation coefficient statistics r between the KME and the fitted values from Model 111 are $r = 0.9953$ for Treatment group 1 and $r = 0.9839$ for Treatment group 2, respectively. Comparing the value 0.9953 with the percentage points in Table C.5 (with $p = 0.6$, since the estimated p is 0.58 for Treatment group 1), we see that 0.9953 is well above any 20% point for $n = 20$. Hence we decide that the Weibull mixture model fits the Treatment group 1 data quite well. Similarly, the value of 0.9839 for Treatment group 2 substantially exceeds any 20% point for $n = 20$ in Table C.3 (with $p = 0.4$, as $\tilde{p} = 0.44$ for Treatment group 2). Thus the fit of the Weibull mixture model with the Treatment group 2 data is also quite good.

6.4 FACTORIAL ANALYSES

Second return to prison: a two-factor analysis

This example is of the analysis of a somewhat larger data set wherein we can assess the effects of two factors jointly. The data consist of the times to reimprisonment, following their *second* release from prison, of 1050 male prisoners serving sentences longer than 3 months. Thus the time to event-of-interest in this example is the time for a prisoner to return to prison, following his second release from it. Of course, the time to second return is censored if in fact a second return has not occurred by the end of follow-up on that releasee.

The prisoners were not homogeneous with respect to their age, so we classify the prisoners' ages according to an age-group factor with three levels:

- Age-group 1: ≤ 20 years old at first imprisonment
- Age-group 2: 20–40 years old at first imprisonment
- Age-group 3: > 40 years old at first imprisonment

The other factor of interest here is the parole status of the prisoners on their second release:

- Parole group 1: no parole (unsupervised release)
- Parole group 2: parole (supervised release)

We are interested mainly in whether these factors affect the probability that recidivism ultimately occurs (in this case, whether a return to prison, for the third time, occurs at any time) and whether they affect the rate at which those prisoners who ultimately return do so. We can ask these questions separately for the parole and age-group factors taken separately, and we can also ask how the factors affect rates of recidivism and probabilities of ultimate recidivism taken *jointly*. In particular, is there any interaction between the factors in the sense that a difference in recidivism between parolees and nonparolees may occur for some but not all age-groups? Previous analyses of recidivism data like this (e.g. Figure 5.3) have demonstrated that a Weibull mixture model is a reasonable description of the distribution of the time to recidivism, and that often a substantial immune component exists — many prisoners appear never to return to prison upon release from it. Accordingly, we will proceed with an analysis like that described in Section 6.3, but with the added feature of considering the two factors jointly.

Before doing this, however, we need to determine which of the models, 000, 100, etc., provides the best fit to the data, treating the two factors separately at this stage. The method of doing this is exactly as outlined in Section 6.3, so we omit the details. We find that for both the parole and age-group factors (and indeed for certain combined models also) the 110 model is adequate and also necessary for modelling the data. In other words, we need separate p's and λ's for each parole group, but the shape parameter α can be taken to be the same with no significant loss of information. The same is true for the age-groups, so we will work with the 110 model from now on.

As in an analysis of variance for normally distributed data, or an analysis of deviance for a generalised linear model, we begin by testing for interaction between the factors. If we find a significant interaction, it means that the difference between parole groups, either in the rate of recidivism or in the probability of ultimate recidivism, varies significantly over the age-groups — and likewise, differences between age-groups are not the same for each parole group. If this occurs we need to consider the parole $\times$ age categories separately to determine which differences are significant. If on the other hand there is no significant interaction between the factors, we are justified in ignoring the effect of one when assessing the effect of the other. In that case the data analysis reduces to examining the factors separately as we did in Section 6.3.

To see how this works out for the current data set, refer to Table 6.7 for the relevant deviance values. (Note that the deviances in Table 6.7 refer to the model fitted with time measured in 'months', or 30-day periods, as is common in criminology. The same is true for the deviances and estimates of λ reported in Tables 6.8 to 6.10. To change the scale back to years, we would multiply the estimates of λ by 365/30, and subtract from the deviance the value $2\log(365/30)\Sigma_{i=1}^{n} c_i$, that is, 4.997 times the number of failures in the data set.)

The values in Table 6.7 were obtained from the Fortran program described in Section 5.2, which also fits factorial Weibull mixture models to the censored data (see Section 8.1 for the parametric description of general factorial models). Recall that each of the deviances in the table refers to the 110 model, so the same α parameter — the shape parameter of the Weibull — applies to each level of a factor in the model, but the p and λ parameters are allowed to vary.

We begin by testing for interaction between the parole (P) and age-group (A) factors. Row 3 of Table 6.7 gives a deviance ($-2\log L$) value of 4929.22 for the 110 model which allows separate parameters for the P and A factors. Row 4 allows for extra parameters in the model so as to describe the P.A interaction, and gives a deviance of 4925.25 for the fitted model. The difference, 3.97, between these deviances is listed in row 6 of the table, together with the difference in the number of parameters fitted between the P + A and P + A + P.A models, which in this case (for the 110 model) is 4. Considered as an observation on a χ^2 random variable with four degrees of freedom, the deviance difference of 3.97 is far from significant, so we conclude that there is no evidence of interaction between the parole and age-group factors. This simplifies the analysis in that we can now test for the main effects of parole and age-group separately.

Table 6.7 Parole × age analysis, recidivism data. P = parole group effect, A = age-group effect, NS = nonsignificant ($p > 0.05$).

Row	Effect	Parameters	Deviance
1	parole	5	4950.61
2	age-group	7	4944.06
3	P + A	9	4929.22
4	P + A + P.A	13	4925.25
5	000	3	4964.12
		Degrees of freedom	Deviance difference
6	P.A	4	3.97 NS
7	P after A	2	14.84***
8	A after P	4	21.39***
9	parole	2	13.51**
10	age-group	4	20.06***

** $p < 0.01$
*** $p < 0.001$

Next, comparing the model which contains parameters for both P and A, i.e. row 3 of Table 6.7 again, with the model which contains parameters only for A, in row 2, gives a test for the effect of parole, having adjusted for, or 'controlled for', age-group. This is probably the most important comparison in this analysis: it tells us whether supervision of prisoners on release via a parole system, after allowing for possible differences in the ages of the prisoners, has any effect on recidivism, which is, it is hoped, positive, i.e., reduces it! Here we difference the deviances in rows 2 and 3, and the numbers of parameters contained in the fitted models, to obtain the values in row 7 of Table 6.7. The deviance difference of 14.84, as an observation on χ_2^2, is highly significant ($p < 0.001$), so we conclude that there is indeed an effect of parole, after controlling for age-group.

In a similar way we can test for the effect of age-groups, after allowing for parole, in row 8 of Table 6.6, obtaining again a highly significant deviance difference. In rows 9 and 10 of the table we give also the 'marginal' effects of P and A (the same as were obtained in Section 6.3 for the factors involved there) which are simply the differences of the deviances in rows 1 and 2 with the deviance for the 000 model given in row 5. These are little different from the values in rows 7 and 8, as we would expect in view of the insignificant interaction between P and A.

So far we have obtained significant effects of parole and age-group, but what form do they take? Table 6.8 shows the parameter estimates, with 95% confidence intervals, for the levels of the two factors.

In Table 6.8, NF denotes the number of 'failures' (the number returning to prison before the cutoff date, i.e., the number of uncensored observations), and N is the total number in the group under consideration. The table reveals that the ultimate probability of recidivism is estimated to be lower for parolees than for those with no parole (0.57 compared with 0.63) and that rates of fail are also lower for parolees (0.037 compared with 0.052). Both these differences are significant, as is shown by the fact that the 110 model is necessary. The estimated shape parameter $\tilde{\alpha}$ is 0.94, close to 1, suggesting that the failure distribution of the susceptibles is exponential-like. Precision of estimation of α is indicated by a confidence interval of (0.86, 1.01) around the 000 estimate of 0.93. We have not

Table 6.8 Parameter estimates for parole and age-group levels, recidivism data.

Row	Group	Parameter estimate (CI)				
		$\tilde{p}$ (CI)	$\tilde{\lambda}$ (CI)	$\tilde{\alpha}$	NF	N
1	no parole	0.63 (0.58, 0.68)	0.052 (0.044, 0.061)	0.94	286	539
2	parole	0.57 (0.50, 0.64)	0.037 (0.029, 0.047)	0.94	207	511
3	age ≤ 20	0.75 (0.66, 0.82)	0.050 (0.039, 0.064)	0.94	118	190
4	$20 <$ age ≤ 40	0.57 (0.52, 0.62)	0.044 (0.037, 0.052)	0.94	343	778
5	age > 40	0.52 (0.37, 0.67)	0.041 (0.023, 0.072)	0.94	32	82
6	overall (000)	0.60 (0.56, 0.64)	0.045 (0.039, 0.051)	0.93	493	1050

listed the confidence intervals on α in Table 6.8—they are of about that order of magnitude throughout.

For age-groups we note a relatively high probability of recidivism at 0.75 for the youngest reoffenders, whereas the 20–40 and 40+ age-groups seem similar in this respect, but lower than for those 20 or under. The same appears to be true for the rates of failure (λ) in the three groups. This suggests a simplification of the model by pooling together age-groups 2 and 3 into a 20+ age-group. To test whether we lose significant information by this pooling, we construct an analysis of deviance to test that contrast in Table 6.9. (This again is for the 110 model.)

In Table 6.9, row 1 is simply the effect of combining age-groups 2 and 3 (into a new age-group 2) and refitting the age-group factor, now with two levels rather than 3. Row 2 of Table 6.9 is simply row 2 of Table 6.7, and the difference between it and row 1 is in row 4. This gives the required contrast: the deviance difference due to allowing extra parameters for age-groups 2 and 3. The χ_2^2 value of 0.68 is very small, confirming what we suspected from Table 6.8, that there is little difference between age-groups 2 and 3. Finally, row 5 of Table 6.9 compares the new two-level factor for age-groups with the 000 model, giving a highly significant deviance difference, thus reinforcing the highly significant difference between young reoffenders (20 years and under) and those over 20.

The parameter estimates for the new age-group factor are given in Table 6.10. It shows that recidivism rates for young offenders, and estimated ultimate probabilities of recidivism, are higher than for older offenders. Having achieved this simplification, we ask further whether the 110 model is now necessary for the age-group effect: the two $\tilde{\lambda}$ values in Table 6.10 are quite close. The deviance difference between the 110 model and the 100 model for the two-level age-group factor is only 0.80, which is nonsignificant as an observation on χ_1^2. So we can even combine the λ values in this model to obtain an overall value of $\tilde{\lambda} = 0.045$. However, the probabilities of ultimate recidivism are significantly different between young and old offenders, so the 100 model is necessary for this comparison.

We now summarise the conclusions which can be drawn from this analysis.

(a) There are significant differences between both p (probability of ultimate recidivism) and λ (related to rate of recidivism) for parole groups, and

Table 6.9 Age 1 versus Ages 2 and 3, recidivism data.

Row	Effect	No. of parameters	Deviance
1	Age 1 v. Ages 2 + 3	5	4944.74
2	Age-groups	7	4944.06
3	000	3	4964.12
		Degrees of freedom	Deviance difference
4	Age 2 v. Age 3	2	0.68 NS
5	Age 1 v. Ages 2 + 3	2	19.38***

*** $p < 0.001$.

Table 6.10 Parameter estimates for Age-groups 1 and 2.

Row	Group	Parameter estimate (CI)				
		$\tilde{p}$ (CI)	$\tilde{\lambda}$ (CI)	$\tilde{\alpha}$	NF	N
1	Age $\leq$ 20	0.75 (0.67, 0.83)	0.050 (0.039, 0.064)	0.93	118	190
2	Age > 20	0.56 (0.51, 0.60)	0.044 (0.037, 0.051)	0.93	375	860

these persist when age is controlled for (parole decreases recidivism by comparison with no parole).

(b) Only two age-groups — age 20 years and under and age over 20 years — are necessary for an adequate description of the age effect in the data, and these groups differ significantly in their estimated probabilities of ultimate recidivism only (being higher for younger offenders).

(c) There is no evidence of any significant interaction between parole and age.

To complete the analysis, we should, as always, check for goodness of fit of the parametric distribution fitted (the Weibull in this case), and for sufficient follow-up to be confident that the survival curves have levelled. Figure 6.4 shows the

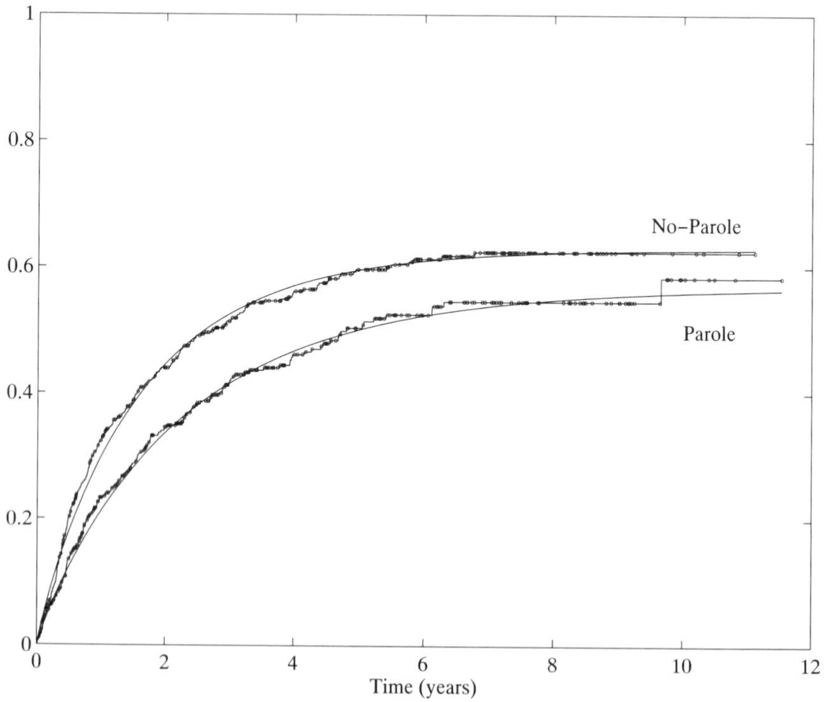

Figure 6.4 Recidivism times, second return to prison: parole and no-parole groups.

KMEs for each of the parole groups separately, together with the fitted Weibulls. Visually, the fit is reasonable but not excellent — there appears to be some systematic departure, especially in recidivism in the first 2 years or so — but we should remember that these curves 'average over' other significant subgroups in the data, for example, the age-groups.

A goodness of fit test of the kind described in Section 5.4 gives

$$r = 0.9979 \text{ for Group 1 (no parole)}$$

$$r = 0.9985 \text{ for Group 2 (parole)}$$

where r is the correlation coefficient between the KME and the fitted values from the Weibull mixture model. For Group 1, the sample size is $n = 539$, the estimated value of p is 0.63 and the proportion of censored observations is 0.47. We expect a uniform censoring distribution in data like this (see the discussion in Section 9.4), so we look in Table C.5 and compare the observed r value of 0.9979 with the percentage points corresponding to $n = 500$ and $B = 2$ or 4. The 20% points are 0.9959 for $B = 2$ and 0.9975 for $B = 4$. Both are less than $r = 0.9979$, hence we conclude that the Weibull mixture model fits fairly well the Group 1 data. For Group 2, $n = 511$, $\tilde{p} = 0.57$ and the proportion of censored observations is 0.59. So we again look at Table C.5, where the 20% point for $n = 500$ and $B = 2$ is 0.9959, well below $r = 0.9985$. Thus the Weibull mixture model provides a good fit for Group 2 data as well, as judged by the correlation statistic.

Testing for sufficient follow-up by the method of Section 4.3 gives for $q_n = N_n/n$ the values $46/539 = 0.0853$ for Group 1 (no parole) and $1/511 = 0.0019$ for Group 2 (parole). From Table B.5 (with $p = 0.6$) we see that 0.0853 is well above the 95% point for $B = 4$ and $n = 500$. Thus there is strong evidence that the follow-up in the no-parole group compares well with the case of $B = 4$, which corresponds to reasonably sufficient follow-up, and so we conclude that the follow-up in the no-parole group is sufficient. On the other hand, 0.0019 is around the 10% point for $B = 2$ and $n = 500$, hence there is substantial evidence that follow-up in the parole group is insufficient. But this is largely due to the single failure at about 10 years, as we discussed in Section 5.5. If we treat this late failure as an outlier and exclude it, the value of q_n for the remaining 510 observations is $94/510 = 0.1843$, which exceeds the 90% point of 0.1340 for $B = 6$ and $n = 500$. Hence there is strong evidence that follow-up is sufficient for this group too, if we exclude the outlying observation.

Parole × marital status analysis: recidivism data

This smaller data set consists of the recidivism times (return to prison after first release) of 216 male non-Aboriginal Western Australian releasees, cross-classified by their parole status at release (1 denotes a parolee, 2 denotes release to freedom, i.e. no parole) and marital status at imprisonment (1 denotes unmarried, 2 denotes married). We analysed the (2,1) group (no parole, unmarried) in

Section 5.3, and found that the data for that group was, according to the likelihood ratio test, as well fitted by an ordinary Weibull as by a Weibull mixture model; thus there was no significant evidence of immunes. However, follow-up for that group was judged to be insufficient to be conclusive.

In this section we fit the covariate model to test for the effects of parole status and marital status, jointly for the four groups in the cross-classification. We easily determine by the method of Section 6.3 that Model 100 is adequate (and necessary) for the data, so we judge that there are significant differences in immune proportions between the four groups—$(1, 1) =$ (parole, unmarried), $(1, 2) =$ (parole, married), $(2, 1) =$ (no parole, unmarried) and $(2, 2) =$ (no parole, married)—but no significant differences in λ or in α between the four groups. Despite this simplification we will still tabulate (Table 6.11) the estimated values of p, λ and α for the four groups separately (that is, the estimates for the 111 model) so as to illustrate an interesting aspect of the analysis.

In row 2 of Table 6.11, the estimates for the (2,1) group are the same as those given in Section 5.3. The Model 100 estimates for the same group, in row 6, are somewhat different, and we note that the estimated susceptible proportion for the group, $\tilde{p}$, is now 0.72, lower than the original 0.84, and with a shorter confidence interval of $(0.54, 0.85)$, suggesting with somewhat more confidence that there is some evidence of immunes for this group. The confidence interval for the estimate of 0.84 in row 2 is (0.12, 0.91), which almost contains 1.

How do we explain the discrepancy between the 111 and 100 model estimates for this group? Have we magically obtained more precision in the second analysis? The answer, in effect, is yes. Provided we are prepared to 'borrow strength' from the other three groups, by allowing their data values to contribute to the estimated parameters for the (2,1) group, we do get an enhancement from the effectively larger sample size obtained by including the other groups in the analysis. Remember that, for the 100 model, common values of the λ and α parameters are fitted to all four groups. This extra information has the effect here of forcing the estimate of p for the (2,1) group down, away from the boundary at 1. Of course this is only a reasonable procedure if we are prepared to believe

Table 6.11 Parameter estimates for parole (P) and marital (M) effects, recidivism data.

Row	Model	P	M	p (95% CI)	λ	α	$-2\log L$	NF	N
1	111	1	1	0.46 (0.30, 0.62)	0.062	1.25	206.8	20	52
2		2	1	0.84 (0.12, 0.91)	0.036	0.79	310.7	34	62
3		1	2	0.40 (0.26, 0.57)	0.053	0.69	259.6	24	68
4		2	2	0.54 (0.34, 0.73)	0.050	0.95	163.5	16	34
							940.6		
5	100	1	1	0.53 (0.36, 0.69)	0.048	0.86			
6		2	1	0.72 (0.54, 0.85)	0.048	0.86			
7		1	2	0.38 (0.27, 0.50)	0.048	0.86	945.6		
8		2	2	0.59 (0.41, 0.74)	0.048	0.86			

what our analysis suggests: that the λ and α parameters are the same for the groups, allowing us to report their common values.

The analysis also tells us that, inasmuch as the 100 model is an adequate description of the data for all four groups, we can detect no significant difference in λ and α between the four groups; this is only true to within the accuracy afforded by the data.

Given the above findings, we believe that there is reasonable evidence for an immune proportion in the (2,1) population. The analysis of Section 5.3 told us only that we could not reject the hypothesis of no immunes present, and that follow-up was insufficient to be conclusive. From the present analysis, we deduce that, *taking into account the information from closely related groups*, there is some evidence for a levelling of the distribution of the (2,1) group. Note that the late failures for this group are sporadic, and that the estimate of the asymptotic level of the (2,1) group under the 111 model, i.e. 0.84, represents a big extrapolation from the body of the data. This is illustrated in Figure 6.5, where we plot the 111 fits to the four groups.

In conclusion, we mention briefly why we think that the goodness of fit of the Weibull mixture distribution should be assessed from the Model 111 fit, rather

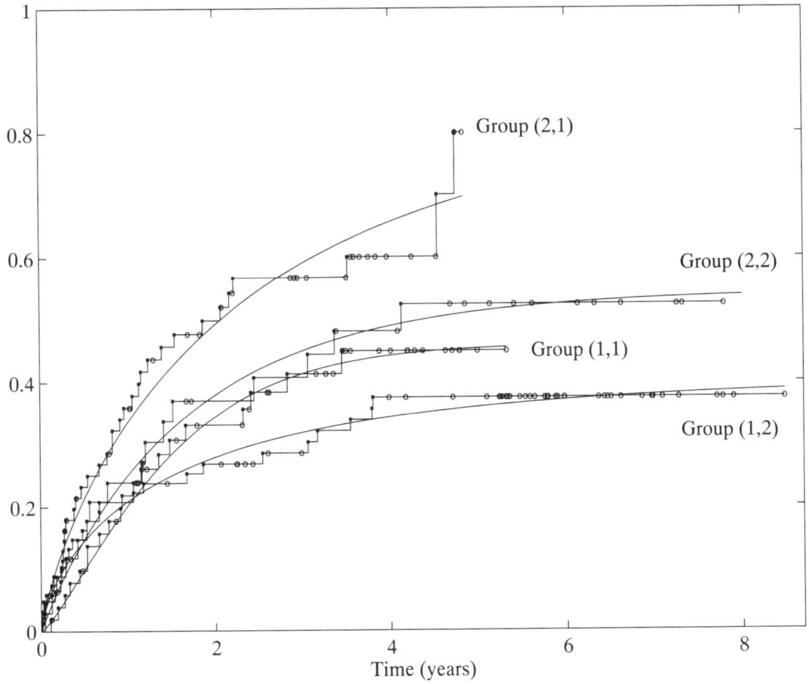

Figure 6.5 Recidivism times, first return to prison: parole × marital status groups, model 111 fit. Group $(1, 2)$ = parole, married; Group $(1, 1)$ = parole, unmarried; Group $(2, 1)$ = no parole, unmarried; Group $(2, 2)$ = No parole, married.

than from the Model 100 fit, which was judged adequate by the likelihood ratio (i.e. the deviance) tests we have performed. (A similar issue arises in regard to the other data analyses in this chapter.) We should keep in mind the following basic principle of hypothesis testing: acceptance of the null hypothesis only indicates insufficient evidence against, rather than strong evidence in support of, the hypothesis. The likelihood ratio tests are directed not to the goodness of fit of the distribution *per se* so much as to an adequate representation of the likelihood surface, in terms of the parameters available, in the region of the maximum likelihood estimates. These are of course related but distinct concepts. More specifically, Model 111 always provides a better fit than any of those models (Models 110, 101, etc.) which have restrictions on the parameters, since it allows the maximum degrees of freedom in terms of available parameters. Our acceptance of the simpler model, Model 100, as a result of the deviance test does not change this fact, because it is based on the ground that such a simplification does not *significantly* increase the value of the deviance. Nevertheless, it does increase the deviance somewhat, and so provides a worse fit than Model 111. This difference can be substantial from the viewpoint of goodness of fit, even if not deemed significant in terms of the deviance.

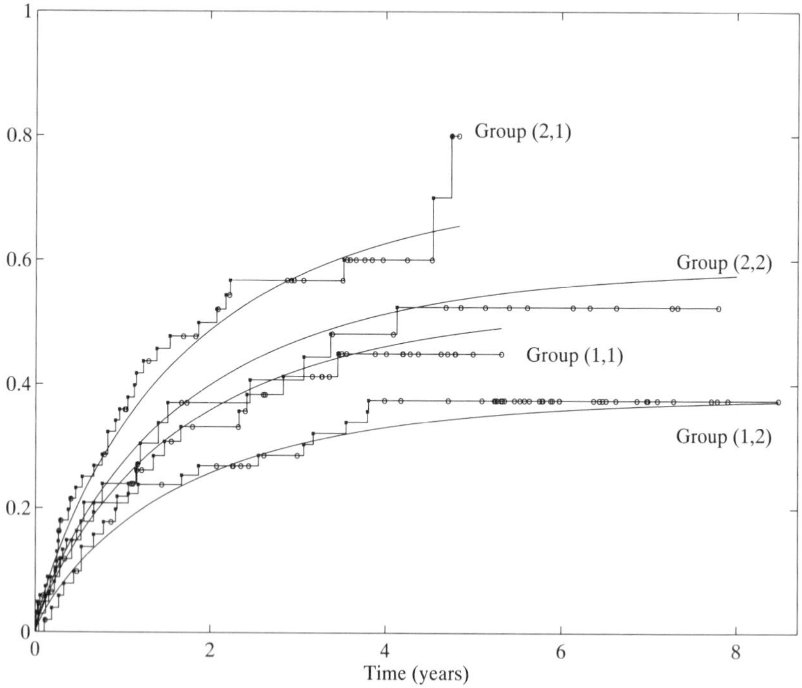

Figure 6.6 Recidivism times, first return to prison: parole × marital status groups, model 100 fit. Group (1, 2) = parole, married; Group (1, 1) = parole, unmarried; Group (2, 1) = no parole, unmarried; Group (2, 2) = no parole, married.

To compare the approaches, in Figure 6.6 we plot the 100 model fits for the parole $\times$ marital status recidivism data. Inspection of the graph reveals that the fits are noticeably different in detail from those displayed in Figure 6.5, even though, according to our analysis, the parameters λ and α do not differ significantly between the four groups.

CHAPTER 7

Large-Sample Properties of Parametric Models: Single Samples

7.1 INTERIOR AND BOUNDARY MAXIMUM LIKELIHOOD ESTIMATES

There are many good accounts of the large-sample properties of MLEs and the deviance statistic in general statistical models, such as Cox and Hinkley (1974, Ch. 9), so we will only briefly summarise in this section the main aspects that are needed for the analyses which were demonstrated in Chapters 5 and 6.

Briefly, as a general rule, MLEs of parameters which are *interior* to the parameter space are asymptotically normally distributed, and the corresponding deviance statistics for testing hypotheses specifying *interior* values of the parameters, are asymptotically distributed as chi-square random variables (with certain degrees of freedom), as the sample size becomes large. This is true provided the model(s) fitted possess sufficient mathematical regularity or are sufficiently 'well behaved'.

On the other hand, estimators of parameters in samples from distributions in which the null hypothesis under test (which thereby specifies the 'true' distribution governing the observations) specifies a value on the *boundary* of the parameter space are *not* in general normally distributed, asymptotically, even for 'regular' models; nor is the deviance statistic for testing such a boundary null hypothesis asymptotically distributed as chi-square, in general. Since this is precisely the situation which occurs when we wish to test for the presence or absence of immunes (the hypothesis $H_0 : p = 1$ specifies p to be on the boundary of the parameter space $(0, 1]$ for p), this caution is of particular interest to us.

It is often the case, in fact, when testing boundary hypotheses, that the large-sample distribution of the deviance d_n turns out to be a mixture of a chi-square distribution and a mass at 0 (a value of $d_n = 0$ corresponding to a boundary maximum of the likelihood). Cox and Hinkley (1974, p. 321) provide a discussion of this phenomenon. It is particularly relevant to our situation, since, as we shall show in Section 7.3, under some reasonable conditions, the asymptotic distribution of the deviance statistic d_n designed to test for the presence or absence of immunes in a censored exponential sample, is indeed a 50–50 mixture of a chi-square distribution with one degree of freedom and a point mass at 0, a result that we used in the analysis of the data in Section 5.3.

In order to justify such data analyses, and also the various other data analyses involving hypothesis tests illustrated throughout Chapters 5 and 6, we collect in this chapter the results that have been worked out so far for the large-sample distributions of MLEs, and of deviances, for the various parametric models we are interested in. These are, mainly, the exponential and exponential mixture models, for censored survival data, both with and without immunes. We need to consider both the 'interior' situations and the 'boundary' test for the presence of immunes. We begin with the single-parameter 'ordinary' exponential model, which while remaining simple enough to analyse explicitly, nevertheless brings out some important aspects of the assumptions required on the censoring and survival distributions to produce the results we need.

7.2 THE ORDINARY EXPONENTIAL SURVIVAL MODEL FOR A SINGLE SAMPLE

The basic asymptotic results

Suppose that $t_1, t_2, \ldots, t_n$ and $c_1, c_2, \ldots, c_n$ are survival times and censor indicators generated according to the i.i.d. censoring model of Section 2.1, where the survival distribution is in fact $F(t) = 1 - e^{-\lambda t}$, the censoring distribution G is arbitrary (so far), and the immune proportion is 0 (so $p = 1$). Let us be more precise in formulation. The 'true' distribution of the random variables representing the actual survival times, t_i^*, will be $F(t) = 1 - e^{-\lambda_0 t}$, where λ_0 is a particular (given) value of λ (the 'true' value). We have to maximise the log-likelihood (the logarithm of the likelihood given by (5.12), apart from constants), i.e. the expression

$$l_n(\lambda) = n_u \log \lambda - \lambda n\bar{t}, \tag{7.1}$$

for values of λ in $(0, \infty)$. Here $n_u = \sum_1^n c_i$ is the number of uncensored observations, and $\bar{t} = \sum_1^n t_i/n$ is the mean of all survival times. As we saw in (5.18), this maximisation is easily done, and it results in a unique MLE

$$\tilde{\lambda}_n = n_u/(n\bar{t}). \tag{7.2}$$

We are in an interior situation here since $\lambda_0 > 0$ is in the interior of the parameter space $(0, \infty)$ for λ. The next theorem gives the large sample distribution of $\tilde{\lambda}_n$.

Theorem 7.1 *Assume the i.i.d. censoring model with $F(t) = 1 - e^{-\lambda_0 t}$, $t \geq 0$, and assume that the censoring distribution G is not degenerate at 0. Then as $n \to \infty$, $\tilde{\lambda}_n$ is consistent in probability for λ_0, and is asymptotically normally distributed around λ_0. More specifically,*

$$\sqrt{n}(\tilde{\lambda}_n - \lambda_0) \xrightarrow{D} N(0, \sigma^2), \tag{7.3}$$

where $0 < \sigma^2 < \infty$. Furthermore, σ^2 is consistently estimated by $(n\tilde{\lambda}_n)^2/n_u$ so that as $n \to \infty$

$$\frac{\sqrt{n_u}}{\tilde{\lambda}_n}(\tilde{\lambda}_n - \lambda_0) \xrightarrow{D} N(0, 1). \tag{7.4}$$

In addition, the distribution of the deviance difference

$$d_n = 2(l_n(\tilde{\lambda}_n) - l_n(\lambda_0)) \tag{7.5}$$

approaches a chi-square distribution with one degree of freedom as $n \to \infty$.

Proof Suppose that G is not degenerate at 0. We will first prove (7.3). This can easily be proved directly from the explicit representation (7.2) of $\tilde{\lambda}_n$, but we prefer to present here a method which can be generalised to more complex models. Let

$$S_n(\lambda) = \frac{\mathrm{d}l_n(\lambda)}{\mathrm{d}\lambda} \tag{7.6}$$

be the first derivative of $\log L$, and let

$$F_n(\lambda) = -\frac{\mathrm{d}^2 l_n(\lambda)}{\mathrm{d}\lambda^2}$$

be the negative of the second derivative of $\log L$. Then from (7.1) we obtain

$$S_n(\lambda) = \frac{n_u}{\lambda} - n\bar{t} \text{ and } F_n(\lambda) = \frac{n_u}{\lambda^2} = \frac{1}{\lambda^2}\sum_{i=1}^{n} c_i. \tag{7.7}$$

Make a Taylor expansion of $S_n(\tilde{\lambda}_n)$ around $S_n(\lambda_0)$, and recall that $S_n(\tilde{\lambda}_n) = 0$ since $\tilde{\lambda}_n$ is the MLE. Thus

$$0 = S_n(\tilde{\lambda}_n) = S_n(\lambda_0) + (\tilde{\lambda}_n - \lambda_0)\frac{\mathrm{d}S_n(\dot{\lambda}_n)}{\mathrm{d}\lambda} \tag{7.8}$$

$$= S_n(\lambda_0) - (\tilde{\lambda}_n - \lambda_0)F_n(\dot{\lambda}_n). \tag{7.9}$$

Here $\dot{\lambda}_n$ is some value of λ between λ_0 and $\tilde{\lambda}_n$, i.e. $\dot{\lambda}_n = a\lambda_0 + (1-a)\tilde{\lambda}_n$ for some (random) $a = a_n \in [0, 1]$. (Thomas (1960, p. 796), for example, gives a statement of Taylor's theorem in the form used in (7.8).) From (7.9) we have

$$\sqrt{n}(\tilde{\lambda}_n - \lambda_0) = \frac{S_n(\lambda_0)}{\sqrt{n}}\frac{n}{F_n(\dot{\lambda}_n)}. \tag{7.10}$$

Now notice that

$$\tilde{\lambda}_n = \frac{n_u}{n\bar{t}} = \sum_{i=1}^{n} c_i \Big/ \sum_{i=1}^{n} t_i = \frac{1}{n}\sum_{i=1}^{n} c_i \Big/ \frac{1}{n}\sum_{i=1}^{n} t_i. \tag{7.11}$$

Since the c_i are i.i.d. random variables we have by the strong law of large numbers that, as $n \to \infty$,

$$\frac{1}{n}\sum_{i=1}^{n} c_i \to E(c_1) \qquad \text{a.s.}$$

We can calculate $E(c_1)$ (recall that c_1 is a Bernoulli r.v.) as

$$\begin{aligned} E(c_1) = P(c_1 = 1) = P(t_1^* \le u_1) &= \int_{[0,\infty)} F(y)\,\mathrm{d}G(y) \\ &= \int_{[0,\infty)} (1 - \mathrm{e}^{-\lambda_0 y})\,\mathrm{d}G(y). \end{aligned} \tag{7.12}$$

Similarly we have, as $n \to \infty$,

$$\frac{1}{n}\sum_{i=1}^{n} t_i \to E(t_1) \qquad \text{a.s.}$$

and we can also calculate

$$\begin{aligned} E(t_1) &= E(\min(t_1^*, u_1)) = E(t_1^* I(t_1^* \le u_1) + u_1 I(t_1^* > u_1)) \\ &= \int_{[0,\infty)} \left(\int_{[0,y]} t\,\mathrm{d}F(t) + y \int_{(y,\infty)} \mathrm{d}F(t) \right) \mathrm{d}G(y) \\ &= \int_{[0,\infty)} \left(\int_0^y t\lambda_0 \mathrm{e}^{-\lambda_0 t}\,\mathrm{d}t + y\mathrm{e}^{-\lambda_0 y} \right) \mathrm{d}G(y) \\ &= \int_{[0,\infty)} \left(\frac{1 - \mathrm{e}^{-\lambda_0 y}}{\lambda_0} \right) \mathrm{d}G(y). \end{aligned} \tag{7.13}$$

The integral in (7.13) is positive since G is not degenerate at 0. From (7.11) to (7.13) we obtain

$$\tilde{\lambda}_n \to \frac{E(c_1)}{E(t_1)} = \lambda_0 \qquad \text{a.s.,}$$

showing that $\tilde{\lambda}_n$ is indeed consistent (almost surely) for λ_0. This is an important and useful property of an estimator, as we discussed in Section 3.1.

Returning to (7.10), since $\tilde{\lambda}_n \to \lambda_0$ a.s., we have that $\dot{\lambda}_n = a\lambda_0 + (1-a)\tilde{\lambda}_n$ also converges to λ_0 a.s. as $n \to \infty$, so by (7.7)

$$\frac{F_n(\dot{\lambda}_n)}{n} = \frac{n_u}{n\dot{\lambda}_n^2} = \frac{1}{n\dot{\lambda}_n^2}\sum_{i=1}^{n} c_i \to \frac{E(c_1)}{\lambda_0^2} \qquad \text{a.s.} \tag{7.14}$$

Also note that

$$\frac{S_n(\lambda_0)}{\sqrt{n}} = \frac{1}{\lambda_0\sqrt{n}}\sum_{i=1}^{n}(c_i - \lambda_0 t_i)$$

is a sum of n i.i.d. random variables, normalised by a multiple of $\sqrt{n}$. Furthermore, each summand has expected value $E(c_1) - \lambda_0 E(t_1) = 0$ by (7.12) and (7.13). Consequently by the central limit theorem, $S_n(\lambda_0)/\sqrt{n}$ is asymptotically normally distributed, as $N(0, \gamma^2)$, say, where

$$\gamma^2 = \frac{\text{Var}(c_1 - \lambda_0 t_1)}{\lambda_0^2} = \frac{E(c_1 - \lambda_0 t_1)^2}{\lambda_0^2}, \tag{7.15}$$

provided $0 < \gamma^2 < \infty$. We can evaluate the quantity γ^2 directly, by manipulations such as we used in (7.12) and (7.13), or we can note that

$$E(S_n(\lambda_0))^2 = E\left(\frac{\partial l_n(\lambda_0)}{\partial \lambda}\right)^2 = -E\left(\frac{\partial^2 l_n(\lambda_0)}{\partial \lambda^2}\right) = E(F_n(\lambda_0)) \tag{7.16}$$

(by an identity in likelihood theory which is proved in Theorem 7.2). Either method gives (see (7.7))

$$\gamma^2 = \text{Var}\left(\frac{S_n(\lambda_0)}{\sqrt{n}}\right) = E\left(\frac{F_n(\lambda_0)}{n}\right) = E\left(\frac{1}{n\lambda_0^2}\sum_{i=1}^{n} c_i\right) = \frac{E(c_1)}{\lambda_0^2}. \tag{7.17}$$

We always have $\gamma^2 < \infty$ since $c_1 \le 1$ a.s. and $\lambda_0 > 0$. Note also that $\gamma^2 > 0$ if and only if $E(c_1) > 0$, equivalently, by (7.12), if and only if the censoring distribution G is not degenerate at 0. We assumed that this is so, so we can return to (7.10) to obtain

$$\sqrt{n}(\tilde{\lambda}_n - \lambda_0) \xrightarrow{D} N(0, \gamma^2)\frac{1}{\gamma^2} = N\left(0, \frac{1}{\gamma^2}\right).$$

This proves (7.3) with $\sigma^2 = 1/\gamma^2$. Also since, as in (7.14) and (7.17), $F_n(\tilde{\lambda}_n)/n = n_u/(n\tilde{\lambda}_n^2) \to \gamma^2$ a.s., we have (7.4), too.

Next we prove that d_n in (7.5) is asymptotically χ_1^2. To do this, expand $l_n(\lambda_0)$ around $l_n(\tilde{\lambda}_n)$ by Taylor's theorem to get

$$\begin{aligned} d_n = 2(l_n(\tilde{\lambda}_n) - l_n(\lambda_0)) &= 2(\tilde{\lambda}_n - \lambda_0)S_n(\tilde{\lambda}_n) + (\tilde{\lambda}_n - \lambda_0)^2 F_n(\dot{\lambda}_n) \\ &= (\tilde{\lambda}_n - \lambda_0)^2 F_n(\dot{\lambda}_n) \end{aligned} \tag{7.18}$$

(recall that $S_n(\tilde{\lambda}_n) = 0$). Here $\dot{\lambda}_n$ is between λ_0 and $\tilde{\lambda}_n$; it is not the same as the $\dot{\lambda}_n$ in (7.14), but (7.14) still holds for this $\dot{\lambda}_n$. Thus $F_n(\dot{\lambda}_n)/n \to E(c_1)/\lambda_0^2 = \gamma^2 = 1/\sigma^2$, a.s. Also by (7.3), $\sqrt{n}(\tilde{\lambda}_n - \lambda_0)/\sigma \xrightarrow{D} N(0, 1)$. Consequently

$$d_n = \frac{n(\tilde{\lambda}_n - \lambda_0)^2}{\sigma^2}\frac{\sigma^2 F_n(\dot{\lambda}_n)}{n} \xrightarrow{D} (N(0,1))^2 \stackrel{D}{=} \chi_1^2, \tag{7.19}$$

as required. ■

Using the asymptotic results

What is the use of an asymptotic result like (7.3) or (7.4)? Actually it is the 'Studentised' result (7.4) that is of use in practice. It says that the distribution of $(\sqrt{n_u}/\tilde{\lambda}_n)(\tilde{\lambda}_n - \lambda_0)$ is close to the standard normal distribution in large samples, provided that G is not degenerate at 0, the approximation improving as the sample size increases. We can alternatively express this as '$\tilde{\lambda}_n$ is approximately distributed as $N(\lambda_0, \tilde{\lambda}_n^2/n_u)$ in large samples, providing the censoring is not too heavy'.

Thus, although we do not know the exact distribution of $\tilde{\lambda}_n$ for a finite n (it depends on the unknown λ_0 and on the unknown censoring distribution G), we can use the properties of the normal distribution to assess the precision of estimation of λ_0 by $\tilde{\lambda}_n$ in the sample, by using the normal approximation to calculate confidence intervals on the unknown parameter λ_0. Similarly, (7.5) can be used for testing the hypothesis $H_0 : \lambda = \lambda_0$, which thereby specifies that the 'true' value of λ is λ_0, as required in Theorem 7.1. If d_n as calculated from (7.5) is too large plausibly to be an observation on a χ_1^2 random variable, we will reject H_0.

What about the requirement that G not be degenerate at 0? This is not at all limiting. In fact, it is obviously a minimal assumption under the independent censoring model, since if G were degenerate at 0 then all the u_i would equal 0 a.s., so all the $t_i(= \min(t_i^*, u_i))$ would also equal 0 a.s. In other words, no nonzero observations could be observed, and of course no consistent estimates of λ_0 could then be constructed. We know that G is not degenerate at 0 if we have observed some nonzero observations. (A data set without any such is obviously not interesting at all!)

The general principle illustrated here carries over to more complex models. We will find it necessary to restrict censoring variables in some way so that they cannot get too small — in other words, the censoring must not be too heavy, relative to the failure time distribution. We have already met this idea in various places in looking at properties of nonparametric estimates in Chapters 3 and 4, and it will occur, in some form, in each of our analyses of parametric models, too. This is logical because, as the above simple example illustrates, with extremely heavy censoring we cannot expect estimators to behave reasonably. On the other hand, the restrictions required will not in general be as simple and minimal as merely requiring that the censoring distribution not degenerate at 0, either. Nevertheless, we strive as usual for minimal assumptions governing our results — we want them to hold for as wide a variety of censoring distributions as possible, since the censoring distribution is usually unknown.

Note that there is no problem in the above analysis with allowing G to be degenerate at some point $u_0 > 0$; this is the case of Type I censoring (see Section 2.1), and it results in consistent and asymptotically normal estimation of λ_0 by $\tilde{\lambda}_n$, according to Theorem 7.1.

The method of proof of Theorem 7.1, though somewhat simplified because of the simple form of the exponential distribution (and the lack of immunes) is fairly typical also of the more complicated results we will quote in later sections. Crucial steps in the proof are to obtain the first derivative, $S_n(\lambda)$, and the negative second derivative, $F_n(\lambda)$, of the log-likelihood, then to 'find' an MLE (ideally *the* MLE) by solving the estimating equation $S_n(\tilde{\lambda}_n) = 0$. In Theorem 7.1 we were lucky enough to be able to solve this estimating equation explicitly for an 'interior' value $\tilde{\lambda}_n (> 0)$. Asymptotic properties of $\tilde{\lambda}_n$ then followed from two main properties:

(i) $S_n(\lambda_0)$ is a sum of independent, mean zero, random variables, hence when suitably normalised (by $\sqrt{n}$ in the above case) it is asymptotically normal.

(ii) $F_n(\lambda_0)$ is a sum of independent random variables, hence when normalised (by n in the above case) it approaches a constant in large samples, in probability. This constant is, under conditions on the censoring and survival distributions, not 0 or ∞. Furthermore, $F_n(\dot{\lambda}_n)$ is sufficiently close to $F_n(\lambda_0)$, for $\dot{\lambda}_n$ sufficiently close to λ_0, so that $F_n(\dot{\lambda}_n)$ also converges, when normalised by n, to a constant.

Two major problems which occur in more complicated analyses virtually disappeared in the above proof. We had no problem demonstrating the *existence* of a (*unique*) MLE $\tilde{\lambda}_n$; and furthermore proof of its consistency for λ_0 was quite straightforward. In some models, establishing these properties represents the major part of the proof of a theorem like Theorem 7.1: we can seldom find explicit formulae for the MLEs. Note also that in Theorem 7.1 we had to consider only one parameter; most interesting models are higher-dimensional. Fortunately, the tools developed for the one-dimensional case generalise, though the need and opportunities for more sophisticated analyses multiply.

A likelihood identity

We used in the proof of Theorem 7.1 an identity in likelihood theory (7.16) which holds in the censored data case, but not for the same reasons as in the case of ordinary (uncensored) data, where it is a well-known property (Cox and Hinkley 1974, p. 281). Since this result is crucial in many of our proofs, we will quote and prove the result here.

Let L_n be any likelihood of the form

$$L_n(\boldsymbol{\theta}_1, \ldots, \boldsymbol{\theta}_n) = \prod_{i=1}^{n} (f_{\theta_i}(t_i))^{c_i} (1 - F_{\theta_i}(t_i))^{1-c_i} \tag{7.20}$$

where $F_{\theta_i}(t)$ is any survival distribution (possibly improper) and $f_{\theta_i}(t)$ is its density. The vectors $\boldsymbol{\theta}_i$, as usual, are parameter vectors, possibly depending on observation number, i, with 'true' value $\boldsymbol{\theta}_{i0}$, say, and c_i is the censor indicator associated with survival time t_i. Let

$$l_n(\boldsymbol{\theta}_1, \ldots, \boldsymbol{\theta}_n) = \log L(\boldsymbol{\theta}_1, \ldots, \boldsymbol{\theta}_n).$$

Then we have Theorem 7.2.

Theorem 7.2 *Suppose $l_n(\boldsymbol{\theta}_1, \ldots, \boldsymbol{\theta}_n)$ is twice continuously differentiable with respect to $\boldsymbol{\theta}_i$ for $\boldsymbol{\theta}_i$ in some open neighbourhood N_i of $\boldsymbol{\theta}_{i0}$, and suppose*

$$E\left(\frac{\partial^2 l_n(\boldsymbol{\theta}_1, \ldots, \boldsymbol{\theta}_n)}{\partial \boldsymbol{\theta}_i^2}\right)_{\boldsymbol{\theta}_i=\boldsymbol{\theta}_{i0}}$$

is a finite matrix. Suppose also that for each i, the right extreme of F_{θ_i}, say τ_i, does not depend on the value of $\boldsymbol{\theta}_i$, and $f_{\theta_i}(t) > 0$ for all $0 < t < \tau_i$. Then

$$E\left(\frac{\partial l_n(\boldsymbol{\theta})}{\partial \boldsymbol{\theta}_i}\right)_{\boldsymbol{\theta}_i=\boldsymbol{\theta}_{i0}} = 0 \tag{7.21}$$

and

$$\left[E\left(\frac{\partial l_n(\boldsymbol{\theta})}{\partial \boldsymbol{\theta}_i}\right) E\left(\frac{\partial l_n(\boldsymbol{\theta})}{\partial \boldsymbol{\theta}_i}\right)^T\right]_{\boldsymbol{\theta}_i=\boldsymbol{\theta}_{i0}} = -E\left(\frac{\partial^2 l_n(\boldsymbol{\theta})}{\partial \boldsymbol{\theta}^2}\right)_{\boldsymbol{\theta}_i=\boldsymbol{\theta}_{i0}}. \tag{7.22}$$

Proof The proof does not depend crucially on the value of n or on the vector nature of the $\boldsymbol{\theta}_i$, so we simply prove it for $n = 1$ and $\boldsymbol{\theta}_i = \theta$, a scalar parameter. In this case we drop the subscripts i and n on the quantities c_i, $\boldsymbol{\theta}_i$, L_n and l_n, take $t_i = t_1$ and simply write

$$L(\theta) = f_\theta^c(t_1)(1 - F_\theta(t_1))^{1-c}. \tag{7.23}$$

Let $l(\theta) = \log L(\theta)$. Let t_1^* have (true) survival distribution $F_{\theta_0}(t)$, corresponding to the true value θ_0 of θ, and let t_1^* be censored by censor variable u with distribution G, resulting in a censored survival time $t_1 = \min(t_1^*, u)$. Keep θ in the neighbourhood N of θ_0 for which $f_\theta(t) > 0$ for all $t > 0$. We have

$$\frac{\mathrm{d}l(\theta)}{\mathrm{d}\theta} = \frac{1}{L(\theta)}\frac{\mathrm{d}L(\theta)}{\mathrm{d}\theta} \tag{7.24}$$

and

$$\frac{\mathrm{d}^2 l(\theta)}{\mathrm{d}\theta^2} = -\frac{1}{L^2(\theta)}\left(\frac{\mathrm{d}L(\theta)}{\mathrm{d}\theta}\right)^2 + \frac{1}{L(\theta)}\frac{\mathrm{d}^2 L(\theta)}{\mathrm{d}\theta^2}. \tag{7.25}$$

From (7.24) we can calculate

$$E\left(\frac{\mathrm{d}l(\theta)}{\mathrm{d}\theta}\right)_{\theta=\theta_0} = E\left(\frac{1}{L(\theta)}\frac{\mathrm{d}L(\theta)}{\mathrm{d}\theta}(I(c=1) + I(c=0))\right)_{\theta=\theta_0}.$$

(As usual, $I()$ denotes the indicator function.) At this stage we also drop the subscript for θ_0. Then

$$E\left(\frac{\mathrm{d}l(\theta)}{\mathrm{d}\theta}\right) = E\left(\frac{1}{f_\theta(t_1^*)}\frac{\mathrm{d}f_\theta(t_1^*)}{\mathrm{d}\theta} I(t_1^* \leq u)\right)$$

$$
\begin{aligned}
&+E\left(\frac{1}{1-F_\theta(u)}\frac{\mathrm{d}(1-F_\theta(u))}{\mathrm{d}\theta}I(t_1^*>u)\right)\\
&=E\left(\int_0^u \frac{1}{f_\theta(y)}\frac{\mathrm{d}f_\theta(y)}{\mathrm{d}\theta}f_\theta(y)\,\mathrm{d}y\right)-E\left(\frac{\mathrm{d}F_\theta(u)}{\mathrm{d}\theta}\right)\\
&=E\left(\frac{\mathrm{d}}{\mathrm{d}\theta}\int_0^u f_\theta(y)\,\mathrm{d}y\right)-E\left(\frac{\mathrm{d}F_\theta(u)}{\mathrm{d}\theta}\right)\\
&=E\left(\frac{\mathrm{d}F_\theta(u)}{\mathrm{d}\theta}\right)-E\left(\frac{\mathrm{d}F_\theta(u)}{\mathrm{d}\theta}\right)=0.
\end{aligned}
$$

This proves (7.21).

From (7.24) and (7.25) we see that (7.22) holds, i.e., in this case,

$$
E\left(\frac{\mathrm{d}l(\theta)}{\mathrm{d}\theta}\right)^2_{\theta=\theta_0}=-E\left(\frac{\mathrm{d}^2l(\theta)}{\mathrm{d}\theta^2}\right)_{\theta=\theta_0}
$$

if and only if

$$
E\left(\frac{1}{L(\theta)}\frac{\mathrm{d}^2L(\theta)}{\mathrm{d}\theta^2}\right)_{\theta=\theta_0}=0. \qquad (7.26)
$$

However, this is true because (writing θ for θ_0 again)

$$
\begin{aligned}
E\left(\frac{1}{L(\theta)}\frac{\mathrm{d}^2L(\theta)}{\mathrm{d}\theta^2}\right)&=E\left(\frac{1}{L(\theta)}\frac{\mathrm{d}^2L(\theta)}{\mathrm{d}\theta^2}(I(c=1)+I(c=0))\right)\\
&=E\left(\frac{1}{f_\theta(t_1^*)}\frac{\mathrm{d}^2f_\theta(t_1^*)}{\mathrm{d}\theta^2}I(t^*\le u)\right)+E\left(\frac{1}{1-F_\theta(u)}\frac{\mathrm{d}^2(1-F_\theta(u))}{\mathrm{d}\theta^2}I(t_1^*>u)\right)\\
&=E\left(\int_0^u\frac{1}{f_\theta(y)}\frac{\mathrm{d}^2f_\theta(y)}{\mathrm{d}\theta^2}f_\theta(y)\,\mathrm{d}y\right)-E\left(\frac{1}{1-F_\theta(u)}\frac{\mathrm{d}^2F_\theta(u)}{\mathrm{d}\theta^2}(1-F_\theta(u))\right)\\
&=E\left(\int_0^u\frac{\mathrm{d}^2f_\theta(y)}{\mathrm{d}\theta^2}\,\mathrm{d}y\right)-E\left(\frac{\mathrm{d}^2F_\theta(u)}{\mathrm{d}\theta^2}\right)\\
&=E\left(\frac{\mathrm{d}^2}{\mathrm{d}\theta^2}\int_0^u f_\theta(y)\,\mathrm{d}y\right)-E\left(\frac{\mathrm{d}^2F_\theta(u)}{\mathrm{d}\theta^2}\right)=0.
\end{aligned}
$$

■

Theorem 7.2 is essential for the asymptotic analysis of parametric models for censored data. In fact the relation (7.21), that the expected value of the first derivative of the log-likelihood, evaluated at the true parameter value (and under the true distribution of the random variables), is equal to 0, is necessary for consistent estimation in a wide class of (or perhaps all) parametric models. We want to emphasise, too, that the 'true' distribution function under which the expectations in Theorem 7.2 are evaluated, may be improper. This is important to our later results. Next we will set out in detail an asymptotic analysis for the simplest such case we are interested in — the exponential mixture model (for a single sample).

7.3 THE EXPONENTIAL MIXTURE MODEL FOR A SINGLE SAMPLE

Likelihood properties

Consider the likelihood for the exponential mixture model, which has the form (see (5.15))

$$L_n(\lambda, p) = \prod_{i=1}^{n} (p\lambda e^{-\lambda t_i})^{c_i} (1 - p + pe^{-\lambda t_i})^{1-c_i}$$

(apart from constants). For this, we have the log-likelihood

$$l_n(\lambda, p) = \sum_{i=1}^{n} \{c_i (\log p + \log \lambda) - c_i \lambda t_i + (1 - c_i) \log(1 - p + pe^{-\lambda t_i})\}, \tag{7.27}$$

with its derivatives

$$\frac{\partial l_n(\lambda, p)}{\partial \lambda} = \sum_{i=1}^{n} \left\{ c_i \left(\frac{1}{\lambda} - t_i \right) - \frac{(1 - c_i) p t_i e^{-\lambda t_i}}{1 - p + pe^{-\lambda t_i}} \right\}, \tag{7.28}$$

and

$$\frac{\partial l_n(\lambda, p)}{\partial p} = \sum_{i=1}^{n} \left\{ \frac{c_i}{p} - \frac{(1 - c_i)(1 - e^{\lambda t_i})}{1 - p + pe^{-\lambda t_i}} \right\}. \tag{7.29}$$

Suppose the 'true' parameters are λ_0 and p_0; by this we mean that the c.d.f. of the true (unobserved) survival times t_i^* is $p_0(1 - e^{-\lambda_0 t})$, $t \geq 0$.

Form the vector of first derivatives

$$S_n(\lambda, p) = \begin{bmatrix} \dfrac{\partial l_n(\lambda, p)}{\partial \lambda} \\ \dfrac{\partial l_n(\lambda, p)}{\partial p} \end{bmatrix}.$$

Then (7.21) tells us that

$$E(S_n(\lambda_0, p_0)) = 0. \tag{7.30}$$

(This is also quite easy to verify directly here; the reader may care to try it for an exercise.) In the language of the theory of estimating equations, (7.30) means that the estimating equation given by equating the first derivative of the log-likelihood to zero, i.e. the equation

$$S_n(\lambda, p) = 0, \tag{7.31}$$

is *unbiased*, and therefore that consistent estimation of λ_0 and p_0 by the solution of (7.31)—by the MLEs of λ_0 and p_0—is possible.

We did not assume, in the above discussion, that the parameter p_0 is *interior* to its parameter space (0, 1]; in fact (7.30) remains true even if $p_0 = 1$, so that the true survival distribution is the ordinary exponential model $1 - \mathrm{e}^{-\lambda_0 t}$, and there are no immunes in the population. We are still, in (7.27), fitting a model which *allows* for an immune proportion in the population, and will estimate this proportion by $1-\tilde{p}_n$, where $\tilde{p}_n$ is the MLE of p_0. This value of $\tilde{p}_n$ may no longer satisfy (7.31), however, in that it may not have $\partial l_n(\tilde{\lambda}_n, \tilde{p}_n)/\partial p = 0$, although it will still be true that $\partial l_n(\tilde{\lambda}_n, \tilde{p}_n)/\partial\lambda = 0$. If $\tilde{p}_n$ is consistent for $p_0 = 1$, we should obtain a value of $\tilde{p}_n$ close to 1, or even equal to 1, at least in large samples.

We will see below that this is precisely what happens in the 'boundary' case when $p_0 = 1$. But for now let us restrict ourselves to the 'interior' case, when $0 < p_0 < 1$, and show, in the next section, that, under very mild conditions, the MLEs of $\tilde{\lambda}_n$ and $\tilde{p}_n$ given by solving (7.31) are consistent for λ_0 and p_0. Before turning to this task in the next section, we will first write down the formulae for the observed and expected information matrices for the exponential mixture model.

To do this, simply differentiate (7.28) and (7.29) once more to get

$$-\frac{\partial^2 l_n(\lambda, p)}{\partial\lambda^2} = \sum_{i=1}^{n} \left\{ \frac{c_i}{\lambda^2} - \frac{(1-c_i)p(1-p)t_i^2\mathrm{e}^{-\lambda t_i}}{(1-p+p\mathrm{e}^{-\lambda t_i})^2} \right\}, \tag{7.32}$$

$$-\frac{\partial^2 l_n(\lambda, p)}{\partial\lambda\partial p} = \sum_{i=1}^{n} \frac{(1-c_i)t_i\mathrm{e}^{-\lambda t_i}}{(1-p+p\mathrm{e}^{-\lambda t_i})^2}, \tag{7.33}$$

and

$$-\frac{\partial^2 l_n(\lambda, p)}{\partial p^2} = \sum_{i=1}^{n} \left\{ \frac{c_i}{p^2} + \frac{(1-c_i)(1-\mathrm{e}^{-\lambda t_i})^2}{(1-p+p\mathrm{e}^{-\lambda t_i})^2} \right\}. \tag{7.34}$$

The above constitute the elements of the symmetric matrix $F_n(\lambda, p)$ which is the negative second derivative matrix of $l_n(\lambda, p)$. Taking expectations of them (using the idea of the proof of Theorem 7.2, looking at cases where $c_i = 1$ or $c_i = 0$ separately; or see Appendix Note 1) we obtain the elements of the expected information matrix $D_n = (d_{ij})$ as

$$d_{11} = p_0 E \left\{ \frac{1-\mathrm{e}^{-\lambda_0 u}}{\lambda_0^2} - \frac{(1-p_0)u^2\mathrm{e}^{-\lambda_0 u}}{1-p_0+p_0\mathrm{e}^{-\lambda_0 u}} \right\}, \tag{7.35}$$

$$d_{12} = E \left\{ \frac{u\mathrm{e}^{-\lambda_0 u}}{(1-p_0+p_0\mathrm{e}^{-\lambda_0 u})} \right\}, \tag{7.36}$$

and

$$d_{22} = E \left\{ \frac{1-\mathrm{e}^{-\lambda_0 u}}{p_0(1-p_0+p_0\mathrm{e}^{-\lambda_0 u})} \right\}. \tag{7.37}$$

These are finite quantities for all λ_0, $\lambda_0 > 0$, and p_0, $0 < p_0 < 1$, by the properties of the exponential function.

Asymptotics of the exponential mixture model: interior case

Theorem 7.1, for the ordinary exponential model, was used to illustrate the kinds of results we look for in an asymptotic analysis. Very similar results hold for the two-parameter exponential *mixture* model, though there are some points of difference. We will consider in the present section the case when there are indeed immunes present in the population, so the 'true' value p_0 of the parameter p in the mixture model:

$$F(t) = p(1 - e^{-\lambda t}), \qquad t \geq 0, \tag{7.38}$$

lies strictly between 0 and 1. (Likewise λ_0, the true value of λ, lies in $(0, \infty)$; we always assume this.) Together they specify a value of $\boldsymbol{\theta}_0 = (p_0, \lambda_0)$ *interior* to the parameter space for $\boldsymbol{\theta}$.

To estimate $\boldsymbol{\theta}_0$, we maximise the log-likelihood $l_n(\boldsymbol{\theta})$ (given in (7.27)) to obtain an MLE $\tilde{\boldsymbol{\theta}}_n$. As explained previously, there is no explicit formula for $\tilde{\boldsymbol{\theta}}_n$; it must be obtained numerically for each data set. Nevertheless, we can deduce some of the properties of $\tilde{\boldsymbol{\theta}}_n$ by an asymptotic analysis. This analysis parallels to some extent the Newton–Raphson procedure for finding $\tilde{\boldsymbol{\theta}}_n$, in that it proceeds via Taylor expansion of the derivative of the log-likelihood. This is followed by the application of two results from probability theory—the law of large numbers and the central limit theorem—to the linearised equations. The method was exemplified in Theorem 7.1, and we will not give detailed proofs for the results stated here; interested readers may consult the notes at the end of this chapter.

For the exponential mixture model we have the following theorem.

Theorem 7.3 *Assume the i.i.d. censoring model of Section 2.1, with $F(t) = p_0(1 - e^{-\lambda_0 t})$, $t \geq 0$. Suppose $0 < \lambda_0 < \infty$ and $0 < p_0 < 1$, and that the censoring distribution G does not degenerate at 0. Then an MLE $\tilde{\boldsymbol{\theta}}_n = (\tilde{p}_n, \tilde{\lambda}_n)$ of $\boldsymbol{\theta}_0 = (p_0, \lambda_0)$ exists and is locally uniquely defined with probability approaching 1, is consistent for $\boldsymbol{\theta}_0$, and is asymptotically normally distributed.*

The idea of a sequence of events A_n occurring 'with probability approaching 1' simply means that $P(A_n) \to 1$ as $n \to \infty$ (see also Section 3.1). Practically, it means that we will observe the occurrence of the event in question—in the case of Theorem 7.3, the existence of the MLE—with high probability in large samples. In Theorem 7.3, the event in question also specifies that the MLE exists *uniquely* in a neighbourhood of the true value $\boldsymbol{\theta}_0$ (this is what we mean by 'locally uniquely') so we can be confident 'with high probability' of this property in any large enough sample.

To apply Theorem 7.3, we need to know the normalisation under which $\tilde{\boldsymbol{\theta}}_n$ converges to normality as $n \to \infty$. This is provided by the square root of minus the second derivative matrix of $l_n(\theta)$. In fact, let

$$D_n = E\left(-\frac{\partial^2 l_n(\boldsymbol{\theta})}{\partial \boldsymbol{\theta}^2}\right)_{\theta=\theta_0}$$

which in this case is a 2×2 matrix, and let $D_n^{1/2}$ denote the symmetric square root of D_n. (See Appendix Note 3 for a discussion of square roots of a matrix.) Then, under the conditions of Theorem 7.3 and as $n \to \infty$, we have

$$D_n^{1/2}(\tilde{\boldsymbol{\theta}}_n - \boldsymbol{\theta}_0) \xrightarrow{D} N(0, I_2) \tag{7.39}$$

where $N(0, I_2)$ denotes the bivariate standard normal distribution (I_2 is the 2×2 identity matrix). Equation (7.39) indeed establishes the approximate normality of the MLE $\tilde{\boldsymbol{\theta}}_n$ in large samples, but is not practically useful since D_n is unknown. However, the matrix

$$F_n(\tilde{\boldsymbol{\theta}}_n) = \left(-\frac{\partial^2 \log l_n(\boldsymbol{\theta})}{\partial \boldsymbol{\theta}^2}\right)_{\theta=\tilde{\theta}_n}, \tag{7.40}$$

i.e. minus the second derivative matrix of the log-likelihood evaluated at the MLE $\tilde{\boldsymbol{\theta}}_n$, is a consistent estimator of D_n in the sense that

$$D_n^{-1/2} F_n(\tilde{\boldsymbol{\theta}}_n) D_n^{-1/2} \xrightarrow{P} I_2. \tag{7.41}$$

Thus, as well as (7.39), we have

$$(F_n(\tilde{\boldsymbol{\theta}}_n))^{1/2}(\tilde{\boldsymbol{\theta}}_n - \boldsymbol{\theta}_0) \xrightarrow{D} N(0, I_2) \tag{7.42}$$

and this is a result we can use since the quantities involved can be calculated from the sample. (To justify replacing D_n in (7.39) by $F_n(\tilde{\boldsymbol{\theta}}_n)$ to obtain (7.42) when (7.41) holds, see Appendix Note 3.) The relation (7.42) says that, approximately in large samples, the MLE $\tilde{\boldsymbol{\theta}}_n$ is bivariate normally distributed with mean vector $\boldsymbol{\theta}_0$ and covariance matrix $(F_n(\tilde{\boldsymbol{\theta}}_n))^{-1}$. This property can be used for calculating confidence intervals around the components $\tilde{p}_n$, $\tilde{\lambda}_n$, of $\hat{\theta}_n$ separately, or a confidence ellipse around $(\tilde{p}_n, \tilde{\lambda}_n)$, jointly.

Also useful is a test that $\theta = (p, \lambda)$ takes on a particular value $\boldsymbol{\theta}_0$ specified by

$$H_0 : \boldsymbol{\theta} = \boldsymbol{\theta}_0. \tag{7.43}$$

This then becomes the 'true' value of $\boldsymbol{\theta}$ in Theorem 7.3. As usual we calculate the deviance statistic

$$d_n = 2(l_n(\tilde{\boldsymbol{\theta}}_n) - l_n(\tilde{\boldsymbol{\theta}}_{H_0})) \tag{7.44}$$

to test H_0, and we have the 'standard' asymptotic result that d_n is asymptotically distributed as χ^2 with degrees of freedom equal to the number of parameters constrained under H_0. This is true under the conditions of Theorem 7.3.

The exponential mixture model: boundary case

We discussed earlier the fact that a test of the boundary hypothesis

$$H_0 : p_0 = 1 \tag{7.45}$$

(no immunes present) in the exponential mixture model lies outside 'standard' asymptotic theory due to the boundary nature of the hypothesis. However, as we indicated in Section 7.1, only a fairly minor modification of the theory is needed. We still use the deviance statistic to test H_0, but it turns out that this test statistic is, in large samples, approximately distributed as a random variable which is a 50–50 mixture of χ_1^2 and a point mass at 0, rather than as a chi-square random variable. In other words, it has the distribution of a random variable X, where

$$P(X \le x) = \tfrac{1}{2} + \tfrac{1}{2}P(\chi_1^2 \le x), \qquad x \ge 0. \tag{7.46}$$

In this section we will state and discuss the conditions under which such a result holds, though again proofs are omitted.

Besides p there is also the rate parameter λ of the exponential distribution to be estimated, and since this is not specified by H_0 in (7.45), it is free to vary. Our approach to testing H_0 is, as usual, to compare the likelihood evaluated at the unrestricted MLE with the likelihood evaluated at the restricted estimator obtained under H_0. Here the unrestricted MLE $(\tilde{\lambda}_n, \tilde{p}_n)$ is the same one we discussed in detail in the previous section: it exists, uniquely, with probability approaching 1, and we emphasised also in the previous section that it may occur on the boundary—we may have $\tilde{p}_n = 1$. Under H_0, we again maximise the likelihood, now for variations in the only parameter left to estimate, namely λ. But under H_0 the exponential mixture model reduces to the ordinary exponential model, and we can explicitly find the MLE for λ; it is $\tilde{\lambda}_{H_0} = \tilde{\lambda}_n$, where $\tilde{\lambda}_n$ is given by (7.2). Of course we have $\tilde{p}_{H_0} = 1$, so the restricted MLE is $\tilde{\theta}_{H_0} = (\tilde{\lambda}_{H_0}, 1)$. To test H_0, we form the deviance difference

$$d_n = 2(l_n(\tilde{\theta}_n) - l_n(\tilde{\theta}_{H_0})) \tag{7.47}$$

and note that, if a boundary maximum of the likelihood occurs, i.e. if $\tilde{p}_n = 1$, then $\tilde{\theta}_n = \tilde{\theta}_{H_0}$ and $d_n = 0$. In this case we will not reject H_0, so the test is rather simple for a boundary estimate of p.

If H_0 is in fact true, then we expect this boundary situation to occur in about 50% of samples, for n large. In the other 50%, approximately, we will find a value of $\tilde{p}_n$ in (0, 1), and a corresponding value of $d_n > 0$. To test whether this value is too 'large', we compare it with the 95th percentile of the distribution of the random variable X defined in (7.46), rejecting H_0 if d_n exceeds this value. We illustrated this test on some data sets in Section 5.3 and Chapter 6.

The above practice is based on the following theoretical result.

Theorem 7.4 *Assume the i.i.d. censoring model, and suppose the true distribution of the survival times is, as specified by H_0 in (7.45), the ordinary exponential* *distribution* $F(t) = 1 - e^{-\lambda_0 t}$, $t \geq 0$. *Let* $\tilde{\boldsymbol{\theta}}_n = (\tilde{\lambda}_n, \tilde{p}_n)$ *be estimates obtained by maximising the log-likelihood* $l_n(\boldsymbol{\theta})$ *in (7.27) for the exponential mixture model. Suppose the censoring random variable* u *is not degenerate at 0 and satisfies*

$$E(e^{\lambda_0 u}) < \infty. \tag{7.48}$$

Then $\tilde{\boldsymbol{\theta}}_n$ *is uniquely determined in a neighbourhood of* $\boldsymbol{\theta}_0$, *with probability approaching 1, and* $\tilde{\boldsymbol{\theta}}_n$ *is consistent for* $\boldsymbol{\theta}_0$. *Let* D *be the* 2×2 *matrix*

$$D = \begin{bmatrix} \lambda_0^{-2} E(1 - e^{-\lambda_0 u}) & E(u) \\ E(u) & E(e^{\lambda_0 u} - 1) \end{bmatrix}, \tag{7.49}$$

which is finite under (7.48), and let (X, Y) *be a 2-vector which has the bivariate normal distribution with mean vector 0 and covariance matrix* D^{-1}. *Suppose in addition that*

$$E(e^{(\lambda_0+\eta)u}) < \infty \qquad \text{for some } \eta > 0. \tag{7.50}$$

Then for each real x *and all* $y < 0$ *the joint asymptotic distribution of* $\sqrt{n}(\hat{\lambda}_n - \lambda_0)$ *and* $\sqrt{n}(\hat{p}_n - 1)$ *is given by*

$$\lim_{n\to\infty} P\{\sqrt{n}(\hat{\lambda}_n - \lambda_0) \leq x, \sqrt{n}(\hat{p}_n - 1) \leq y\} = P\{X \leq x, Y \leq y\}, \tag{7.51}$$

and the limiting distribution of the deviance as given by (7.47) is

$$\lim_{n\to\infty} P\{d_n \leq x\} = \tfrac{1}{2} + \tfrac{1}{2}P\{\chi_1^2 \leq x\}, \qquad \text{for } x \geq 0. \tag{7.52}$$

How do we use all this information in practice? We must begin by assuming that an exponential mixture distribution is a good fit to the data — this can be assessed by the method of Section 5.4, for example. Supposing this is so, we next test for the presence of immunes.

Consider firstly the case when $d_n \leq c_\alpha$, the $100(1 - \alpha)$th percentile of the distribution on the right-hand side of (7.52), so that H_0 is accepted at the $\alpha\%$ significance level ($\alpha = 0.05$, say). Then we do not have significant evidence of immunes, and we can treat the data as if it comes from an ordinary (not mixture) exponential distribution. The large-sample distribution of the MLE $\tilde{\lambda}_n$ is then as given for that model by the discussion in Section 7.2. In fact, $\sqrt{n}(\tilde{\lambda}_n - \lambda_0)$ is approximately normal with mean 0 and variance $\lambda_0^2/E(1 - e^{-\lambda_0 u})$. In practice, we will take $\sqrt{n}(\tilde{\lambda}_n - \lambda_0)$ as approximately normal with mean 0 and variance $1/f^{11}(\tilde{\boldsymbol{\theta}}_n)$, where $f^{11}(\tilde{\boldsymbol{\theta}}_n)$ is the (1, 1) element of the matrix $F_n(\tilde{\boldsymbol{\theta}}_n)$.

What of the other case, when d_n exceeds c_α, so we will reject H_0 and decide that there is evidence of immunes in the population? In fact, the proportion of immunes is then estimated to be $1 - \tilde{p}_n$, and the rate of failure is likewise estimated to be $\tilde{\lambda}_n$, where $\tilde{\theta}_n = (\tilde{\lambda}_n, \tilde{p}_n)$ is the MLE. Now the results of Theorem 7.4 can be used to calculate confidence intervals on the true (but unknown) p_0 and λ_0. From (7.51) we can also deduce the marginal limiting distribution

$$\lim_{n\to\infty} P\{\sqrt{n}(\tilde{\lambda}_n - \lambda_0) \le x\} = P\{X \le x, Y < 0\} + \tfrac{1}{2}P\{Z \le x\},$$
$$\text{for } -\infty < x < \infty, \qquad (7.53)$$

where Z is univariate normal with mean 0 and variance $\lambda_0^2/E(1 - e^{-\lambda_0 u})$; and from (7.51) we also obtain

$$\lim_{n\to\infty} P\{\sqrt{n}(\tilde{p}_n - 1) \le y\} = P\{Y \le y\} \qquad \text{for } y < 0. \qquad (7.54)$$

These results are interesting but not directly useful as they stand since the distributions of X and Y depend on the elements of D, and hence on the unknown censoring distribution. But as usual we surmount this problem by Studentising. The estimator

$$F_n(\tilde{\theta}_n) = \left.\frac{-\partial^2 \log L}{\partial \theta^2}\right|_{\theta=\tilde{\theta}_n}$$

is a consistent estimator of the matrix D in the sense that $F_n(\tilde{\theta}_n)/n \xrightarrow{P} D$ elementwise; hence we can substitute for D the sample value $F_n(\tilde{\theta}_n)/n$, and use (7.53) and (7.54) to obtain approximations for the distributions of $\tilde{\lambda}_n$ and $\tilde{p}_n$. Note that the necessity of estimating p in addition to λ modifies the asymptotic distribution of $\tilde{\lambda}_n$, showing that incorrect inference would be made if p were ignored when in fact immunes are judged to be present.

We will omit the proof of Theorem 7.4, since it is technically complicated and the details are set out fully in Zhou and Maller (1995). However, a much simpler proof can be given for the case when λ is *known* (hence, need not be estimated) and this illustrates the essential ideas. In particular, it shows clearly why a boundary estimate of p occurs with probability approaching 1/2 in large samples, and explains the asymptotic mixture distribution (7.52) for the deviance statistic d_n. So we will give this result here.

Theorem 7.5 *Assume the i.i.d. censoring model, with $F_0(t) = 1 - e^{-\lambda_0 t}$, and assume that λ_0 is known. Suppose the censoring random variable u is not degenerate at 0. Then a unique MLE $\tilde{p}_n$ of p exists in (0, 1]. If in addition*

$$E(e^{\lambda_0 u}) < \infty, \qquad (7.55)$$

then $P\{\tilde{p}_n = 1\} \to 1/2$ as $n \to \infty$, $P\{d_n = 0\} \to 1/2$ as $n \to \infty$, and d_n has the limiting mixture distribution specified by (7.52).

Proof When λ_0 is known, we can write the log-likelihood of a sample $t_1, \ldots, t_n$, apart from constants, as

$$l_n(p) = \sum_{i=1}^{n} \{c_i \log p + (1 - c_i)\log(1 - pF_i)\}, \qquad 0 < p \le 1;$$

this is derived from (7.27). Here for brevity we let $F_i = 1 - e^{-\lambda_0 t_i}$. Define

$$S_n(p) = \mathrm{d}l_n(p)/\mathrm{d}p.$$

It is easy to show by differentiation that the random function of p defined by

$$X_n(p) = pS_n(p) = \sum_{i=1}^{n} \left(\frac{c_i - pF_i}{1 - pF_i} \right) \tag{7.56}$$

is nonincreasing in p on [0,1]. Also $X_n(0) > 0$ with probability approaching 1 as $n \to \infty$ since

$$\frac{1}{n} X_n(0) = \frac{1}{n} \sum_{i=1}^{n} c_i \to E(c_1) \qquad \text{as } n \to \infty,$$

and $E(c_1) > 0$ because c_1 is not degenerate at 0. Thus there is a p^* in (0,1) such that $X_n(p^*) = 0$, equivalently $S_n(p^*) = 0$, if and only if $X_n(1) < 0$ (see Figure 7.1). It follows that

$$\tilde{p}_n = \begin{cases} p^* & \text{if } X_n(1) < 0 \\ 1 & \text{if } X_n(1) \geq 0 \end{cases}$$

is the unique MLE of p in (0,1].

By assumption (7.55),

$$E(\mathrm{e}^{\lambda u}) - 1 = \int_{[0,\infty)} \frac{\mathrm{d}G(y)}{1 - F(y)} = \sigma^2, \text{ say,}$$

is finite. As an application of Theorem 7.2, $X_n(1)$ has mean zero and variance $n\sigma^2$. Since the t_i and c_i are i.i.d., we can apply the central limit theorem to obtain,

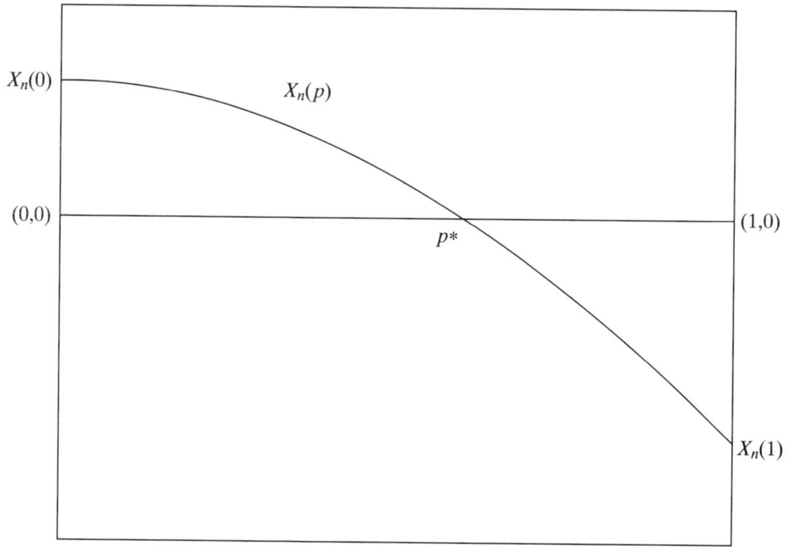

Figure 7.1 Schematic illustration of interior maximum of $X_n(p)$.

as $n \to \infty$,

$$\frac{X_n(1)}{\sqrt{n}} \xrightarrow{D} N(0, \sigma^2).$$

Thus $P\{X_n(1) \geq 0\} \to 1/2$, equivalently $P\{\tilde{p}_n = 1\} \to 1/2$, and so $\tilde{p}_n$ is not asymptotically normal. However, using a Taylor expansion we can show that for $x < 0$,

$$\begin{aligned} P\{\sigma\sqrt{n}(\tilde{p}_n - 1) \leq x | \tilde{p}_n < 1\} &= P\{\sigma\sqrt{n}(\tilde{p}_n - 1) \leq x | X_n(1) < 0\} \\ &\to P\{N(0,1) \leq x | N(0,1) \leq 0\}, \end{aligned}$$

so $\sigma\sqrt{n}(\tilde{p}_n - 1)$ has a truncated normal distribution, asymptotically.

To test the null hypothesis $H_0 : p = 1$, we use d_n as defined by (7.47), i.e.

$$d_n = 2\{l_n(\tilde{p}_n) - l_n(1)\}.$$

When $\tilde{p}_n = 1$, $d_n = 0$, so

$$P\{d_n = 0\} \to 1/2. \tag{7.57}$$

When $\tilde{p}_n < 1$, $S_n(\tilde{p}_n) = 0$, so a Taylor expansion shows that

$$\begin{aligned} l_n(1) - l_n(\tilde{p}_n) &= (1 - \tilde{p}_n)S_n(\tilde{p}_n) + \frac{1}{2}(1 - \tilde{p}_n)^2 \frac{\mathrm{d}S_n(\dot{p}_n)}{\mathrm{d}p} \\ &= -\tfrac{1}{2}(1 - \tilde{p}_n)^2 F_n(\dot{p}_n) \end{aligned}$$

for some $\dot{p}_n$ between $\tilde{p}_n$ and 1. Here $F_n(p) = -\mathrm{d}^2 l_n(p)/\mathrm{d}p^2$ and we can use the weak law of large numbers to show that $F_n(\dot{p}_n)/n \xrightarrow{P} \sigma^2$. Thus, as $n \to \infty$, for $x > 0$,

$$\begin{aligned} P\{d_n \leq x | \tilde{p}_n < 1\} &= P\left\{\sigma^2 n(\tilde{p}_n - 1)^2 \frac{F_n(\dot{p}_n)}{\sigma^2 n} \leq x \middle| \tilde{p}_n < 1\right\} \\ &\to P\{[N(0,1)]^2 \leq x | N(0,1) \leq 0\} \\ &= \frac{1}{2}\frac{P\{[N(0,1)]^2 \leq x\}}{P\{N(0,1) \leq 0\}} = P\{\chi_1^2 \leq x\}. \end{aligned} \tag{7.58}$$

Then the fact that d_n has limiting distribution given by (7.52) follows from (7.57) and (7.58) and the fact that $P\{\tilde{p}_n = 1\} \to 1/2$ as $n \to \infty$. ■

In Theorem 7.5, condition (7.55) represents the current manifestation of something that we mentioned earlier: there must be some measure of the censoring distribution not being too 'heavy'. In the present case, condition (7.55) rules out censoring distributions which do not have moment generating functions.

Asymptotic uniqueness of the estimators

In very general nonlinear models, an analysis of likelihood estimation cannot hope to establish *global* uniqueness WPA1 of a maximum of the likelihood, as opposed to uniqueness in relatively small neighbourhoods of the true parameter, WPA1. This latter property is all that we have established in Theorems 7.3 to 7.5. Global uniqueness is a useful property to know in practice, since it tells us that there is *a single* maximum of the likelihood to be found (by Newton—Raphson or some other numerical procedure). It does not, however, rule out the possibility of local maxima of the likelihood, but so far in practice (in all the data analyses we have done, such as those in Chapters 5 and 6 using exponential and Weibull mixture models), we have not found a situation in which the likelihood had more than one local maximum.

The special structure of the exponential mixture model allows us to prove a version of global uniqueness of the maximum likelihood estimator $\boldsymbol{\theta} = (\lambda, p)$, although we need to develop our own methods. We will establish that a unique global maximum of the likelihood for the exponential mixture model exists with probability approaching 1 (WPA1) as $n \to \infty$. Unlike for the ordinary exponential model, we cannot do this by establishing overall positive definiteness of minus the second derivative matrix of the log-likelihood, $-\partial^2 l_n(\boldsymbol{\theta})/\partial\boldsymbol{\theta}^2$, since this does not hold; it is not hard to find an example in which the diagonal element of $-\partial^2 l_n(\boldsymbol{\theta})/\partial\boldsymbol{\theta}^2$ in (7.32) is negative when $p < 1$. Instead we proceed by direct examination of $l_n(\boldsymbol{\theta})$.

From (7.27) we have the log-likelihood as

$$l_n(\boldsymbol{\theta}) = \sum_{i=1}^{n} \{c_i(\log p + \log \lambda - \lambda t_i) + (1 - c_i)\log(1 - p + pe^{-\lambda t_i})\}. \qquad (7.59)$$

Let $\tilde{\boldsymbol{\theta}}_n$ be any maximiser of $l_n(\boldsymbol{\theta})$, i.e. any random variable satisfying $l_n(\tilde{\boldsymbol{\theta}}_n) = \sup_{\boldsymbol{\theta}} l_n(\boldsymbol{\theta})$. Define a neighbourhood of θ_0 by

$$N(\boldsymbol{\theta}_0, \delta) = \{\boldsymbol{\theta} : |\boldsymbol{\theta} - \boldsymbol{\theta}_0| < \delta,\ p \le 1\}.$$

We will now prove Theorem 7.6.

Theorem 7.6 *For any sufficiently small $\delta > 0$ and $\varepsilon > 0$, there is an N such that*

$$P\{\tilde{\boldsymbol{\theta}}_n \text{ exists uniquely in } (0, \infty) \times (0, 1] \text{ and } \tilde{\boldsymbol{\theta}}_n \in N(\boldsymbol{\theta}_0, \delta)\} > 1 - \varepsilon$$

for all $n > N$.

Proof We already know from Theorem 7.3 that $\tilde{\boldsymbol{\theta}}_n$ is uniquely defined in a neighbourhood $N(\boldsymbol{\theta}_0, \delta)$, WPA1 when δ is sufficiently small. So to prove the theorem, we need only show that when $\eta > 0$ is sufficiently small and $R > 0$ is sufficiently large, as $n \to \infty$,

$$P\{\tilde{\boldsymbol{\theta}}_n \in [0, \eta] \times [0, 1]\} \to 0, \qquad (7.60)$$

$$P\{\tilde{\boldsymbol{\theta}}_n \in [R, \infty] \times [0, 1]\} \to 0, \qquad (7.61)$$

$$P\{\tilde{\boldsymbol{\theta}}_n \in [0, \infty] \times [0, \eta]\} \to 0, \qquad (7.62)$$

$$P\{\tilde{\boldsymbol{\theta}}_n \in [\eta, R] \times [\eta, 1-\eta] - N(\boldsymbol{\theta}_0, \delta)\} \to 0, \tag{7.63}$$

$$P\{\tilde{\boldsymbol{\theta}}_n \in [\eta, R] \times [1-\eta, 1] - N(\boldsymbol{\theta}_0, \delta)\} \to 0. \tag{7.64}$$

We begin with (7.60). First suppose $\tilde{\boldsymbol{\theta}}_n \in [0, \eta] \times [0, 1]$, where $0 < \eta < 1$. Then the maximum of $l_n(\boldsymbol{\theta})$ in this region must at least equal its value outside this region, in particular, its value at (1,1). Thus

$$\sup_{0<p\le 1} \sup_{\lambda \le \eta} l_n(\boldsymbol{\theta}) \ge l_n(1, 1). \tag{7.65}$$

Now by (7.59),

$$l_n(1, 1) = -\sum_{i=1}^{n} t_i = -n\bar{t}_n. \tag{7.66}$$

Also by (7.59),

$$l_n(\boldsymbol{\theta}) \le \sum_{i=1}^{n} c_i(\log\lambda - \lambda t_i) \le \sum_{i=1}^{n} c_i \log\lambda = n\bar{c}_n \log\lambda \tag{7.67}$$

because $\log p \le 0$ and $\log(1 - p + pe^{-\lambda t_i}) \le 0$ for all $p \le 1$, $\lambda > 0$, and $i = 1, \ldots, n$. By the weak law of large numbers, $\bar{c}_n \xrightarrow{P} E(c_1) > 0$. It follows that when η is so small that $\log\eta < -2E(t_1)/E(c_1)$, we have

$$2 \sup_{0<\lambda\le\eta} \bar{c}_n \log\lambda \le E(c_1)\log\eta < -2E(t_1) \tag{7.68}$$

WPA1. Combine (7.65) to (7.68) to get, when $\log\eta < -2E(t_1)/E(c_1)$, that

$$\begin{aligned} P\{\tilde{\boldsymbol{\theta}}_n \in [0, \eta] \times [0, 1]\} &\le P\left\{ \sup_{0<p\le 1} \sup_{0<\lambda\le\eta} l_n(\boldsymbol{\theta}) \ge -n\bar{t}_n \right\} \\ &\le P\left\{ \sup_{0<\lambda\le\eta} \bar{c}_n \log\lambda \ge -\bar{t}_n \right\} \to 0. \end{aligned}$$

This proves (7.60).

The proofs for (7.61) and (7.62) are much the same. For (7.61), use a similar argument to (7.65), (7.66) and the first inequality in (7.67) to get

$$\begin{aligned} P\{\tilde{\boldsymbol{\theta}}_n \in [R, \infty) \times [0, 1]\} &\le P\left\{ \sup_{0<p\le 1} \sup_{\lambda\ge R} l_n(\boldsymbol{\theta}) \ge l_n(1, 1) = -n\bar{t}_n \right\} \\ &\le P\left\{ \sup_{\lambda\ge R} \left(\sum_{i=1}^{n} c_i(\log\lambda - \lambda t_i) \right) \ge -n\bar{t}_n \right\} \\ &= P\left\{ \sup_{\lambda\ge R} \left(\bar{c}_n \log\lambda - \frac{\lambda}{n} \sum_{i=1}^{n} c_i t_i \right) \ge -\bar{t}_n \right\}. \end{aligned} \tag{7.69}$$

But since $E(c_1t_1) > 0$, $2E(c_1)\log\lambda - \frac{\lambda}{2}E(c_1t_1) \to -\infty$ as $\lambda \to \infty$, and so when R is sufficiently large,

$$\sup_{\lambda\geq R}\left\{2E(c_1)\log\lambda - \frac{\lambda}{2}E(c_1t_1)\right\} < -2E(t_1).$$

It follows that

$$P\left\{\sup_{\lambda\geq R}\left(\bar{c}_n\log\lambda - \frac{\lambda}{n}\sum_{i=1}^{n}c_it_i\right) < -\bar{t}_n\right\}$$

$$\geq P\left\{\bar{c}_n < 2E(c_1), \frac{1}{n}\sum_{i=1}^{n}c_it_i > \frac{1}{2}E(c_1t_1), \bar{t}_n < 2E(t_1)\right\} \to 1$$

as $n \to \infty$, which, together with (7.69), proves (7.61).

For (7.62), we use the inequality (see (7.59))

$$\frac{1}{n}l_n(\boldsymbol{\theta}) \leq \frac{1}{n}\sum_{i=1}^{n}c_i(\log p + \log\lambda - \lambda t_i)$$

$$= \bar{c}_n\log p + \bar{c}_n\log\lambda - \lambda\frac{1}{n}\sum_{i=1}^{n}c_it_i$$

$$\leq \bar{c}_n\left\{\log p + \log\left(\frac{n\bar{c}_n}{\sum_{i=1}^{n}c_it_i}\right)\right\} \qquad \text{(for all } \boldsymbol{\theta}\text{)}.$$

The last inequality holds because when $a \geq 0$ and $b \geq 0$, the function $a\log\lambda - b\lambda$ takes its maximum at $\lambda = a/b$, so

$$a\log\lambda - b\lambda \leq a\log\frac{a}{b} - a \leq a\log\frac{a}{b}.$$

Consequently, by again comparing the value of $l_n(\boldsymbol{\theta})$ with that of $l_n(1,1) = -n\bar{t}_n$, we obtain

$$P\{\tilde{\boldsymbol{\theta}}_n \in [0,\infty)\times[0,\eta]\} \leq P\left\{\sup_{p\leq\eta}\frac{1}{n}l_n(\boldsymbol{\theta}) \geq -\bar{t}_n\right\}$$

$$\leq P\left\{\log\eta + \log\left(\frac{n\bar{c}_n}{\sum_{i=1}^{n}c_it_i}\right) \geq -\frac{\bar{t}_n}{\bar{c}_n}\right\}.$$

The last probability converges to 0 as $n \to \infty$ provided that η is so small that $\log\eta < -\log(E(c_1)/E(c_1t_1)) - (E(t_1)/E(c_1))$. Thus (7.62) holds.

To show (7.63), we will need the following inequality. We can write the log-likelihood in the form (see (7.59))

$$l_n(\boldsymbol{\theta}) = \sum_{i=1}^{n} \log(f(t_i, \boldsymbol{\theta})^{c_i}[1 - F(t_i, \boldsymbol{\theta})]^{1-c_i}) \tag{7.70}$$

where $f(t, \boldsymbol{\theta}) = p\lambda e^{-\lambda t}$ and $F(t, \boldsymbol{\theta}) = p(1 - e^{-\lambda t})$. Define the function

$$z(\boldsymbol{\theta}) = E_{\boldsymbol{\theta}_0}\left\{\frac{1}{n} l_n(\boldsymbol{\theta})\right\} = E_{\boldsymbol{\theta}_0}\{\log(f(t_1, \boldsymbol{\theta})^{c_1}[1 - F(t_1, \boldsymbol{\theta})]^{1-c_1})\}. \tag{7.71}$$

Then, under the i.i.d. censoring model, we will show that

$$z(\boldsymbol{\theta}_0) \geq z(\boldsymbol{\theta}) \tag{7.72}$$

for all $\boldsymbol{\theta}$, with equality if and only if

$$X(\boldsymbol{\theta}) = \frac{f(t_1, \boldsymbol{\theta})^{c_1}[1 - F(t_1, \boldsymbol{\theta})]^{1-c_1}}{f(t_1, \boldsymbol{\theta}_0)^{c_1}[1 - F(t_1, \boldsymbol{\theta}_0)]^{1-c_1}} \tag{7.73}$$

is a constant, a.s.

The proof of this is a version of a standard argument in likelihood theory. We include it here for ease of reference. Since the function $\log x$ is concave (second derivative negative), Jensen's inequality tells us that

$$z(\boldsymbol{\theta}_0) - z(\boldsymbol{\theta}) = E\{-\log X(\boldsymbol{\theta})\} \geq -\log(E\{X(\boldsymbol{\theta})\}).$$

Now evaluate the right-hand side of this to get

$$\begin{aligned} -\log(E\{X(\boldsymbol{\theta})\}) &= -\log\left\{E\left[\int_0^u \frac{f(t, \boldsymbol{\theta})}{f(t, \boldsymbol{\theta}_0)} f(t, \boldsymbol{\theta}_0)\, dt\right]\right. \\ &\qquad \left. + E\left[\frac{1 - F(u, \boldsymbol{\theta})}{1 - F(u, \boldsymbol{\theta}_0)}(1 - F(u, \boldsymbol{\theta}_0))\right]\right\} \\ &= -\log\{E[F(u, \boldsymbol{\theta})] + E[1 - F(u, \boldsymbol{\theta})]\} = -\log 1 = 0. \end{aligned}$$

(We used equations (7.80) and (7.81) of Appendix Note 1 to evaluate the expectations.) Thus we have shown that $z(\boldsymbol{\theta}_0) \geq z(\boldsymbol{\theta})$, and equality holds if and only if $X(\boldsymbol{\theta})$ is a constant a.s., since this is the case when equality holds in Jensen's inequality. Furthermore, since $X(\boldsymbol{\theta})$ is not constant a.s. for $\boldsymbol{\theta} \neq \boldsymbol{\theta}_0$ and $z(\boldsymbol{\theta})$ is continuous in $\boldsymbol{\theta}$, we can conclude that, for any given $\delta > 0$, there exists $\varepsilon > 0$ such that

$$z(\boldsymbol{\theta}_0) - z(\boldsymbol{\theta}) \geq \varepsilon \qquad \text{for all } \boldsymbol{\theta} \notin N(\boldsymbol{\theta}_0, \delta). \tag{7.74}$$

Let $\varepsilon > 0$, $\delta > 0$ satisfy (7.74). Then we prove the following: for any $\boldsymbol{\theta}_1 \in [\eta, R] \times [\eta, 1 - n] - N(\boldsymbol{\theta}_0, \delta)$, there exists a $\delta_1 > 0$ such that

$$P\left\{\sup_{\boldsymbol{\theta} \in N_1} \frac{1}{n} l_n(\boldsymbol{\theta}) \geq \frac{1}{n} l_n(\boldsymbol{\theta}_0) - \frac{\varepsilon}{2}\right\} \to 0 \qquad \text{as } n \to \infty \tag{7.75}$$

where

$$N_1 = N(\boldsymbol{\theta}_1, \delta_1) = \{\boldsymbol{\theta} : |\boldsymbol{\theta} - \boldsymbol{\theta}_1| < \delta_1\}.$$

To see this, let $\boldsymbol{\theta}_1 \in [\eta, R] \times [\eta, 1-\eta] - N(\boldsymbol{\theta}_0, \delta)$. Then by (7.74) we have $z(\boldsymbol{\theta}_0) \geq z(\boldsymbol{\theta}_1) + \varepsilon$ and

$$\begin{aligned}
&P\left\{\sup_{\boldsymbol{\theta}\in N_1} \frac{1}{n} l_n(\boldsymbol{\theta}) \geq \frac{1}{n} l_n(\boldsymbol{\theta}_0) - \frac{\varepsilon}{2}\right\} \\
&\quad \leq P\left\{\sup_{\boldsymbol{\theta}\in N_1} \frac{1}{n}[l_n(\boldsymbol{\theta}) - l_n(\boldsymbol{\theta}_1)] \geq \frac{1}{n} l_n(\boldsymbol{\theta}_0) - \frac{1}{n} l_n(\boldsymbol{\theta}_1) - \frac{\varepsilon}{2}\right\} \\
&\quad \leq P\left\{\sup_{\boldsymbol{\theta}\in N_1} \frac{1}{n}[l_n(\boldsymbol{\theta}) - l_n(\boldsymbol{\theta}_1)] \geq \varepsilon - \left|\frac{1}{n} l_n(\boldsymbol{\theta}_0) - z(\boldsymbol{\theta}_0)\right| \right. \\
&\qquad \left. - \left|\frac{1}{n} l_n(\boldsymbol{\theta}_1) - z(\boldsymbol{\theta}_1)\right| - \frac{\varepsilon}{2}\right\} \\
&\quad \leq P\left\{\sup_{\boldsymbol{\theta}\in N_1} \frac{1}{n}[l_n(\boldsymbol{\theta}) - l_n(\boldsymbol{\theta}_1)] \geq \frac{\varepsilon}{4}\right\} + P\left\{\left|\frac{1}{n} l_n(\boldsymbol{\theta}_0) - z(\boldsymbol{\theta}_0)\right| > \frac{\varepsilon}{8}\right\} \\
&\qquad + P\left\{\left|\frac{1}{n} l_n(\boldsymbol{\theta}_1) - z(\boldsymbol{\theta}_1)\right| > \frac{\varepsilon}{8}\right\} \\
&\quad = P\left\{\sup_{\boldsymbol{\theta}\in N_1} \frac{1}{n}[l_n(\boldsymbol{\theta}) - l_n(\boldsymbol{\theta}_1)] \geq \frac{\varepsilon}{4}\right\} + o(1)
\end{aligned}$$

where the last equality is due to the law of large numbers. Thus it suffices to show that for sufficiently small $\delta_1 > 0$,

$$P\left\{\sup_{\boldsymbol{\theta}\in N_1} \frac{1}{n}[l_n(\boldsymbol{\theta}) - l_n(\boldsymbol{\theta}_1)] \geq \frac{\varepsilon}{4}\right\} \to 0 \qquad \text{as } n \to \infty. \tag{7.76}$$

Let $\boldsymbol{\theta}_1 = (\lambda_1, p_1)$. If $\bar{t}_n < 2E(t_1)$, then

$$\begin{aligned}
&\frac{1}{n}[l_n(\boldsymbol{\theta}) - l_n(\boldsymbol{\theta}_1)] \\
&= \frac{1}{n}\sum_{i=1}^{n}\left(c_i\left\{\log\frac{p}{p_1} + \log\frac{\lambda}{\lambda_1} - (\lambda - \lambda_1)t_i\right\} + (1 - c_i)\log\frac{1 - p + p\mathrm{e}^{-\lambda t_i}}{1 - p_1 + p_1\mathrm{e}^{-\lambda_1 t_i}}\right) \\
&\leq \left|\log\frac{p}{p_1}\right| + \left|\log\frac{\lambda}{\lambda_1}\right| + |\lambda - \lambda_1|\bar{t}_n + \frac{1}{n}\sum_{i=1}^{n}\left|\log\frac{1 - p + p\mathrm{e}^{-\lambda t_i}}{1 - p_1 + p_1\mathrm{e}^{-\lambda_1 t_i}}\right| \\
&\leq \frac{1}{\tilde{p}}|p - p_1| + \frac{1}{\tilde{\lambda}}|\lambda - \lambda_1| + |\lambda - \lambda_1| 2E(t_1) \\
&\quad + \frac{1}{n}\sum_{i=1}^{n}\left\{\frac{t_i\mathrm{e}^{-\tilde{\lambda} t_i}}{1 - \tilde{p} + \tilde{p}\mathrm{e}^{-\tilde{\lambda} t_i}}|\lambda - \lambda_1| + \frac{1 - \mathrm{e}^{-\tilde{\lambda} t_i}}{1 - \tilde{p} + \tilde{p}\mathrm{e}^{-\tilde{\lambda} t_i}}|p - p_1|\right\}
\end{aligned} \tag{7.77}$$

where $\tilde{p}$ is between p and p_1, and $\tilde{\lambda}$ is between λ and λ_1. Since $\boldsymbol{\theta}_1 \in [\eta, R] \times [\eta, 1-\eta]$, λ is bounded away from 0 and ∞, while p is bounded away from 0 and 1. Thus (7.77) leads to

$$\begin{aligned}\frac{1}{n}[l_n(\boldsymbol{\theta}) - l_n(\boldsymbol{\theta}_1)] &\leq M_1(|p - p_1| + |\lambda - \lambda_1|) + 2E(t_1)|\lambda - \lambda_1| \\ &\quad + \frac{1}{1-\tilde{p}}\{2E(t_1)|\lambda - \lambda_1| + |p - p_1|\} \\ &\leq M_2(|p - p_1| + |\lambda - \lambda_1|) \qquad (\text{if } \bar{t}_n < 2E(t_1))\end{aligned}$$

for all $\boldsymbol{\theta} \in N_1$, where M_1, M_2 are two constants dependent on η, R and $E(t_1)$ only. Thus when δ_1 is sufficiently small, $\bar{t}_n < 2E(t_1)$ implies

$$\sup_{\boldsymbol{\theta} \in N_1} \frac{1}{n}[l_n(\boldsymbol{\theta}) - l_n(\boldsymbol{\theta}_1)] < \frac{\varepsilon}{4}.$$

It follows that

$$P\left\{\sup_{\boldsymbol{\theta} \in N_1} \frac{1}{n}[l_n(\boldsymbol{\theta}) - l_n(\boldsymbol{\theta}_1)] \geq \frac{\varepsilon}{4}\right\} \leq P\{\bar{t}_n \geq 2E(t_1)\} \to 0$$

as $n \to \infty$. Thus (7.75) holds.

Now because $J = [\eta, R] \times [\eta, 1-\eta] - N(\boldsymbol{\theta}_0, \delta)$ is a compact subset of $\mathbb{R}^2$, (7.75) together with the finite cover theorem implies the existence of a finite number of subsets $N_1, N_2, \ldots, N_k$ of $\mathbb{R}^2$ such that (7.75) holds with N_1 replaced with each of $N_1, \ldots, N_k$ and $J \subseteq \cup_{i=1}^k N_i$. It follows that

$$P\left\{\sup_{\boldsymbol{\theta} \in J} \frac{1}{n} l_n(\boldsymbol{\theta}) \geq l_n(\boldsymbol{\theta}_0) - \frac{\varepsilon}{2}\right\} \to 0 \tag{7.78}$$

and this clearly implies (7.63).

It remains to prove (7.64). Let $J = [\eta, R] \times [\eta, 1-\eta] - N(\boldsymbol{\theta}_0, \delta)$ as before and let $J_1 = [\eta, R] \times [1-\eta, 1] - N(\boldsymbol{\theta}_0, \delta)$. Note $1 - p + p\mathrm{e}^{-\lambda t_i} \leq \eta + (1-\eta)\mathrm{e}^{-\lambda t_i}$ for $p \geq 1 - \eta$, hence $\boldsymbol{\theta} \in J_1$ implies by (7.59) that

$$\begin{aligned}l_n(\boldsymbol{\theta}) &\leq \sum\{c_i(\log\lambda + \log p - \lambda t_i) + (1 - c_i)\log[\eta + (1-\eta)\mathrm{e}^{-\lambda t_i}]\} \\ &= l_n(\lambda, 1-\eta) + n\bar{c}_n(\log p - \log(1-\eta)).\end{aligned} \tag{7.79}$$

But

$$|\log p - \log(1-\eta)| = \frac{1}{\tilde{p}}|p - (1-\eta)| \leq \frac{\eta}{(1-\eta)} < \frac{\varepsilon}{2}$$

if $p \geq 1-\eta$ and η is sufficiently small (where $\tilde{p} \in [1-\eta, p]$). Hence (7.79) gives

$$\frac{1}{n} l_n(\boldsymbol{\theta}) \leq \frac{1}{n} l_n(\lambda, 1-\eta) + \frac{\varepsilon}{2} \qquad \text{for all } \boldsymbol{\theta} \in J_1.$$

It follows that

$$P\left\{\sup_{\theta\in J_1}\frac{1}{n}l_n(\theta)\geq l_n(\theta_0)\right\}\leq P\left\{\sup_{\theta\in J_1}\frac{1}{n}l_n(\lambda,1-\eta)+\frac{\varepsilon}{2}\geq l_n(\theta_0)\right\}$$

$$\leq P\left\{\sup_{\theta\in J}\frac{1}{n}l_n(\theta)\geq l_n(\theta_0)-\frac{\varepsilon}{2}\right\}\to 0$$

as $n\to\infty$, by (7.78), on noting that $\theta=(\lambda,1-\eta)\in J$ if $\theta\in J_1$. This in turn implies

$$P\{\tilde{\theta}_n\in J_1\}\to 0 \qquad \text{as } n\to\infty$$

and (7.64) follows. This completes the proof of Theorem 7.6.

APPENDIX NOTES TO CHAPTER 7

1. Calculating expectations of functions of censored random variables

Suppose a generalised independent censoring model holds, so that the t_i^* are independent survival times with cumulative distribution $p_iF_{i0}(t)$, the u_i are independent nonnegative random variables with distribution $G_i(t)$, and the t_i^* and u_i are independent. As usual, let $t_i=\min(t_i^*,u_i)$. Then if Q is any measurable real-valued function, we have

$$E(c_iQ(t_i))=p_iE\left\{\int_{[0,u_i]}Q(t)\,\mathrm{d}F_{i0}(t)\right\} \tag{7.80}$$

and

$$E((1-c_i)Q(t_i))=E((1-p_iF_{i0}(u_i))Q(u_i)), \tag{7.81}$$

provided the expectations on the right exist. These equations are proved in Ghitany *et al.* (1994) using a method similar to the proof of Theorem 7.2.

2. Boundary hypothesis tests

It was understood early in the theory of hypothesis testing that an estimator of a parameter which lies on a boundary of the parameter space for that parameter will not behave, distributionally, like the estimator of a parameter which lies interior to its parameter space. When the estimator or hypothesis testing procedure is maximum likelihood, and the sample size is large, the estimator in the interior case will be approximately normally distributed (under regularity conditions, of course); whereas in the boundary case we recognise, informally, that the range of variation of the estimator must be restricted, and we do not expect a limiting normal distribution to result.

Chernoff (1954) seems to have been the first to give a rigorous asymptotic result for a boundary estimator. He worked with a very general setup, dealing with the maximum likelihood estimator of a parameter restricted to lie in or on the boundary of what is (at least local to the true parameter) a cone. This setup is general enough to cover most practical situations. Chernoff assumed various regularity conditions (such as uniform integrability of the third derivative of the log-likelihood in a neighbourhood of the true parameters) and obtained, as the limiting distribution of the deviance statistic for testing the hypothesis that θ_0 lies on the boundary of the cone, a mixture of chi-square distributions.

Later writers (Feder 1968; Moran 1971) gave variants or extensions of Chernoff's results. Cox and Hinkley (1974, pp. 303–321) gave a discussion of the issue too. More recently, Self and Liang (1987) gave a precise specification of the problem and obtained rigorous asymptotic results, proving also the *existence* and *uniqueness* of (an) MLE in large samples, which, as we pointed out above, are major aspects of the theory, often overlooked by less formal treatments. Our own contribution is Theorem 7.3, which deals with the hypothesis $H_0 : p_0 = 1$, and is thus a test for the presence of immunes. Theorem 7.3 is proved in detail in Zhou and Maller (1995), where we extend the method of Fahrmeir and Kaufmann (1985) for generalised linear models (interior case only) to the exponential mixture model for immunes (boundary case).

The Self-Liang setup is too restrictive for this model, but the special properties of the exponential distribution allow the very general result given in Theorem 7.3. Even more recently, Vu (1995) has generalised our results to a wide class of distributions in the exponential family, and Vu and Zhou (1997) have given some very general results for boundary hypothesis tests.

3. 'Studentisation' by 'matrix square roots'

In working out the large-sample behaviour of an estimator $\tilde{\theta}_n$ (MLE or otherwise) derived from a sample of size n, we are often able to prove results like

$$D_n^{1/2}(\tilde{\theta}_n - \theta_0) \xrightarrow{D} N(0, I_k) \tag{7.82}$$

as $n \to \infty$ (where $N(0, I_k)$ denotes the k-variate standard normal distribution and I_k is the $k \times k$ identity matrix), and also

$$D_n^{-1/2} F_n(\tilde{\theta}_n) D_n^{-1/2} \xrightarrow{P} I_k. \tag{7.83}$$

The matrix D_n is often (for the MLE, in particular) the expected information matrix given by

$$D_n = E\left(-\frac{\partial^2 l_n(\theta)}{\partial \theta^2}\right)_{\theta=\theta_0},$$

but when maximising the likelihood, we usually take $F_n(\tilde{\theta}_n)$ to be minus the second derivative matrix of the log-likelihood evaluated at the MLE:

$$F_n(\tilde{\theta}_n) = \left(-\frac{\partial^2 \log l_n(\theta)}{\partial \theta^2}\right)_{\theta=\tilde{\theta}_n}. \tag{7.84}$$

The matrix $D_n^{1/2}$ is a 'square root' of D_n, that is, a matrix which satisfies

$$D_n^{1/2}(D_n^{1/2})^T = D_n. \tag{7.85}$$

Such a matrix square root is not uniquely defined, and there may be various 'versions' for which (7.85) holds. Following Theorem 7.3, for example, we used the 'symmetric square root' of the matrix D_n, and mentioned that (7.82) and (7.83) hold for the MLE (see (7.39) and (7.41)). Fahrmeir and Kaufmann (1985) mention the use of the 'Cholesky' square root of a matrix in a similar context concerning the asymptotic normality of the regression estimator in generalised linear models.

Both types of matrix were considered by Vu *et al.* (1995) who show that (7.82) and (7.83) imply that the 'Studentised' result corresponding to (7.82) holds: we have, in general,

$$(F_n(\tilde{\theta}_n))^{1/2}(\tilde{\theta}_n - \theta_n) \xrightarrow{D} N(0, I_k). \tag{7.86}$$

This is the version that is practically useful: D_n is unknown but can be consistently estimated by $F_n(\tilde{\theta}_n)$ in the sense of (7.83), and the latter matrix can be calculated from the sample. Equation (7.86) then establishes the approximate normality of the MLE $\tilde{\theta}_n$ in large samples, with a variance matrix which can be estimated from the sample. With this result, we can calculate confidence ellipsoids on the unknown value of θ_0.

CHAPTER 8

Large-Sample Properties of Parametric Models with Covariates

In this chapter we will formalise the method of dealing with survival data with covariates that we used in the data analyses of Chapter 6. There we used the Weibull mixture model for the distribution of the lifetimes of susceptibles, and here we restrict our theoretical development to the Weibull or even to the exponential mixture model, which is general enough to reveal the main features of the method without getting too abstract or complicated. The exponential mixture model, in particular, is very amenable to a rigorous asymptotic analysis, which we will state and prove precisely. This analysis suggests ways in which more general distributions can be handled.

8.1 A WEIBULL COVARIATE MODEL

In this section our setup will be that the survival time distribution of a susceptible individual is a Weibull distribution, but potentially specific to that individual, in that the parameters describing the Weibull distribution may depend on i. The survival time t_i^*, for susceptible individual i, will be assumed to have distribution

$$F_{i0}(t) = 1 - e^{-(\lambda_i t)^{\alpha_i}}, \qquad t \geq 0,$$

where λ_i is related to the covariate information $\boldsymbol{x}_i$ for individual i by

$$\lambda_i = e^{\boldsymbol{\beta}^T \boldsymbol{x}_i}, \tag{8.1}$$

and $\boldsymbol{\beta}$ is a vector of parameters to be estimated (where the superscript 'T' denotes the transpose of a vector or matrix).

A formulation like (8.1) occurs in the theory of *generalised linear models*, from which it is borrowed. Indeed, if there were no question of immunes then the above Weibull model, with allowance for censoring, though not strictly speaking a generalised linear model, can be fitted directly on the computer package GLIM (Baker and Nelder 1978), which is designed to fit a wide class of generalised linear models. Examples of this were given in Chapters 5 and 6. The Weibull model with covariates affecting the λ parameter as above is formulated similarly in Elandt-Johnson and Johnson (1980, p. 358). Aitkin *et al.* (1989) and Collett (1994) give examples of survival data modelling using GLIM, especially with exponential and Weibull models, and Crowder *et al.* (1991) and Ansell and Phillips (1994) similarly model reliability data with the Weibull distribution. In the generalised linear model context, equation (8.1) is called a *log-linear link* of the covariates to the parameters λ_i, because the logarithm of λ_i is the linear function $\boldsymbol{\beta}^T \boldsymbol{x}_i$ of the components of $\boldsymbol{x}_i$. More general links than those involving the exponential function can be considered, but (8.1) is a simple one which keeps the λ_i positive, as of course is essential.

To allow for the effect of covariates on the probability that an individual is immune, we also let the immune probability vary from individual to individual. In effect, we are associating with each individual a distinct probability of being immune, which depends on the covariate information specific to that individual. We shall use a *logistic–linear* link to the covariates, so that the probability that individual i is immune is modelled by

$$p_i = \frac{e^{\boldsymbol{\delta}^T \boldsymbol{x}_i}}{1 + e^{\boldsymbol{\delta}^T \boldsymbol{x}_i}}, \tag{8.2}$$

where $\boldsymbol{\delta}$ is another vector of parameters to be estimated. The logistic–linear link keeps each p_i strictly between 0 and 1, so we are at this stage restricting ourselves to the ‘interior’ case — possible boundary values of some of the p_i’s will be considered later (see Section 8.4).

Again the logistic–linear link is common in the theory of generalised linear models, in which it is used to relate a binomial proportion to a vector of covariates, and such models can be fitted directly on GLIM or some similar packages. However, the exponential model with immunes as formulated by (8.1) and (8.2) is *not* a generalised linear model and cannot be fitted directly on GLIM; we have to use some other numerical maximisation procedure, such as the Newton–Raphson procedure discussed in Section 5.2, for this.

Finally, in the Weibull mixture model we can also allow the shape parameters α to vary between individuals. We again adapt a log-linear link of the form

$$\alpha_i = e^{\boldsymbol{\gamma}^T \boldsymbol{x}_i} \tag{8.3}$$

which keeps each α_i strictly positive, as is required. (We saw in Chapter 6 an example where the shape parameters differed between groups.)

Before discussing the large-sample properties of the mixture models, let us see how the above covariate formulation describes the one- and two-group situations modelled in Chapter 6. When the covariate vector $\boldsymbol{x}_i$ is of length one, and all its values are identical, i.e. it does not depend on i, we can simply write it as a scalar variable $x_i = 1$. (Any other value for x_i would simply be absorbed into the parameters β, δ and γ). We correspondingly take β, δ and γ to be of length

1, and then the formulations (8.1), (8.2) and (8.3) for λ_i, p_i and α_i reduce to

$$\lambda = e^{\beta}, \qquad p = \frac{e^{\delta}}{1+e^{\delta}}, \qquad \alpha = e^{\gamma}. \tag{8.4}$$

This is simply a transformation of the three parameters λ, p and α describing the Weibull into three equivalent parameters β, δ, and γ. This transformation changes the ranges of variation of λ, p and α (which are $(0, \infty)$, $(0, 1]$ and $(0, \infty)$) into the whole real line ($(-\infty, \infty)$ for each), and is a convenient re-representation for the purposes of computer programming, even for the single-group situation we are currently discussing. In fact, this transformation is used in the Fortran program mentioned in Chapters 5 and 6.

To obtain the two-group Weibull model from the general covariate setup, as an example, we take the covariate vector $\boldsymbol{x}_i$ associated with individual i to be two-dimensional, and of the form

$$\boldsymbol{x}_i = (\,1 \quad x_i\,)^T \tag{8.5}$$

where x_i is, as indicated in the Gehan–Freireich data of Table 1.1, a treatment group indicator: $x_i = 1$ for Group 1, and $x_i = 2$ for Group 2. The extra first component 1 for $\boldsymbol{x}_i$ in (8.5) is introduced to allow for a constant or intercept term. We correspondingly take the parameter vectors $\boldsymbol{\beta}$, $\boldsymbol{\delta}$, and $\boldsymbol{\gamma}$ to be two-dimensional:

$$\boldsymbol{\beta} = (\,\beta_1 \quad \beta_2\,)^T, \qquad \boldsymbol{\delta} = (\,\delta_1 \quad \delta_2\,)^T, \qquad \boldsymbol{\gamma} = (\,\gamma_1 \quad \gamma_2\,)^T. \tag{8.6}$$

The above setup is sufficient to describe the two-group mixture model, but it is customary to rewrite the parameter and covariate vectors for the case when the covariates are discrete 'factors' with two levels as

$$\boldsymbol{x}_i = \begin{cases} (\,1 \quad 1\,) & \text{when } x_i = 1 \text{ (Group 1)} \\ (\,1 \quad -1\,) & \text{when } x_i = 2 \text{ (Group 2),} \end{cases}$$

and let

$$\boldsymbol{\beta} = (\,\beta_0 \quad \beta_1\,)^T, \qquad \boldsymbol{\delta} = (\,\delta_0 \quad \delta_1\,)^T, \qquad \text{and } \boldsymbol{\gamma} = (\,\gamma_0 \quad \gamma_1\,)^T. \tag{8.7}$$

Then we have

$$\boldsymbol{\beta}^T \boldsymbol{x}_i = \begin{cases} \beta_0 + \beta_1 & \text{when } x_i = 1 \text{ (Group 1)} \\ \beta_0 - \beta_1 & \text{when } x_i = 2 \text{ (Group 2)} \end{cases} \tag{8.8}$$

and similarly for $\boldsymbol{\delta}^T \boldsymbol{x}_i$ and $\boldsymbol{\gamma}^T \boldsymbol{x}_i$.

From (8.1) and (8.8) we see then that the survival rate parameter λ_i for susceptibles is given by

$$\lambda_i = \begin{cases} e^{\beta_0+\beta_1} & \text{when } x_i = 1 \text{ (Group 1)} \\ e^{\beta_0-\beta_1} & \text{when } x_i = 2 \text{ (Group 2).} \end{cases} \tag{8.9}$$

Thus the ratio of the λ parameters for the two groups is $e^{2\beta_1}$, and this represents, approximately, a measure of the ratio of survival rates for Groups 1 and 2.

For the Weibull distribution, the parameter λ is approximately inversely proportional to the median of the distribution of the survival times of the susceptibles (see Section 5.6), so the model can equivalently be interpreted as saying that the effect of Group 2 is to (increase or decrease, according to whether $\delta_1 > 0$ or $\delta_1 < 0$) the median survival time of an individual by a factor of approximately $e^{2\beta_1}$. Likewise from (8.2) and (8.7) we have for the probability that individual i is immune:

$$p_i = \begin{cases} \dfrac{e^{\delta_0+\delta_1}}{1+e^{\delta_0+\delta_1}} & \text{if } x_i = 1 \text{ (Group 1)} \\ \dfrac{e^{\delta_0-\delta_1}}{1+e^{\delta_0-\delta_1}} & \text{if } x_i = 2 \text{ (Group 2),} \end{cases} \tag{8.10}$$

and again p_i takes only two values, with the difference between them representing the difference in immune proportions between the two treatment groups. Finally, we have by (8.3) and (8.7) a similar representation for the shape parameters:

$$\alpha_i = \begin{cases} e^{\gamma_0+\gamma_1} & \text{when } x_i = 1 \text{ (Group 1)} \\ e^{\gamma_0-\gamma_1} & \text{when } x_i = 2 \text{ (Group 2).} \end{cases} \tag{8.11}$$

Equations (8.9) to (8.11) show clearly that under the Weibull mixture model, the hypothesis that the treatment has no effect on the survival time distribution has three components: there may or may not be an effect on the survival rate of susceptibles, there may or may not be a change in the proportion of immunes, and there may or may not be a difference in the shapes of the Weibull distributions, as measured by the parameter α. In the two-group setup these hypotheses can be written as

$$\begin{aligned} &H_0 : \beta_1 = 0 && \text{(no treatment effect on survival rate of susceptibles)} \\ &H_0' : \delta_1 = 0 && \text{(no treatment effect on proportion of immunes)} \\ &H_0'' : \gamma_1 = 0 && \text{(no treatment effect on shape of survival distribution).} \end{aligned}$$

As shown in Chapter 6, in a data analysis we will be interested in all three of these aspects of the treatment effect. As alternative hypotheses to the above we will simply take $H_1 : \beta_1 \neq 0$, $H_1' : \delta_1 \neq 0$, and $H_1'' : \gamma_1 \neq 0$.

We showed in Chapter 6 how to fit these models to data and test for the effects of a treatment. The models 000, 100, etc., with corresponding null hypotheses H_{000}, H_{100}, etc., listed in Section 6.2, can be expressed parametrically here as

$$\begin{aligned} &H_{000} : \beta_1 = \delta_1 = \gamma_1 = 0; \\ &H_{100} : \beta_1 = \gamma_1 = 0,\ \delta_1 \text{ arbitrary;} \end{aligned}$$

etc., In Chapter 6 we gave detailed examples of testing them using deviance differences, and interpreting the results, so we now move on to discussing the large-sample properties of the MLEs.

8.2 LEAST SQUARES ESTIMATION AND GENERALISED LINEAR MODELS

Before discussing the large-sample properties of survival models with covariates, it is useful to review the properties of some simpler models, as a guide to the sorts of conditions we must expect to impose in order to obtain the results we need. These results, as indicated in Section 7.2, will take the form of ensuring (with high probability) the existence and consistency of a (possibly vector) estimator, and the calculation of its large-sample distribution, which may be the normal distribution as in the case of 'interior' estimation, or a mixture of distributions when the parameter is on the boundary of the parameter space. Finally, we will also need the asymptotic distribution of the likelihood ratio statistic, or more conveniently, the asymptotic distribution of the deviance difference which is used to test a hypothesis of interest about the parameters.

Many of the features of these sorts of analyses are present in the simplest of all statistical models with covariates: the linear regression model for normally distributed data. In this model, observations $y_1, y_2, \ldots, y_n$ are related to covariate vectors $\boldsymbol{x}_1, \boldsymbol{x}_2, \ldots, \boldsymbol{x}_n$ by the linear relationship

$$y_i = \boldsymbol{\beta}_0^T \boldsymbol{x}_i + \varepsilon_i, \qquad 1 \leq i \leq n, \tag{8.12}$$

where the ε_i are i.i.d. r.v.'s, assumed to be normally distributed with mean 0 and finite variance σ^2. Suppose that the $\boldsymbol{x}_i$ are nonrandom vectors of length k. Under the assumption that the matrix

$$B_n = \sum_{i=1}^{n} \boldsymbol{x}_i \boldsymbol{x}_i^T \tag{8.13}$$

is nonsingular, the maximum likelihood estimator $\tilde{\boldsymbol{\beta}}_n$ of $\boldsymbol{\beta}$ can be calculated as

$$\tilde{\boldsymbol{\beta}}_n = B_n^{-1} \sum_{i=1}^{n} y_i \boldsymbol{x}_i. \tag{8.14}$$

From this explicit expression it is easy to deduce the large-sample properties of $\tilde{\boldsymbol{\beta}}_n$. We have that $\tilde{\boldsymbol{\beta}}_n$ is consistent for $\boldsymbol{\beta}_0$ in probability, i.e. that $\tilde{\boldsymbol{\beta}}_n \xrightarrow{P} \boldsymbol{\beta}_0$ as $n \to \infty$, if and only if the minimum eigenvalue of the matrix B_n in (8.13) tends to infinity as $n \to \infty$. We will express this as

$$\lambda_{\min}(B_n) \to \infty \qquad (n \to \infty), \tag{8.15}$$

the symbol 'λ' being a typical designator of an eigenvalue.

Under the assumption of the normality of the ε_i, $\tilde{\boldsymbol{\beta}}_n$ is always normally distributed with covariance matrix

$$\mathrm{Var}(\tilde{\boldsymbol{\beta}}_n) = \sigma^2 B_n^{-1}. \tag{8.16}$$

(Note that the inverse matrix B_n^{-1} exists since B_n is assumed nonsingular.) This certainly implies that

$$(\sigma^2 B_n)^{1/2}\tilde{\boldsymbol{\beta}}_n \tag{8.17}$$

has, asymptotically, a standard normal distribution in k dimensions, where k is the rank of B_n; in fact, the expression in (8.17) is distributed as a k-dimensional standard normal distribution for each n ($\geq k$). (For a discussion of the 'square root' of a matrix as used in (8.17), see Appendix Note 3 to Chapter 7.)

When the ε_i are not necessarily normally distributed, but are merely assumed to have mean 0 and variance σ^2, $\tilde{\boldsymbol{\beta}}_n$ remains consistent for $\boldsymbol{\beta}_0$ under (8.15). Since no distribution is then specified for the y_i in (8.12), we cannot refer to $\tilde{\boldsymbol{\beta}}_n$ as defined in (8.14) as being the MLE of $\boldsymbol{\beta}_0$, but it is the least squares estimator (LSE) of $\boldsymbol{\beta}_0$ in the sense of being the value of $\boldsymbol{\beta}$ which minimises the sum of squares $\sum_{i=1}^{n}(y_i - \boldsymbol{\beta}^T\boldsymbol{x}_i)^2$. It turns out that for this more general setup, $\tilde{\boldsymbol{\beta}}_n$ is asymptotically normal if and only if the extra condition

$$\max_{1\leq i\leq n} \boldsymbol{x}_i^T B_n^{-1}\boldsymbol{x}_i \to 0 \qquad (n \to \infty) \tag{8.18}$$

is satisfied. This is a kind of 'uniform asymptotic negligibility' condition on the covariates, which roughly states that no single one of the covariate vectors $\boldsymbol{x}_i$ dominates the others. It arises when applying the central limit theorem to the weighted sum $\sum_{i=1}^{n} y_i\boldsymbol{x}_i$ of independent (but not necessarily identically distributed) random variables.

Condition (8.15), on the other hand, tells us that the 'information' contributed by the values of the regressor variables $\boldsymbol{x}_i$ increases without bound as the number of observations increases — as we would expect to be necessary for consistent estimation. It is implied by condition (8.18); see Appendix Note 1.

The nonsingularity of the matrix B_n, for a particular value of n, reflects the fact that the covariate vectors $\boldsymbol{x}_i, \ldots, \boldsymbol{x}_n$ do not lie in a space of lower than k dimensions; in other words, none of the $\boldsymbol{x}_i$ is a linear combination of the others. Although this might seem a natural condition in a regression context, in an analysis of variance setup it will not be true when the model is 'overparametrised', meaning that some of the parameters can be written as linear combinations of the others. But we can always eliminate these linear dependencies and assume that, by some modification of the model, B_n is nonsingular for each n. We will do this to simplify some of our formulations.

Perhaps the next simplest class of models for data with covariates, after the linear regression models, are the *generalised linear models*. Their introduction has revolutionised the practice of statistics in the last two decades by providing a natural method of analysis of data which consists of observations on independent random variables, not necessarily normally distributed, but having a distribution belonging to a certain exponential family of distributions. The name comes about because the dependence of the distribution of each observation on its covariate is assumed to be by way of some function, usually nonlinear, of the linear combination $\boldsymbol{\beta}^T\boldsymbol{x}_i$. Here, as usual, $\boldsymbol{\beta}$ is a vector of parameters to be estimated, and $\boldsymbol{x}_i$ is the covariate vector associated with individual i.

This special 'generalised linear' structure of the model allows relatively easy data analytical methods, based on maximum likelihood, to be implemented for many of the distributions commonly met in practice, especially the binomial, Poisson, exponential and (as a simple but basic special case) the normal. Furthermore, the special 'generalised linear' structure of the model also allows the large-sample theory of linear regression models to be carried over, for the most part, to the generalised linear models. In particular, the consistency and asymptotic normality of the MLE $\tilde{\boldsymbol{\beta}}_n$ of $\boldsymbol{\beta}_0$ can be shown to hold under conditions analogous to the 'uniform asymptotic negligibility' of covariates condition (8.18).

These kinds of results provide us with the clue as to how to analyse survival models with covariates. Following the example of linear regression models, we formulated a model in Section 8.1 in which the dependence of the failure distributions on the covariates was of the 'generalised linear' type. In other words, the covariates always enter the model via a linear combination of the form $\boldsymbol{\beta}^T \boldsymbol{x}_i$, where $\boldsymbol{\beta}$ is a vector of parameters to be estimated. In seeking to draw out the large-sample properties of an MLE $\tilde{\boldsymbol{\beta}}_n$ of $\boldsymbol{\beta}$, we are guided by similar analyses for generalised linear models. In particular, we will expect consistency and asymptotic normality of $\tilde{\boldsymbol{\beta}}_n$ to hold under some sort of 'uniform asymptotic negligibility' condition on the covariate vectors. In the next section we set out some results like this for the survival models we are interested in.

8.3 ASYMPTOTICS OF THE ORDINARY EXPONENTIAL MODEL WITH COVARIATES

Before attempting to understand the large-sample behaviour of the maximum likelihood estimates of the parameters in mixture models, it is useful to consider the simpler setup, when the survival distributions of the susceptibles are exponential and there are no immunes in the population. We now wish to allow the survival distributions to differ between individuals in a way which depends on the value of an individual's covariate vector. We incorporate this dependence into the exponential distribution by expressing the parameter λ of the exponential as a log-linear function of the covariate, as in (8.1).

Thus the ('true') distribution of t_i^* is given by

$$P\{t_i^* \leq t\} = 1 - e^{-\lambda_{i0} t}, \qquad t \geq 0, \tag{8.19}$$

where

$$\lambda_{i0} = e^{\boldsymbol{\beta}_0^T \boldsymbol{x}_i}. \tag{8.20}$$

Also associated with individual i, as usual, will be a censoring time u_i with distribution G_i, and a censor indicator $c_i = I(t_i^* \leq u_i)$. The random variables t_i^* and u_i are assumed to be mutually independent but we stress that now neither the t_i^* not the u_i are assumed to be identically distributed. We will refer to this model as the 'independent censoring model with covariates'.

Our brief review of the large-sample properties of linear and generalised linear models in Section 8.2 and our analysis of single-sample exponential models in Section 7.2 suggest the kinds of assumptions on covariates and censoring distributions that we expect will be needed. For the covariates, we will assume that, for all n, the matrix

$$B_n = \sum_{i=1}^{n} \boldsymbol{x}_i \boldsymbol{x}_i^T \qquad \text{is positive definite.} \tag{8.21}$$

For the censor variables, we will assume that, for all i,

$$P(u_i = 0) = 0. \tag{8.22}$$

This is a little stronger than the requirement that the u_i not be degenerate at 0 (all that was needed in Theorem 7.1 for the single-sample exponential model), but it simplifies the analysis and leads to no real loss of generality, we expect, since it is hard to envisage a practical situation in which censor variables take the value 0 with positive probability, thus producing some zero, censored failure times with positive probability.

The likelihood of the sample $t_1, t_2, \ldots, t_n$ for the above model is

$$L_n(t_1, \ldots, t_n, \boldsymbol{\beta}) = \prod_{i=1}^{n} (\lambda_i \mathrm{e}^{-\lambda_i t_i})^{c_i} (\mathrm{e}^{-\lambda_i t_i})^{1-c_i}$$

where $\lambda_i = \mathrm{e}^{\boldsymbol{\beta}^T \boldsymbol{x}_i}$, and so the log-likelihood has the form

$$l_n(\boldsymbol{\beta}) = \sum_{i=1}^{n} (c_i \boldsymbol{\beta}^T \boldsymbol{x}_i - t_i \mathrm{e}^{\boldsymbol{\beta}^T \boldsymbol{x}_i}). \tag{8.23}$$

As usual we take $\boldsymbol{\beta}$ and $\boldsymbol{x}_i$ to be of length k, so there are k parameters (the components of $\boldsymbol{\beta}$) to be estimated.

We now carry out a 'vector differentiation' of $l_n(\boldsymbol{\beta})$ with respect to $\boldsymbol{\beta}$, so that $l_n(\boldsymbol{\beta})$ is differentiated with respect to each component of $\boldsymbol{\beta}$, then the resulting derivatives are collected together into a vector of length k. Due to the 'generalised linear' structure of the model, this is easy to do. It is not hard to verify from (8.23) that we then obtain the vector equation

$$\frac{\partial l_n(\boldsymbol{\beta})}{\partial \boldsymbol{\beta}} = \sum_{i=1}^{n} (c_i - t_i \mathrm{e}^{\boldsymbol{\beta}^T \boldsymbol{x}_i}) \boldsymbol{x}_i. \tag{8.24}$$

As in Chapter 7, we denote $\partial l_n(\boldsymbol{\beta})/\partial \boldsymbol{\beta}$ by $S_n(\boldsymbol{\beta})$. $S_n(\boldsymbol{\beta})$ is a random vector, the randomness being contributed in this instance by the c_i and the t_i.

Next we carry out a vector differentiation of the vector $S_n(\boldsymbol{\beta})$ by differentiating each component of $S_n(\boldsymbol{\beta})$ with respect to each component of $\boldsymbol{\beta}$, and collecting the derivatives into a $k \times k$ matrix. We denote the negative of this matrix by

$F_n(\boldsymbol{\beta})$. Thus, by (8.24),

$$F_n(\boldsymbol{\beta}) = -\frac{\partial S_n(\boldsymbol{\beta})}{\partial \boldsymbol{\beta}} = -\frac{\partial^2 l_n(\boldsymbol{\beta})}{\partial \boldsymbol{\beta}^2} = \sum_{i=1}^{n} t_i \mathrm{e}^{\boldsymbol{\beta}^T \boldsymbol{x}_i} \boldsymbol{x}_i \boldsymbol{x}_i^T. \tag{8.25}$$

Note that $F_n(\boldsymbol{\beta})$ is a random matrix, the randomness being contributed in this instance by the t_i (the $\boldsymbol{x}_i$ are 'fixed' — nonrandom — vectors).

To fit the exponential model to a data set, $S_n(\boldsymbol{\beta})$ is evaluated for the data points at hand and equated to 0. This vector equation represents k nonlinear equations which can be solved in principle, i.e. numerically under (8.21) and (8.22), as we will show, to obtain a *unique* MLE of $\boldsymbol{\beta}$ which we denote by $\tilde{\boldsymbol{\beta}}_n$. Before discussing the large-sample properties of $\tilde{\boldsymbol{\beta}}_n$, let us derive some fundamental properties of the model which are useful. Define, as usual, the expected information matrix D_n by

$$D_n = E(F_n(\boldsymbol{\beta}_0)).$$

This is a nonrandom matrix. We have from (8.25) that

$$D_n = \sum_{i=1}^{n} E(t_i) \mathrm{e}^{\boldsymbol{\beta}_0^T \boldsymbol{x}_i} \boldsymbol{x}_i \boldsymbol{x}_i^T. \tag{8.26}$$

Theorem 8.1 *Suppose conditions (8.21) and (8.22) hold. Then for each $n \geq 1$, D_n is positive definite, and $F_n(\boldsymbol{\beta})$ is positive definite a.s. for all $\boldsymbol{\beta}$. Also the equation*

$$S_n(\boldsymbol{\beta}) = 0$$

has a unique solution $\tilde{\boldsymbol{\beta}}_n$, for each $n \geq 1$.

Proof Take any $\boldsymbol{\beta}$ and $n \geq 1$. To prove that the symmetric matrix $F_n(\boldsymbol{\beta})$ is positive definite, we look at the quadratic form $\boldsymbol{u}^T F_n(\boldsymbol{\beta})\boldsymbol{u}$. If this is positive for all nonzero vectors $\boldsymbol{u}$ (or equivalently, for all unit vectors $\boldsymbol{u}$), then $F_n(\boldsymbol{\beta})$ is positive definite. Now by (8.25)

$$\boldsymbol{u}^T F_n(\boldsymbol{\beta})\boldsymbol{u} = \sum_{i=1}^{n} t_i \mathrm{e}^{\boldsymbol{\beta}^T \boldsymbol{x}_i} (\boldsymbol{u}^T \boldsymbol{x}_i)^2. \tag{8.27}$$

The random variable $t_i = \min(t_i^*, u_i)$ has probability 0 of taking the value 0, because this is true of the exponentially distributed t_i^*, obviously, and it is true of u_i by assumption (8.22). Consequently (8.27) can be 0 only if $\boldsymbol{u}^T \boldsymbol{x}_i = 0$ for $1 \leq i \leq n$, which implies that $\sum_{i=1}^{n} (\boldsymbol{u}^T \boldsymbol{x}_i)^2 = 0$. This contradicts our assumption (8.21) that B_n is positive definite, and shows that $F_n(\boldsymbol{\beta})$ is positive definite with probability 1 for each $n > 1$ and $\boldsymbol{\beta}$.

A similar proof shows that D_n is positive definite for each $n \geq 1$. The positive definiteness of $F_n(\boldsymbol{\beta})$ proves that $l_n(\boldsymbol{\beta})$ is a concave function of $\boldsymbol{\beta}$, a.s., (second derivative strictly negative for all $\boldsymbol{\beta}$) and hence has a global maximum $\tilde{\boldsymbol{\beta}}_n$ which must be the solution of $\partial l_n(\boldsymbol{\beta})/\partial\boldsymbol{\beta} = S_n(\boldsymbol{\beta}) = 0$. This vector $\tilde{\boldsymbol{\beta}}_n$ is of course the unique MLE of $\boldsymbol{\beta}_0$. ■

Note that we cannot display the MLE $\tilde{\boldsymbol{\beta}}_n$ explicitly; we can only infer its existence from Theorem 8.1. Also, as we remarked above, the assumption (8.21) that the matrix B_n is positive definite for a particular value of n may represent a real restriction in some models, but we can always reparametrise the model (in a perhaps less convenient fashion) so as to make B_n nonsingular. If B_n is positive definite, not for all $n \geq 1$, but for all n sufficiently large, and (8.22) also holds, then the conclusions of Theorem 8.1 hold for all n sufficiently large, and this is all that is needed in an asymptotic analysis.

We turn now to the asymptotic behaviour of the MLE $\tilde{\boldsymbol{\beta}}_n$ defined in Theorem 8.1. Let I_k be the $k \times k$ identity matrix, and let $N(0, I_k)$ be the k-dimensional standard normal distribution.

Theorem 8.2 *Assume the independent censoring model with covariates, suppose that B_n as defined by (8.21) is nonsingular for all sufficiently large n, and that (8.22) holds. Suppose also that*

$$\max_{1\leq i\leq n} \boldsymbol{x}_i^T D_n^{-1} \boldsymbol{x}_i \to 0 \qquad \text{as } n \to \infty. \tag{8.28}$$

Then the unique MLE $\tilde{\boldsymbol{\beta}}_n$ defined in Theorem 8.1 is consistent for $\boldsymbol{\beta}_0$, and is asymptotically normally distributed as $n \to \infty$ in the sense that

$$D_n^{1/2}(\tilde{\boldsymbol{\beta}}_n - \boldsymbol{\beta}_0) \xrightarrow{D} N(0, I_k). \tag{8.29}$$

The matrix D_n is consistently estimated by $F_n(\tilde{\boldsymbol{\beta}}_n)$ in the sense that, as $n \to \infty$,

$$D_n^{-1/2} F_n(\tilde{\boldsymbol{\beta}}_n) D_n^{-1/2} \xrightarrow{P} I_k, \tag{8.30}$$

and we also have the Studentised result

$$F_n^{1/2}(\tilde{\boldsymbol{\beta}}_n)(\tilde{\boldsymbol{\beta}}_n - \boldsymbol{\beta}_0) \xrightarrow{D} N(0, I_k). \tag{8.31}$$

The deviance statistic

$$d_n = 2(l_n(\tilde{\boldsymbol{\beta}}_n) - l_n(\boldsymbol{\beta}_0)) \tag{8.32}$$

is asymptotically distributed as χ^2 with k degrees of freedom.

The proof of Theorem 8.2, which is quite technical if we write out all the details, is given in Appendix Note 2. Let us illustrate here some consequences of the theorem in special cases.

When all the covariate vectors $\boldsymbol{x}_i$ are the same, the t_i^* are i.i.d., and we might as well take $\boldsymbol{x}_i = 1$, a scalar, and $\boldsymbol{\beta}_0^T \boldsymbol{x}_i = \beta_0$, also a scalar. In this case all individuals have the same survival time distribution $1 - \exp(-e^{\beta_0} t)$, $t \geq 0$. However, Theorem 8.2 does not reduce in this case to Theorem 7.1, which describes the single-sample case, because there we assumed in addition that the censoring random variables u_i were i.i.d., whereas in the present section they may have different distributions, G_i. So we can get some information here on how the censoring affects the estimation of $\boldsymbol{\beta}$ when the censoring is not i.i.d.

For this situation, D_n as given by (8.26) reduces to the scalar sequence

$$D_n = e^{\beta_0} \sum_{i=1}^{n} E(t_i), \tag{8.33}$$

or equivalently, by (8.67) in Appendix Note 2, to

$$D_n = \sum_{i=1}^{n} E(c_i).$$

Thus the uniform asymptotic negligibility criterion (8.28) reduces here to the requirement that the series $\sum_{i=1}^{\infty} E(c_i)$ should diverge. This is another version of the principle that the asymptotic properties we desire will hold when the censoring is not too heavy—we must have $E(c_i) = P(c_i = 1)$ not too small. It can also be shown that the divergence of $\sum_{i=1}^{n} E(c_i)$ is also necessary for the consistency of $\tilde{\boldsymbol{\beta}}_n$ when the t_i^* are i.i.d.

A special case of the above result, when the t_i^* are i.i.d exponential random variables and the u_i are degenerate at nonstochastic values, was obtained by Kalbfleisch and Prentice (1980, p. 49), who discuss its ramifications further. One important question raised by them is how results dependent on properties of the censoring random variables—which remain unobserved for individuals who failed—are to be interpreted and applied. In the end, of course, we must make further assumptions, such as that the u_i are i.i.d. (in which case $\sum_{i=1}^{\infty} E(c_i)$ diverges as long as the u_i are not degenerate at 0), or we must endeavour to estimate some of the censoring distribution(s) and then check whether the required conditions seem approximately to be satisfied. We will return to this point in Section 9.4.

8.4 ASYMPTOTICS OF THE EXPONENTIAL MIXTURE MODEL WITH COVARIATES

The exponential mixture model: interior case

The analysis of the ordinary exponential model given in the previous section can be carried over in much the same form to the exponential mixture model. There are special properties of the ordinary exponential model which simplify the asymptotic analysis, compared with the more general mixture model, and the results we obtain for the general mixture model are not quite as good, in the sense that slightly more restrictive conditions are required, at least on the covariates, to obtain the results we want to use in practice.

The reason for this is the extra complication created by the immune probabilities associated with each individual. These will be modelled, as discussed in Section 8.1, by a logistic linear function of the covariate vectors, but for extra generality we allow the covariate vector to differ from that which is linked to the rate parameters. Specifically, we assume that susceptibles have independent failure times t_i^* whose distributions have probability density functions

$$f_{i0}(t) = \lambda_{i0} e^{-\lambda_{i0} t}$$

where

$$\lambda_{i0} = e^{\boldsymbol{\beta}_0^T \boldsymbol{x}_i},$$

while individual i also has a probability p_{i0} of being immune to failure, where

$$p_{i0} = \frac{e^{\boldsymbol{\delta}_0^T \boldsymbol{y}_i}}{1 + e^{\boldsymbol{\delta}_0^T \boldsymbol{y}_i}}.$$

Here the subscript zero refers to 'true' values of the parameters $\boldsymbol{\beta}$ and $\boldsymbol{\delta}$, which as usual we must estimate, whereas $\boldsymbol{x}_i$ and $\boldsymbol{y}_i$ are vectors containing covariate information for individual i. We suppose $\boldsymbol{x}_i$ is of length k_1 and $\boldsymbol{y}_i$ is of length k_2, so that $\boldsymbol{\beta}$ and $\boldsymbol{\beta}_0$ are of length k_1, and $\boldsymbol{\delta}$ and $\boldsymbol{\delta}_0$ are of length k_2. In general, $\boldsymbol{x}_i$ and $\boldsymbol{y}_i$ may differ but they may be thought of as subvectors of an overall vector of covariate information, of length $k = k_1 + k_2$, for individual i.

As usual, we observe survival times t_i which are the minima of t_i^* and u_i for independent censoring random variables u_i, which are also independent of the t_i^*, and we observe $c_i = 1$ if $t_i^* \leq u_i$ and $c_i = 0$ otherwise. Given n such observations, the log-likelihood for the exponential mixture model is, apart from constants (see (7.27)),

$$l_n(\boldsymbol{\theta}) = \sum_{i=1}^{n} \{c_i(\log p_i + \log \lambda_i - \lambda_i t_i) + (1 - c_i)\log(1 - p_i + p_i e^{-\lambda_i t_i})\}. \quad (8.34)$$

Here we put the parameter subvectors $\boldsymbol{\beta}$ and $\boldsymbol{\delta}$ together into a parameter vector $\boldsymbol{\theta} = (\boldsymbol{\beta}^T\ \boldsymbol{\delta}^T)^T$, with true value $\boldsymbol{\theta}_0 = (\boldsymbol{\beta}_0^T\ \boldsymbol{\delta}_0^T)^T$.

We have to maximise $l_n(\boldsymbol{\theta})$ for variations in all the components of $\boldsymbol{\beta}$ and $\boldsymbol{\delta}$, so we differentiate with respect to the vector $\boldsymbol{\theta}$ to obtain

$$S_n(\boldsymbol{\theta}) = \frac{\partial l_n(\boldsymbol{\theta})}{\partial \boldsymbol{\theta}} = \begin{bmatrix} \dfrac{\partial\, l_n(\boldsymbol{\theta})}{\partial \boldsymbol{\beta}} \\ \dfrac{\partial l_n(\boldsymbol{\theta})}{\partial \boldsymbol{\delta}} \end{bmatrix}.$$

As a result again of the nice 'generalised linear' structure of our model it is easy to see that we can write $S_n(\boldsymbol{\theta})$ in the form

$$S_n(\boldsymbol{\theta}) = \sum_{i=1}^{n} X_i s_i(\boldsymbol{\theta}) \quad (8.35)$$

where the $s_i(\boldsymbol{\theta})$ are two-dimensional random vectors with components

$$s_{i1}(\boldsymbol{\theta}) = c_i(1-\lambda_i t_i) - (1-c_i)\frac{p_i\lambda_i t_i \mathrm{e}^{-\lambda_i t_i}}{1-p_i+p_i\mathrm{e}^{-\lambda_i t_i}} \tag{8.36}$$

and

$$s_{i2}(\boldsymbol{\theta}) = (1-p_i)\left(c_i - \frac{(1-c_i)p_i(1-\mathrm{e}^{-\lambda t_i})}{1-p_i+p_i\mathrm{e}^{-\lambda_i t_i}}\right), \tag{8.37}$$

and the X_i are nonstochastic $(k_1+k_2)\times 2$ matrices given by

$$X_i = \begin{pmatrix} x_i & 0 \\ 0 & y_i \end{pmatrix}. \tag{8.38}$$

Thus we see that, for this model, we have to contend with a matrix rather than a vector of covariate information, even if $\boldsymbol{x}_i = \boldsymbol{y}_i$.

Differentiating again, we obtain minus the second derivative of the log-likelihood as

$$F_n(\boldsymbol{\theta}) = -\frac{\partial^2 l_n(\boldsymbol{\theta})}{\partial\boldsymbol{\theta}^2} = \begin{bmatrix} \dfrac{\partial^2 l_n(\boldsymbol{\theta})}{\partial\boldsymbol{\beta}^2} & \dfrac{\partial^2 l_n(\boldsymbol{\theta})}{\partial\boldsymbol{\beta}\partial\boldsymbol{\delta}^T} \\ \dfrac{\partial^2 l_n(\boldsymbol{\theta})}{\partial\boldsymbol{\delta}\partial\boldsymbol{\beta}^T} & \dfrac{\partial^2 l_n(\boldsymbol{\theta})}{\partial\boldsymbol{\delta}^2} \end{bmatrix}$$

$$= \sum_{i=1}^{n} X_i \mathcal{F}_i X_i^T \tag{8.39}$$

where $\mathcal{F}_i(\boldsymbol{\theta})$ is a 2×2 symmetric random matrix with diagonal elements

$$f_i^{11}(\boldsymbol{\theta}) = c_i\lambda_i t_i + (1-c_i)\frac{p_i\lambda_i t_i\mathrm{e}^{-\lambda_i t_i}}{1-p_i+p_i\mathrm{e}^{-\lambda_i t_i}}\left\{1-\frac{(1-p_i)\lambda_i t_i}{1-p_i+p_i\mathrm{e}^{-\lambda_i t_i}}\right\}, \tag{8.40}$$

$$f_i^{22}(\boldsymbol{\theta}) = p_i(1-p_i)\left\{1-\frac{(1-c_i)\mathrm{e}^{-\lambda_i t_i}}{(1-p_i+p_i\mathrm{e}^{-\lambda_i t_i})^2}\right\}, \tag{8.41}$$

and off-diagonal elements

$$f_i^{12}(\boldsymbol{\theta}) = f_i^{21}(\boldsymbol{\theta}) = p_i(1-p_i)(1-c_i)\frac{\lambda_i t_i\mathrm{e}^{-\lambda_i t_i}}{(1-p_i+p_i\mathrm{e}^{-\lambda t_i})^2}. \tag{8.42}$$

In this section we restrict ourselves to the interior case, when $0 < p_{i0} < 1$ for $1 \le i \le n$ (implicitly assumed already by taking $\boldsymbol{\delta}$ finite). So it is easy to see that all the s_{ir} and f_i^{rs}, $1 \le r, s \le 2$, have finite expectations, and we can define, for each $\boldsymbol{\theta}$,

$$D_n = -E\left(\frac{\partial^2 l_n(\boldsymbol{\theta}_0)}{\partial\boldsymbol{\theta}^2}\right)$$

$$= E(F_n(\boldsymbol{\theta}_0)) = \sum_{i=1}^{n} X_i E(\mathcal{F}_i(\boldsymbol{\theta}_0))X_i^T$$

$$= \sum_{i=1}^{n} X_i \mathcal{D}_i X_i^T, \qquad \text{say.} \tag{8.43}$$

For each $A \geq 1$ define a neighbourhood $N(\theta_0)$ by

$$N(\theta_0) = \{\theta : (\theta - \theta_0)^T D_n (\theta - \theta_0) \leq A\},$$

and as before let $N(0, I_k)$ denote the standard normal random vector in k dimensions. Then we have the following theorem.

Theorem 8.3 *Suppose the independent censoring model with covariates holds, that D_n^{-1} exists for all n large enough, and that*

$$\lim_{n\to\infty} \sum_{i=1}^{n} (\text{trace } (X_i^T D_n^{-1} X_i))^{3/2} = 0. \tag{8.44}$$

Then an MLE $\tilde{\theta}_n$ of θ_0 exists locally uniquely to θ_0 WPA1 as $n \to \infty$. Also $\tilde{\theta}_n$ is consistent in probability for θ_0, and is asymptotically normal as $n \to \infty$: we have both

$$D_n^{1/2}(\tilde{\theta}_n - \theta_0) \xrightarrow{D} N(0, I_k) \tag{8.45}$$

and the Studentised result

$$(F_n(\tilde{\theta}_n))^{1/2}(\tilde{\theta}_n - \theta_0) \xrightarrow{D} N(0, I_k). \tag{8.46}$$

Also, if d_n denotes the deviance statistic

$$d_n = 2(l_n(\tilde{\theta}_n) - l_n(\theta_0)) \tag{8.47}$$

then d_n has asymptotically a χ_k^2 distribution as $n \to \infty$.

Condition (8.44) expresses the version of asymptotic negligibility of the covariates that we need here. We will omit the proof of Theorem 8.3 since, in principle, it is the same as that of Theorem 8.2 though the details are more complicated to check. They are available in Ghitany *et al.* (1994) (see also Ghitany and Maller 1992) for those interested. However, we will prove another useful result here.

In Section 6.3 a number of 'between-groups' analyses of deviance for Weibull mixture models were demonstrated on various data sets. When the parameters are interior to their parameter spaces — specifically, when the probabilities of individuals being immune are positive — the large sample results needed to justify the analysis we carried out are contained in Theorem 8.3. To see how this comes about, let us consider a 'one-way analysis of deviance' setup in which the null hypothesis states that the immune probabilities of individuals in any one of K groups are the same, but no restriction is put on the rate parameters λ.

This setup can be specified as follows. Suppose $n_k = n_k(n)$ individuals belong to Group k, $1 \leq k \leq K$, and assume that

$$n_k(n) \geq \delta n \tag{8.48}$$

for some $\delta > 0$, where as usual

$$n = n_1 + n_2 + \cdots + n_K$$

is the total number of individuals in the sample. Condition (8.48) means that the numbers of observations in each group must increase roughly proportionally to the sample size. Let $k(i)$ denote the group to which individual i belongs. The covariate vectors $\boldsymbol{x}_i$ and $\boldsymbol{y}_i$ associated with individual i are of length K and have a 1 in the $k(i)$-th position and 0's elsewhere. Assume that, within group k, the survival times t_i^* and censoring random variables u_i are i.i.d., with the cumulative distribution function of the t_i^* being the exponential mixture distribution $p_{(k)}(1 - e^{-\lambda_{(k)}t})$, for some $p_{(k)}$ and $\lambda_{(k)} > 0$. Suppose as usual that the t_i^* and the u_i are independent, and that none of the u_i degenerate at 0. The hypothesis we wish to test is

$$H_0 : 0 < p_{(1)} = p_{(2)} = \cdots = p_{(K)} < 1 \tag{8.49}$$

and for this we will use the deviance statistic

$$d_n = 2(l_n(\tilde{\theta}_n) - l_n(\dot{\theta}_n)) \tag{8.50}$$

where l_n is the log-likelihood, $\tilde{\theta}_n$ is the MLE of $\boldsymbol{\theta} = (\boldsymbol{\beta}^T \boldsymbol{\delta}^T)^T$ and $\dot{\theta}_n$ is the MLE of $\boldsymbol{\theta}$ under H_0. We will show that, in large samples, d_n has approximately a chi-square distribution with $K - 1$ degrees of freedom, as we would expect.

Theorem 8.4 *If H_0 in (8.49) is true and (8.48) holds, then as $n \to \infty$,*

$$d_n \xrightarrow{D} \chi^2_{K-1}. \tag{8.51}$$

The proof of Theorem 8.4 is given in Appendix Note 3. We mention here, what is useful for applications, that the results of Theorem 8.4 can easily be extended to higher-way 'analysis of deviance' layouts, such as we used in Chapter 6. For this we would need a version of (8.48) generalised in the obvious way. The proof in Appendix Note 3 demonstrates the necessity of assuming (8.48) in Theorem 8.4, since if (8.48) failed, there would be the possibility of 'ephemeral' groups which became asymptotically small as $n \to \infty$. Such a group could still be 'substantial' in that $n_k(n) \to \infty$ as $n \to \infty$, but this is ruled out by (8.48).

The exponential mixture model: boundary case

In Section 6.3 we also used hypothesis tests for the equality of λ's between levels of a grouping factor. This can be justified just as in Theorem 8.4, and there is as usual no problem with 'boundary' estimates of the λ's. But such a problem does arise with the p's in that some or all groups may not contain immune components, i.e. they may have $p = 1$. For a single group we showed in Section 7.3 that the deviance statistic for testing $H_0 : p_0 = 1$ has, asymptotically, not a chi-square distribution but the distribution of X where

$$P(X \leq x) = \tfrac{1}{2} + \tfrac{1}{2}P\{\chi_1^2 \leq x\}. \tag{8.52}$$

When there is more than one group, or a factorial structure of groups, we test for differences in immune proportions between groups in two stages.

Stage 1 Test $H_{01} : p_{(1)} = \cdots = p_{(K)} = 1$.

This can be done using the single-sample result of Section 7.3. Let $\hat{\lambda}_{(k)}$ and $\hat{p}_{(k)}$ be MLEs of $\lambda_{(k)}$ and $p_{(k)}$ in group k, let $I(k)$ be the indices of the individuals in group k, and let

$$\tilde{\lambda}_{(k)} = \frac{\sum\limits_{i \in I(k)} c_i}{\sum\limits_{i \in I(k)} t_i}$$

be the estimated parameter for the ordinary exponential model fitted in group k, $1 \leq k \leq K$. Define $\tilde{l}_k = l_n(\tilde{\lambda}_{(k)}, \tilde{p}_{(k)}) - l_n(\tilde{\lambda}_{(k)}, 1)$. Then the $\tilde{l}_k$ are independent for $1 \leq k \leq K$, and hence as $n \to \infty$

$$d_n = 2\sum_{k=1}^{K} \tilde{l}_k \xrightarrow{D} X_1 + \cdots + X_K \tag{8.53}$$

Table 8.1 The percentage points for the distribution of $X_1 + \cdots + X_K$, where each X_k has the same distribution as the X in (8.52).

K	1%	5%	10%
1	5.4119	2.7055	1.6424
2	7.2895	4.2306	2.9524
3	8.7464	5.4345	4.0102
4	10.0186	6.4979	4.9553
5	11.1828	7.4797	5.8351
6	12.2741	8.4070	6.6317
7	13.3127	9.2948	7.4761
8	14.3095	10.1522	8.2567
9	15.2738	10.9855	9.0181
10	16.2109	11.7988	9.7638
11	17.1252	12.5955	10.4962
12	18.0199	13.3778	11.2173
13	18.8976	14.1477	11.9286
14	19.7602	14.9067	12.6313
15	20.6095	15.6560	13.3264
16	21.4468	16.3966	14.0146
17	22.2733	17.1294	14.6966
18	23.0899	17.8551	15.3731
19	23.8976	18.5743	16.0444
20	24.6969	19.2875	16.7110

where the random variables X_k are i.i.d., having the same distribution as the random variable X in (8.52). The percentage points for the distribution of $X_1 + \cdots + X_K$ are tabulated for $K = 2, \ldots, 20$ in Table 8.1, and also for $K = 1$, corresponding to the single-group analysis.

If we accept H_{01} according to the values in Table 8.1, we conclude that there are no immunes in any of the groups and analysis can proceed by 'ordinary' survival analysis techniques. If we reject H_{01} we move on to the next stage.

Stage 2 Test $H_{02} : 0 < p_{(1)} = \cdots = p_{(K)} < 1$.
This is the interior case analysed in the beginning of this section. We use the deviance d_n, approximately distributed as χ^2_{K-1} (see (8.51)), to test H_0. Acceptance of H_{02} implies no difference in immune proportions between groups, rejection means that we must look at the individual group estimates of $p_{(k)}$. Some of these, in turn, may not be significantly different from 1. Examples of this sort of analysis were given in Chapter 6.

APPENDIX NOTES TO CHAPTER 8

1. Large-sample theory of linear and generalised linear regression models

That the consistency of $\tilde{\boldsymbol{\beta}}_n$, the least squares estimator of the vector of regression coefficients in a linear regression, is implied by (and indeed implies) $\lambda_{\min}(B_n) \to \infty (n \to \infty)$ (see (8.13)) was proved by Drygas (1971, 1976), who allowed various more general models than we discussed in Section 8.2. Asymptotic normality of $\tilde{\boldsymbol{\beta}}_n$ was proved to be equivalent to

$$\max_{1 \le i \le n} \boldsymbol{x}_i^T B_n^{-1} \boldsymbol{x}_i \to 0 \qquad (n \to \infty) \tag{8.54}$$

by Eicker (1963). See also Lai *et al.* (1979) for almost sure consistency of $\tilde{\boldsymbol{\beta}}_n$. The fact that (8.54) implies $\lambda_{\min}(B_n) \to \infty$ as $n \to \infty$ is proved for example in Ghitany *et al.* (1994).

Useful large-sample properties of the MLE of the parameter vector $\boldsymbol{\beta}$ in generalised linear models were established under mild conditions, related to (8.28) above, in the path-breaking paper by Fahrmeir and Kaufmann (1985).

2. Proof of Theorem 8.2

The method of proof of Theorem 8.2 relies very heavily on the methods which were developed for generalised linear models. The main step is as follows. Suppose D_n and $F_n(\boldsymbol{\beta})$ are positive definite; this occurs for large enough n and with high probability, by Theorem 8.1. For each $A > 1$ and $n \ge 1$, define a neighbourhood $N(\boldsymbol{\beta}_0)$ of $\boldsymbol{\beta}_0$ by

$$N(\boldsymbol{\beta}_0) = \{\boldsymbol{\beta} : (\boldsymbol{\beta} - \boldsymbol{\beta}_0)^T D_n (\boldsymbol{\beta} - \boldsymbol{\beta}_0) \le A\}. \tag{8.55}$$

We will proceed by considering an arbitrary vector $\boldsymbol{\beta}$ which is on the boundary $\partial N(\boldsymbol{\beta}_0)$ of $N(\boldsymbol{\beta}_0)$; thus $\boldsymbol{\beta}$ is any vector satisfying

$$(\boldsymbol{\beta} - \boldsymbol{\beta}_0)^T D_n (\boldsymbol{\beta} - \boldsymbol{\beta}_0) = A. \tag{8.56}$$

If we can show that $l_n(\boldsymbol{\beta}) < l_n(\boldsymbol{\beta}_0)$ for all such $\boldsymbol{\beta}$, then the (unique) MLE $\tilde{\boldsymbol{\beta}}_n$ of $\boldsymbol{\beta}_0$ must lie in $N(\boldsymbol{\beta}_0)$. The situation for $k = 2$, when $N(\boldsymbol{\beta}_0)$ is an ellipse in two dimensions, is illustrated in Figure 8.1.

Note that, since the t_i and c_i are random variables, $l_n(\boldsymbol{\beta})$ is random and statements like '$l_n(\boldsymbol{\beta}) < l_n(\boldsymbol{\beta}_0)$' must be qualified by 'with probability approaching 1' (WPA1); we will not always mention this explicitly.

To prove that $l_n(\boldsymbol{\beta}) < l_n(\boldsymbol{\beta}_0)$ WPA1, use a Taylor expansion in k dimensions to write, for any $\boldsymbol{\beta}$,

$$l_n(\boldsymbol{\beta}) = l_n(\boldsymbol{\beta}_0) + (\boldsymbol{\beta} - \boldsymbol{\beta}_0)^T S_n(\boldsymbol{\beta}_0) - \tfrac{1}{2}(\boldsymbol{\beta} - \boldsymbol{\beta}_0)^T F_n(\boldsymbol{\beta}')(\boldsymbol{\beta} - \boldsymbol{\beta}_0); \tag{8.57}$$

recall that $S_n(\boldsymbol{\beta})$ is the first derivative (vector) and $F_n(\boldsymbol{\beta})$ is the negative second derivative (matrix) of $l_n(\boldsymbol{\beta})$. In (8.57), $\boldsymbol{\beta}'$ is a vector 'between' $\boldsymbol{\beta}_0$ and $\boldsymbol{\beta}$, i.e. it can be written as

$$\boldsymbol{\beta}' = a\boldsymbol{\beta}_0 + (1 - a)\boldsymbol{\beta}$$

for some a in [0, 1]. Now by (8.57) we have

$$P\{l_n(\boldsymbol{\beta}) < l_n(\boldsymbol{\beta}_0)\} \geq P\{(\boldsymbol{\beta} - \boldsymbol{\beta}_0)^T S_n(\boldsymbol{\beta}_0) < \tfrac{1}{2}(\boldsymbol{\beta} - \boldsymbol{\beta}_0)^T F_n(\boldsymbol{\beta}')(\boldsymbol{\beta} - \boldsymbol{\beta}_0)\}$$

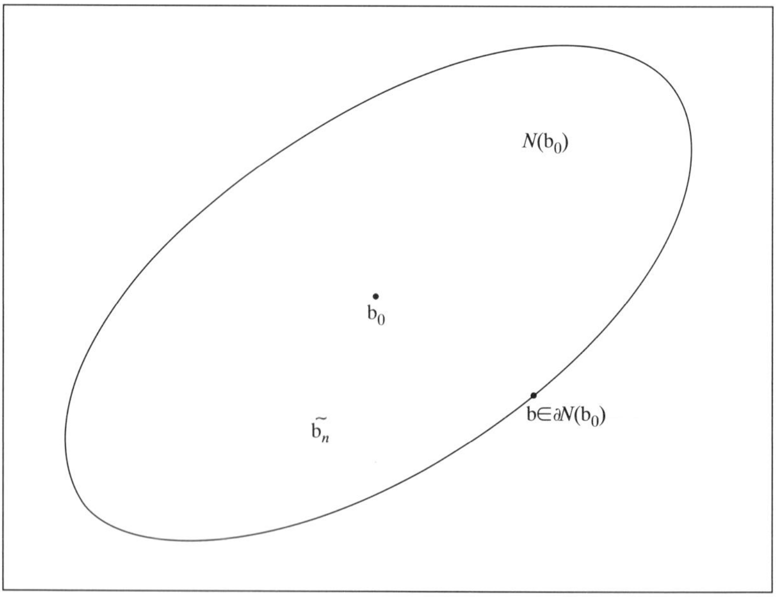

Figure 8.1 Schematic of neighbourhood $N(\boldsymbol{\beta}_0)$ with boundary $\partial N(\boldsymbol{\beta}_0)$ showing the location of the MLE $\tilde{\boldsymbol{\beta}}_n$.

$$\geq P\{(\boldsymbol{\beta}-\boldsymbol{\beta}_0)^T S_n(\boldsymbol{\beta}_0) < \tfrac{1}{4}A, (\boldsymbol{\beta}-\boldsymbol{\beta}_0)^T F_n(\boldsymbol{\beta}')(\boldsymbol{\beta}-\boldsymbol{\beta}_0) > \tfrac{1}{2}A\}$$
$$\geq P\{(\boldsymbol{\beta}-\boldsymbol{\beta}_0)^T F_n(\boldsymbol{\beta}')(\boldsymbol{\beta}-\boldsymbol{\beta}_0) > \tfrac{1}{2}A\} - P\{(\boldsymbol{\beta}-\boldsymbol{\beta}_0)^T S_n(\boldsymbol{\beta}_0) \geq \tfrac{1}{4}A\}. \quad (8.58)$$

We will now be able to deduce the behaviour of these probabilities, as $n \to \infty$, from the general properties of $F_n(\boldsymbol{\beta})$ and $S_n(\boldsymbol{\beta}_0)$ proved in Theorem 7.2. By (7.21) we have

$$E(S_n(\boldsymbol{\beta}_0)) = 0$$

and by (7.22), $S_n(\boldsymbol{\beta}_0)$ has covariance matrix given by

$$\mathrm{Var}(S_n(\boldsymbol{\beta}_0)) = -E\left(\frac{\partial^2 l_n(\boldsymbol{\beta}_0)}{\partial \boldsymbol{\beta}^2}\right) = E(F_n(\boldsymbol{\beta}_0)) = D_n. \quad (8.59)$$

Now $(\boldsymbol{\beta}-\boldsymbol{\beta}_0)^T S_n(\boldsymbol{\beta}_0)$ is a sum of independent random variables, by (8.24), with mean 0. So we can apply Chebychev's inequality to get

$$P\left\{(\boldsymbol{\beta}-\boldsymbol{\beta}_0)^T S_n(\boldsymbol{\beta}_0) \geq \frac{1}{4}A\right\} \leq \frac{16(\boldsymbol{\beta}-\boldsymbol{\beta}_0)^T D_n(\boldsymbol{\beta}-\boldsymbol{\beta}_0)}{A^2}. \quad (8.60)$$

At this stage we assume that $\boldsymbol{\beta} \in \partial N(\boldsymbol{\beta}_0)$. Then by (8.56), the right-hand side of (8.60) has the value $16/A$, and this can be made arbitrarily small by choosing A large enough.

To deal with the other probability in (8.58), we will prove that, uniformly in $\boldsymbol{\beta} \in N(\boldsymbol{\beta}_0)$, we have

$$D_n^{-1/2} F_n(\boldsymbol{\beta}) D_n^{-1/2} \xrightarrow{P} I_k \qquad (n \to \infty). \quad (8.61)$$

This will prove that, when $\boldsymbol{\beta} \in \partial N(\boldsymbol{\beta}_0)$ and $\boldsymbol{\beta}' \in N(\boldsymbol{\beta}_0)$,

$$\begin{aligned}(\boldsymbol{\beta}-\boldsymbol{\beta}_0)^T F_n(\boldsymbol{\beta}')(\boldsymbol{\beta}-\boldsymbol{\beta}_0) &= (\boldsymbol{\beta}-\boldsymbol{\beta}_0)^T D_n^{1/2}\left(D_n^{-1/2} F_n(\boldsymbol{\beta}') D_n^{-1/2}\right) D_n^{1/2}(\boldsymbol{\beta}-\boldsymbol{\beta}_0)\\ &\geq \left((\boldsymbol{\beta}-\boldsymbol{\beta}_0)^T D_n(\boldsymbol{\beta}-\boldsymbol{\beta}_0)\right)\lambda_{\min}(D_n^{-1/2} F_n(\boldsymbol{\beta}') D_n^{-1/2})\\ &= A\,\lambda_{\min}(D_n^{-1/2} F_n(\boldsymbol{\beta}') D_n^{-1/2}) \xrightarrow{P} A.\end{aligned}$$

Consequently, the first probability in (8.58) tends to 1 under (8.61), and as we mentioned, this will prove that $\tilde{\boldsymbol{\beta}}_n$ is in $N(\boldsymbol{\beta}_0)$ with probability approaching 1 for large enough A and n.

In order to prove (8.61), let $\boldsymbol{u}$ be any unit vector in $\mathbb{R}^k$, and use (8.25) to write

$$\begin{aligned}\boldsymbol{u}^T D_n^{-1/2}(F_n(\boldsymbol{\beta}) - F_n(\boldsymbol{\beta}_0)) D_n^{-1/2}\boldsymbol{u} &= \sum_{i=1}^n t_i(e^{\boldsymbol{\beta}^T \boldsymbol{x}_i} - e^{\boldsymbol{\beta}_0^T \boldsymbol{x}_i})(\boldsymbol{u}^T D_n^{-1/2}\boldsymbol{x}_i)^2\\ &= \sum_{i=1}^n t_i e^{\boldsymbol{\beta}_0^T \boldsymbol{x}_i}\Delta_i(\boldsymbol{u}^T D_n^{-1/2}\boldsymbol{x}_i)^2, \quad (8.62)\end{aligned}$$

say, where $\Delta_i = e^{(\boldsymbol{\beta}-\boldsymbol{\beta}_0)^T x_i} - 1$. Now since $\boldsymbol{\beta} \in N(\boldsymbol{\beta}_0)$ we have for each i, $1 \le i \le n$,

$$\begin{aligned}((\boldsymbol{\beta}-\boldsymbol{\beta}_0)^T \boldsymbol{x}_i)^2 &= (\boldsymbol{\beta}-\boldsymbol{\beta}_0)^T D_n^{1/2} D_n^{-1/2} \boldsymbol{x}_i \boldsymbol{x}_i^T D_n^{-1/2} D_n^{1/2} (\boldsymbol{\beta}-\boldsymbol{\beta}_0) \\ &\le (\boldsymbol{\beta}-\boldsymbol{\beta}_0)^T D_n (\boldsymbol{\beta}-\boldsymbol{\beta}_0)(\boldsymbol{x}_i^T D_n^{-1} \boldsymbol{x}_i) \\ &\le A \max_{1\le i\le n} \boldsymbol{x}_i^T D_n^{-1} \boldsymbol{x}_i.\end{aligned}$$

By assumption (8.28), given $\delta > 0$, we can choose $\max_{1\le i\le n} \boldsymbol{x}_i^T D_n^{-1} \boldsymbol{x}_i \le \delta/A$ if n is large enough, so we can choose $|(\boldsymbol{\beta}-\boldsymbol{\beta}_0)^T \boldsymbol{x}_i| \le \delta$ if $1 \le i \le n$ and n is large enough. This in turn means, for such i and n,

$$|\Delta_i| = |e^{(\boldsymbol{\beta}-\boldsymbol{\beta}_0)^T x_i} - 1| \le (\boldsymbol{\beta}-\boldsymbol{\beta}_0)^T x_i e^{|(\boldsymbol{\beta}-\boldsymbol{\beta}_0)^T x_i|} \le \delta e^{\delta}. \tag{8.63}$$

Now return to (8.62) and write the right-hand side as

$$\begin{aligned}\sum_{i=1}^n t_i e^{\boldsymbol{\beta}_0^T x_i} \Delta_i (\boldsymbol{u}^T D_n^{-1/2} \boldsymbol{x}_i)^2 &= \sum_{i=1}^n (t_i - E(t_i)) e^{\boldsymbol{\beta}_0^T \boldsymbol{x}_i} \Delta_i (\boldsymbol{u}^T D_n^{-1/2} \boldsymbol{x}_i)^2 \\ &\quad + \sum_{i=1}^n E(t_i) e^{\boldsymbol{\beta}_0^T \boldsymbol{x}_i} \Delta_i (\boldsymbol{u}^T D_n^{-1/2} \boldsymbol{x}_i)^2 \\ &= a_n + b_n, \text{ say.}\end{aligned} \tag{8.64}$$

We need to show that $a_n \xrightarrow{P} 0$ and $b_n \to 0$ as $n \to \infty$. The following bound on the moments of censored exponential random variables will be needed. To save writing, let $w_i = e^{\boldsymbol{\beta}_0^T \boldsymbol{x}_i}$, let $j \ge 1$, and take $H_i(t) = P(t_i \le t)$. We can integrate by parts to get

$$\begin{aligned}E(w_i t_i)^j &= w_i^j \int_{[0,\infty)} t^j \, dH_i(t) = j w_i^j \int_0^\infty t^{j-1}(1 - H_i(t))\, dt \\ &= j w_i^j \int_0^\infty t^{j-1} e^{-w_i t}(1 - G_i(t))\, dt\end{aligned} \tag{8.65}$$

because

$$1 - H_i(t) = P(t_i > t) = P(t_i^* > t) P(u_i > t) = e^{-w_i t}(1 - G_i(t))$$

and t_i^* has c.d.f. $1 - e^{-w_i t}$. When $j \ge 2$ we can integrate by parts again to get

$$\begin{aligned}E(w_i t_i)^j &= j w_i^j \left\{ \frac{1}{w_i} \int_0^\infty e^{-w_i t}[(j-1)t^{j-2}(1 - G_i(t))\, dt - t^{j-1}\, dG_i(t)] \right\} \\ &\le j(j-1) w_i^{j-1} \int_0^\infty e^{-w_i t} t^{j-2}(1 - G_i(t))\, dt \\ &= j E(w_i t_i)^{j-1} \qquad \text{(by (8.65)).}\end{aligned}$$

Iterating this gives

$$E(w_i t_i)^j \leq j!\, E(w_i t_i), \tag{8.66}$$

valid for $j = 1, 2, 3, \ldots$. In addition to this we have, by (8.65) again,

$$E(c_i) = P(t_i^* \leq u_i) = \int_0^\infty (1 - G_i(t-))w_i \mathrm{e}^{-w_i t}\, \mathrm{d}t = E(w_i t_i). \tag{8.67}$$

Now return to (8.64) and consider the random variable a_n. It has mean 0 and variance

$$\mathrm{Var}(a_n) = \sum_{i=1}^n \mathrm{Var}\,(t_i)w_i^2 \Delta_i^2(\boldsymbol{u}^T D_n^{-1/2}\boldsymbol{x}_i)^4 \leq \delta^2 \mathrm{e}^{2\delta} \sum_{i=1}^n w_i^2\, \mathrm{Var}\,(t_i)(\boldsymbol{u}^T D_n^{-1/2}\boldsymbol{x}_i)^4$$

(by (8.63)). Also by (8.28), for $1 \leq i \leq n$ and n large enough, we have

$$(\boldsymbol{u}^T D_n^{-1/2}\boldsymbol{x}_i)^2 = \boldsymbol{x}_i^T D_n^{-1/2}\boldsymbol{u}\boldsymbol{u}^T D_n^{-1/2}\boldsymbol{x}_i \leq \boldsymbol{x}_i^T D_n^{-1}\boldsymbol{x}_i \leq \delta$$

and we have from (8.66) that, for n large enough,

$$w_i^2\, \mathrm{Var}\,(t_i) \leq E(w_i t_i)^2 \leq 2E(w_i t_i).$$

Thus

$$\begin{aligned}
\mathrm{Var}(a_n) &\leq 2\delta^3 \mathrm{e}^{2\delta} \sum_{i=1}^n E(t_i)w_i(\boldsymbol{u}^T D_n^{-1/2}\boldsymbol{x}_i)^2 \\
&= 2\delta^3 \mathrm{e}^{2\delta} \boldsymbol{u}^T D_n^{-1/2} \left(\sum_{i=1}^n E(t_i)w_i \boldsymbol{x}_i \boldsymbol{x}_i^T \right) D_n^{-1/2}\boldsymbol{u} \\
&= 2\delta^3 \mathrm{e}^{2\delta} \boldsymbol{u}^T D_n^{-1/2} D_n D_n^{-1/2}\boldsymbol{u} = 2\delta^3 \mathrm{e}^{2\delta}.
\end{aligned}$$

Since $\delta > 0$ is arbitrary this shows that $\mathrm{Var}(a_n) \to 0$ as $n \to \infty$, and so by Chebychev's inequality, $a_n \xrightarrow{P} 0$ as $n \to \infty$.

Next consider the sequence b_n; this is easier to deal with. If n is large enough, (8.63) gives

$$\begin{aligned}
b_n &= \sum_{i=1}^n E(t_i)\mathrm{e}^{\boldsymbol{\beta}_0^T x_i} \Delta_i (\boldsymbol{u}^T D_n^{-1/2}\boldsymbol{x}_i)^2 \\
&\leq \delta \mathrm{e}^{\delta} \sum_{i=1}^n E(t_i)\mathrm{e}^{\boldsymbol{\beta}_0^T x_i} (\boldsymbol{u}^T D_n^{-1/2}\boldsymbol{x}_i)^2 = \delta \mathrm{e}^{\delta},
\end{aligned}$$

and this shows that $b_n \to 0$ as $n \to \infty$.

Thus the left-hand side of (8.62) converges to 0 as $n \to \infty$, and since

$$\boldsymbol{u}^T D_n^{-1/2}(F_n(\beta_0) - D_n)D_n^{-1/2}\boldsymbol{u} = \sum_{i=1}^n (t_i - E(t_i))\mathrm{e}^{\beta_0^T x_i}(\boldsymbol{u}^T D_n^{-1/2}\boldsymbol{x}_i)^2$$

converges to 0 as $n \to \infty$ just as we showed for a_n in (8.64), we have proved (8.61).

Together with (8.60) and (8.58), this establishes that

$$\liminf_{n\to\infty} P\{l_n(\boldsymbol{\beta}) < l_n(\boldsymbol{\beta}_0)\} \geq 1 - 16/A^2$$

when $\boldsymbol{\beta} \in \partial N(\boldsymbol{\beta}_0)$. Consequently, as indicated by Figure 8.1, we have

$$\liminf_{n\to\infty} P\{\tilde{\boldsymbol{\beta}}_n \in N(\boldsymbol{\beta}_0)\} \geq 1 - 1/A^2. \tag{8.68}$$

When $\tilde{\boldsymbol{\beta}}_n \in N(\boldsymbol{\beta}_0)$ then

$$A \geq (\tilde{\boldsymbol{\beta}}_n - \boldsymbol{\beta}_0)^T D_n (\tilde{\boldsymbol{\beta}}_n - \boldsymbol{\beta}_0) \geq |\tilde{\boldsymbol{\beta}}_n - \boldsymbol{\beta}_0|^2 \lambda_{\min}(D_n)$$

and $\lambda_{\min}(D_n) \to \infty$ as $n \to \infty$ as a result of (8.28). Given an arbitrary $\varepsilon > 0$, we can thus choose n so large that $1/\lambda_{\min}(D_n) \leq \varepsilon^2/A$, and so we have from (8.68) that

$$\liminf_{n\to\infty} P\{|\tilde{\boldsymbol{\beta}}_n - \boldsymbol{\beta}_0| \leq \varepsilon\} \geq 1 - 1/A^2$$

for all $A > 1$ and $\varepsilon > 0$. Letting $A \to \infty$ now shows that $\tilde{\boldsymbol{\beta}}_n \xrightarrow{P} \boldsymbol{\beta}_0$, so $\tilde{\boldsymbol{\beta}}_n$ is consistent in probability for $\boldsymbol{\beta}_0$.

Now we prove that the random vector $D_n^{1/2}(\tilde{\boldsymbol{\beta}}_n - \boldsymbol{\beta}_0)$ has, asymptotically, a standard normal distribution in k dimensions. To do this, recall that $\tilde{\boldsymbol{\beta}}_n$ satisfies $S_n(\tilde{\boldsymbol{\beta}}_n) = 0$. Also $F_n(\boldsymbol{\beta})$ is minus the first derivative (matrix) of $S_n(\boldsymbol{\beta})$, for each $\boldsymbol{\beta}$. So by Taylor expansion we can write

$$0 = S_n^T(\tilde{\boldsymbol{\beta}}_n) = S_n^T(\boldsymbol{\beta}_0) - (\tilde{\boldsymbol{\beta}}_n - \boldsymbol{\beta}_0)^T F_n(\boldsymbol{\beta}') \tag{8.69}$$

where $\boldsymbol{\beta}' = a\boldsymbol{\beta}_0 + (1-a)\tilde{\boldsymbol{\beta}}_n$ $(0 \leq a \leq 1)$ is a vector between $\boldsymbol{\beta}_0$ and $\tilde{\boldsymbol{\beta}}_n$. By what we proved above, $\tilde{\boldsymbol{\beta}}_n$ is in $N(\boldsymbol{\beta}_0)$ with high probability for large n, so the same is true of $\boldsymbol{\beta}'$. Also $F_n(\boldsymbol{\beta}')$ is invertible, by (8.61), for large n, with high probability, so we can solve (8.69) in the form

$$\tilde{\boldsymbol{\beta}}_n - \boldsymbol{\beta}_0 = F_n^{-1}(\boldsymbol{\beta}') S_n(\boldsymbol{\beta}_0).$$

Then we write

$$D_n^{1/2}(\tilde{\boldsymbol{\beta}}_n - \boldsymbol{\beta}_0) = (D_n^{1/2} F_n^{-1}(\boldsymbol{\beta}') D_n^{1/2}) D_n^{-1/2} S_n(\boldsymbol{\beta}_0).$$

By (8.61), the first factor on the right-hand side converges to the identity matrix I_k, in probability, as $n \to \infty$. So to prove (8.29) it suffices to show that $D_n^{-1/2} S_n(\boldsymbol{\beta}_0)$ is asymptotically standard normal.

To do this we use the Cramer–Wold device (e.g. Shorack and Wellner (1986, p. 862)) and show that

$$\boldsymbol{u}^T D_n^{-1/2} S_n(\boldsymbol{\beta}_0) \xrightarrow{D} N(0, 1)$$

for all unit vectors $\boldsymbol{u}$. By (8.24) we have

$$\boldsymbol{u}^T D_n^{-1/2} S_n(\boldsymbol{\beta}_0) = \sum_{i=1}^{n} (c_i - t_i e^{\boldsymbol{\beta}_0^T \boldsymbol{x}_i})(\boldsymbol{u}^T D_n^{-1/2} \boldsymbol{x}_i). \tag{8.70}$$

The right-hand side of this is a sum of independent random variables each of which has mean 0, as we mentioned above. Each summand has variance

$$\sigma_{in}^2 = E(t_i)e^{\boldsymbol{\beta}_0^T x_i}(\boldsymbol{u}^T D_n^{-1/2}\boldsymbol{x}_i)^2. \tag{8.71}$$

To see this, note that $c_i - t_i e^{\beta_0}$ is the derivative with respect to a scalar β_0 of the logarithm of a likelihood component

$$L_i = (e^{\beta_0} e^{-e^{\beta_0} t_i})^{c_i}(e^{-e^{\beta_0} t_i})^{1-c_i},$$

so by Theorem 7.2, $E(c_i - t_i e^{\beta_0})^2$ equals $-E(\partial(c_i - t_i e^{\boldsymbol{\beta}})/\partial\boldsymbol{\beta})_{\beta_0} = E(t_i)e^{\beta_0}$. Note also from (8.26) that

$$\sum_{i=1}^n \sigma_{in}^2 = 1.$$

So to verify asymptotic normality of $\sum_{i=1}^n Z_{in}$, where

$$Z_{in} = (c_i - t_i e^{\boldsymbol{\beta}_0^T x_i})\boldsymbol{u}^T D_n^{-1/2}\boldsymbol{x}_i,$$

it suffices to verify Liapunov's condition, which here takes the form

$$\sum_{i=1}^n E|Z_{in}|^3 \to 0 \qquad (n \to \infty). \tag{8.72}$$

(Chow and Teicher (1988, p. 298) discuss Liapunov's condition.) To prove (8.72), again let $w_i = e^{\boldsymbol{\beta}_0^T \boldsymbol{x}_i}$, and note that

$$\begin{aligned} |Z_{in}|^3 &\le (c_i + w_i t_i)^3 |\boldsymbol{u}^T D_n^{-1/2}\boldsymbol{x}_i|^3 \\ &\le (c_i + 3c_i w_i t_i + 3c_i(w_i t_i)^2 + (w_i t_i)^3)|\boldsymbol{u}^T D_n^{-1/2}\boldsymbol{x}_i|^3. \end{aligned} \tag{8.73}$$

Using the bound in (8.66), and (8.67), in (8.73) gives

$$E|Z_{in}|^3 \le 16E(w_i t_i)|\boldsymbol{u}^T D_n^{-1/2}\boldsymbol{x}_i|^3,$$

so we have

$$\begin{aligned} \sum_{i=1}^n E|Z_{in}|^3 &\le 16\sum_{i=1}^n w_i E(t_i)|\boldsymbol{u}^T D_n^{-1/2}\boldsymbol{x}_i|^3 \\ &\le 16\sqrt{\max_{1\le i\le n}(\boldsymbol{u}^T D_n^{-1/2}\boldsymbol{x}_i)^2} \sum_{i=1}^n w_i E(t_i)(\boldsymbol{u}^T D_n^{-1/2}\boldsymbol{x}_i)^2 \\ &\le 16\sqrt{\max_{1\le i\le n} \boldsymbol{x}_i^T D_n^{-1}\boldsymbol{x}_i} \to 0 \qquad \text{as } n \to \infty. \end{aligned}$$

This establishes Liapunov's condition (8.72), and hence the asymptotic normality of $D_n^{1/2}(\tilde{\boldsymbol{\beta}}_n - \boldsymbol{\beta}_0)$.

Next we prove (8.30) and (8.31). But (8.30) is immediate from (8.61) since $\tilde{\boldsymbol{\beta}}_n$ is in $N(\boldsymbol{\beta}_0)$ with probability approaching 1, whereas (8.31) follows directly from (8.30) by the matrix argument outlined in Appendix Note 3 to Chapter 7.

It remains only to prove that the deviance difference d_n in (8.32) is asymptotically χ_k^2. To see this, substitute $\tilde{\boldsymbol{\beta}}_n$ for $\boldsymbol{\beta}$ in (8.57) to get

$$\tfrac{1}{2}d_n = (\tilde{\boldsymbol{\beta}}_n - \boldsymbol{\beta}_0)^T S_n(\boldsymbol{\beta}_0) - \tfrac{1}{2}(\tilde{\boldsymbol{\beta}}_n - \boldsymbol{\beta}_0)^T F_n(\boldsymbol{\beta}')(\tilde{\boldsymbol{\beta}}_n - \boldsymbol{\beta}_0)$$

where $\boldsymbol{\beta}' = a\boldsymbol{\beta}_0 + (1-a)\tilde{\boldsymbol{\beta}}_n$, for some a in [0, 1]. Thus $\boldsymbol{\beta}' \in N(\boldsymbol{\beta}_0)$ when $\tilde{\boldsymbol{\beta}}_n \in N(\boldsymbol{\beta}_0)$. Now by (8.61) we can write

$$F_n(\boldsymbol{\beta}') = D_n^{1/2}(I + \varepsilon_n)D_n^{1/2} = D_n + D_n^{1/2}\varepsilon_n D_n^{1/2}$$

where ε_n is a matrix, each of whose elements converges to 0 in probability. Thus

$$\begin{aligned}(\tilde{\boldsymbol{\beta}}_n - \boldsymbol{\beta}_0)^T F_n(\boldsymbol{\beta}')(\tilde{\boldsymbol{\beta}}_n - \boldsymbol{\beta}_0) = {} & (\tilde{\boldsymbol{\beta}}_n - \boldsymbol{\beta}_0)^T D_n(\tilde{\boldsymbol{\beta}}_n - \boldsymbol{\beta}_0)^T \\ & + (\tilde{\boldsymbol{\beta}}_n - \boldsymbol{\beta}_0)^T D_n^{1/2}\varepsilon_n D_n^{1/2}(\tilde{\boldsymbol{\beta}}_n - \boldsymbol{\beta}_0).\end{aligned}$$

Since $D_n^{1/2}(\tilde{\boldsymbol{\beta}}_n - \boldsymbol{\beta}_0)$ converges to $N(0, I_k)$, the second matrix on the right-hand side is $o_p(1)$ as $n \to \infty$. Thus

$$\tfrac{1}{2}d_n = (\tilde{\boldsymbol{\beta}}_n - \boldsymbol{\beta}_0)^T S_n(\boldsymbol{\beta}_0) - \tfrac{1}{2}(\tilde{\boldsymbol{\beta}}_n - \boldsymbol{\beta}_0)^T D_n(\tilde{\boldsymbol{\beta}}_n - \boldsymbol{\beta}_0) + o_p(1)$$

and we make yet another Taylor expansion to write

$$\begin{aligned}(\tilde{\boldsymbol{\beta}}_n - \boldsymbol{\beta}_0)^T S_n(\boldsymbol{\beta}_0) &= (\tilde{\boldsymbol{\beta}}_n - \boldsymbol{\beta}_0)^T (S_n(\tilde{\boldsymbol{\beta}}_n) - F_n(\boldsymbol{\beta}')(\boldsymbol{\beta}_0 - \tilde{\boldsymbol{\beta}}_n)) \\ &= (\tilde{\boldsymbol{\beta}}_n - \boldsymbol{\beta}_0)^T F_n(\boldsymbol{\beta}')(\tilde{\boldsymbol{\beta}}_n - \boldsymbol{\beta}_0).\end{aligned}$$

By the same argument used above, the second term equals $(\tilde{\boldsymbol{\beta}}_n - \boldsymbol{\beta}_0)^T D_n(\tilde{\boldsymbol{\beta}}_n - \boldsymbol{\beta}_0)$ plus a term which is $o_p(1)$, so we finally have

$$d_n = (\tilde{\boldsymbol{\beta}}_n - \boldsymbol{\beta}_0)^T D_n(\tilde{\boldsymbol{\beta}}_n - \boldsymbol{\beta}_0) + o_p(1).$$

Since $D_n^{1/2}(\tilde{\boldsymbol{\beta}}_n - \boldsymbol{\beta}_0)$ is asymptotically standard normal, the right-hand side of this converges to χ_k^2 as $n \to \infty$. This completes the proof of Theorem 8.2. ■

3. Proof of Theorem 8.4

Let

$$p_{(1)} = \cdots = p_{(K)} = p_0, \text{ say,}$$

denote the common value of the p's under H_0. The covariate vectors $\boldsymbol{x}_i$ and $\boldsymbol{y}_i$ have the form

$$\boldsymbol{x}_i = \boldsymbol{y}_i = \begin{bmatrix} 0 \\ \vdots \\ 1_{k(i)} \\ \vdots \\ 0 \end{bmatrix}_{K \times 1} \tag{8.74}$$

where $k(i)$ is the group to which individual i belongs. Define X_i by (8.38). We prove that when (8.48) holds we have the conditions required to apply Theorem 8.3, in fact

$$\liminf_{n\to\infty} \frac{\lambda_{\min}(D_n)}{n} > 0 \tag{8.75}$$

and

$$\lim_{n\to\infty} \sum_{i=1}^{n} (\text{trace } \{X_i^T D_n^{-1} X_i\})^{3/2} = 0. \tag{8.76}$$

To see these, let $\boldsymbol{\nu} = (\nu_1, \ldots, \nu_K, \nu_{K+1}, \ldots, \nu_{2K})^T_{2K\times 1}$ be a unit vector in $\mathbb{R}^{2K}$. Here we have

$$X_i = \begin{pmatrix} \boldsymbol{x}_i & 0 \\ 0 & \boldsymbol{x}_i \end{pmatrix}_{2K\times 2}$$

where $\boldsymbol{x}_i$ is defined in (8.74). By (8.43),

$$\boldsymbol{\nu}^T D_n \boldsymbol{\nu} = \sum_{i=1}^{n} \boldsymbol{\nu}^T X_i \mathcal{D}_i X_i^T \boldsymbol{\nu} = \sum_{k=1}^{K} \sum_{i\in I(k)} \boldsymbol{\nu}_k^T \mathcal{D}_{(k)} \boldsymbol{\nu}_k = \sum_{k=1}^{K} n_k \boldsymbol{\nu}_k^T \mathcal{D}_{(k)} \boldsymbol{\nu}_k \tag{8.77}$$

where $\boldsymbol{\nu}_k = (\nu_k \ \nu_{K+k})^T_{2\times 1}$ and $\mathcal{D}_{(k)}$ is the common value of $\mathcal{D}_i$ within group k.

By (8.48), there exist $\delta > 0$ and $n_0 > 0$ such that $n_k/n \geq \delta$ for all k and $n \geq n_0$. Moreover each $\mathcal{D}_{(k)}$ is nonsingular, since the u_i do not degenerate at 0. Also since

$$\sum_{k=1}^{K} (\nu_k^2 + \nu_{K+k}^2) = 1$$

we have from (8.77) that

$$\frac{\boldsymbol{\nu}^T D_n \boldsymbol{\nu}}{n} \geq \delta \sum_{k=1}^{K} \boldsymbol{\nu}_k^T \mathcal{D}_{(k)} \boldsymbol{\nu}_k \geq \delta \left(\min_{1\leq k\leq K} \lambda_{\min}\{\mathcal{D}_{(k)}\} \right) > 0 \tag{8.78}$$

when $n \geq n_0$. Since (8.78) holds for any unit vector $\boldsymbol{\nu}$, (8.75) follows.

Thus, with $\boldsymbol{\nu}$ a two-dimensional unit vector, and if i belongs to group k, then

$$\boldsymbol{\nu}^T X_i^T D_n^{-1} X_i \boldsymbol{\nu} \leq \frac{1}{\lambda_{\min}\{D_n\}} \leq \frac{a}{n}$$

for some $a > 0$ and n large enough. It follows that

$$\sum_{i=1}^{n} (\text{trace } (X_i^T D_n^{-1} X_i))^{3/2} \leq \sum_{i=1}^{n} \left(\frac{a}{n}\right)^{3/2} \to 0$$

as $n \to \infty$, so (8.76) holds.

We can now complete the proof of Theorem 8.4. Conditions (8.75) and (8.76) allow us to deduce from Theorem 8.3 that an MLE $\tilde{\theta}_n$ exists (WPA1) when the covariates satisfy (8.74). Now suppose $\boldsymbol{x}_i$ is still as in (8.74), but $\boldsymbol{y}_i$ is taken as the scalar 1. Let

$$X_i^* = \begin{pmatrix} \boldsymbol{x}_i & \boldsymbol{0} \\ 0 & 1 \end{pmatrix}_{(K+1)\times 2}$$

and define

$$\boldsymbol{\theta}_0^* = (\beta_{10}, \ldots, \beta_{K0}, \delta_0)^T_{(K+1)\times 1}.$$

Let

$$\boldsymbol{\theta}^* = (\beta_1, \ldots, \beta_K, \delta)^T_{(K+1)\times 1}$$

be an arbitrary vector in $\mathbb{R}^{K+1}$. Virtually the same proof as for (8.75) and (8.76) tells us that a maximiser of $l_n(\boldsymbol{\theta}^*)$ exists WPA1 local to $\boldsymbol{\theta}_0^*$. Write such a maximiser as

$$(\dot{\beta}_1, \ldots, \dot{\beta}_K, \dot{\delta})^T_{(K+1)\times 1}. \tag{8.79}$$

Equivalently, we could define

$$\boldsymbol{\theta} = (\beta_1, \ldots, \beta_K, \delta_1, \ldots, \delta_K)_{2K\times 1}$$

and note that H_0 specifies $K-1$ linear restrictions on $\boldsymbol{\theta}$ which we may express as

$$H\boldsymbol{\theta} = \boldsymbol{0} \tag{8.80}$$

where

$$H = \begin{pmatrix} 0 & \ldots & 0 & 1 & -1 & 0 & \ldots & 0 & 0 \\ \vdots & & & 0 & 1 & -1 & \ldots & 0 & 0 \\ & & & \vdots & \vdots & \vdots & & & \\ 0 & \ldots & 0 & 0 & 0 & 0 & \ldots & 1 & -1 \end{pmatrix}_{(K-1)\times 2K}.$$

Thus to maximise $l_n(\boldsymbol{\theta})$ under H_0 we may maximise

$$l_n(\boldsymbol{\theta}) - \boldsymbol{\theta}^T H^T \boldsymbol{\gamma}$$

unrestrictedly for variations in $\boldsymbol{\theta} \in \mathbb{R}^{2K}$ and $\boldsymbol{\gamma} \in \mathbb{R}^{K-1}$, where $\boldsymbol{\gamma}$ is a vector of Lagrange parameters. This will produce

$$\dot{\boldsymbol{\theta}}_n = (\dot{\beta}_1, \ldots, \dot{\beta}_K, \dot{\delta}, \ldots, \dot{\delta})^T_{2K\times 1}$$

where $\dot{\beta}_k$ and $\dot{\delta}$ are the same as those defined in (8.79). Thus $\dot{\theta}_n$ satisfies

$$\boldsymbol{S}_n(\dot{\boldsymbol{\theta}}_n) - H^T\dot{\boldsymbol{\gamma}} = \boldsymbol{0} \tag{8.81}$$

where $\dot{\boldsymbol{\gamma}}$ is some vector in $\mathbb{R}^{K-1}$. Let

$$\boldsymbol{\theta}_0 = (\beta_{10}, \ldots, \beta_{K0}, \delta_0, \ldots, \delta_0)^T_{2K\times 1}.$$

By (8.80) we have

$$H\dot{\boldsymbol{\theta}}_n = \mathbf{0},$$

and thus, since $H\boldsymbol{\theta}_0 = \mathbf{0}$,

$$H(\dot{\boldsymbol{\theta}}_n - \boldsymbol{\theta}_0) = \mathbf{0}. \tag{8.82}$$

We need to solve for $\dot{\boldsymbol{\theta}}_n$ from these equations. By Taylor expansion we obtain

$$D_n^{-1/2}\mathbf{S}_n(\dot{\boldsymbol{\theta}}_n) = D_n^{-1/2}\mathbf{S}_n(\boldsymbol{\theta}_0) - D_n^{1/2}(\dot{\boldsymbol{\theta}}_n - \boldsymbol{\theta}_0) + o_p(1) \tag{8.83}$$

where we used a version of (8.30) for

$$D_n = E\{F_n(\boldsymbol{\theta}_0)\}_{2K\times 2K}$$

and

$$F_n(\boldsymbol{\theta}) = \sum_{i=1}^{n} X_i \mathcal{F}_i(\boldsymbol{\theta}) X_i^T$$

with $\mathcal{F}_i(\boldsymbol{\theta})$ given by (8.40) to (8.42) and X_i given by

$$X_i = \begin{pmatrix} \boldsymbol{x}_i & \mathbf{0} \\ \mathbf{0} & \mathbf{1} \end{pmatrix}_{2K\times 2}.$$

Here $\mathbf{1}$ is a $K \times 1$ vector of 1's. Equation (8.83) now gives

$$D_n^{1/2}(\dot{\boldsymbol{\theta}}_n - \boldsymbol{\theta}_0) + D_n^{-1/2}\mathbf{S}_n(\dot{\boldsymbol{\theta}}_n) = D_n^{-1/2}\mathbf{S}_n(\boldsymbol{\theta}_0) + o_p(1)$$

so by (8.81),

$$D_n^{1/2}(\dot{\boldsymbol{\theta}}_n - \boldsymbol{\theta}_0) + D_n^{-1/2}H^T\dot{\boldsymbol{\gamma}} = D_n^{-1/2}\mathbf{S}_n(\boldsymbol{\theta}_0) + o_p(1).$$

Also by (8.82)

$$(HD_n^{-1/2})D_n^{1/2}(\dot{\boldsymbol{\theta}}_n - \boldsymbol{\theta}_0) = \mathbf{0}.$$

We can write the last two equations in matrix form as

$$\begin{pmatrix} I & D_n^{-1/2}H^T \\ HD_n^{-1/2} & \mathbf{0} \end{pmatrix} \begin{pmatrix} D_n^{1/2}(\dot{\boldsymbol{\theta}}_n - \boldsymbol{\theta}_0) \\ \dot{\boldsymbol{\gamma}} \end{pmatrix} = \begin{pmatrix} D_n^{-1/2}\mathbf{S}_n(\boldsymbol{\theta}_0) + o_p(1) \\ \mathbf{0} \end{pmatrix}. \tag{8.84}$$

Note that H is of full rank $(K-1)$ and D_n is nonsingular, so $HD_n^{-1}H^T$ is of full rank $(K-1)$. Thus the inverse of the matrix on the left side of (8.84) exists and can be written as

$$\begin{pmatrix} I & D_n^{-1/2}H^T \\ HD_n^{-1/2} & \mathbf{0} \end{pmatrix}^{-1} = \begin{pmatrix} I-P & Q \\ Q^T & R \end{pmatrix}$$

where

$$P = D_n^{-1/2} H^T (H D_n^{-1} H^T)^{-1} H D_n^{-1/2},$$
$$Q = D_n^{-1/2} H^T (H D_n^{-1} H^T)^{-1},$$

and

$$R = -(H D_n^{-1} H^T)^{-1}.$$

Thus we can solve (8.84) as

$$\begin{pmatrix} D_n^{1/2}(\dot{\theta}_n - \theta_0) \\ \dot{\gamma} \end{pmatrix} = \begin{pmatrix} I - P & Q \\ Q^T & R \end{pmatrix} \begin{pmatrix} D_n^{-1/2} \boldsymbol{S}_n(\theta_0) + o_p(1) \\ \boldsymbol{0} \end{pmatrix}$$

which gives

$$D_n^{1/2}(\dot{\theta}_n - \theta_0) = (I - P) D_n^{-1/2} \boldsymbol{S}_n(\theta_0) + o_p(1).$$

By a similar argument, the unrestricted MLE $\tilde{\theta}_n$ satisfies

$$D_n^{1/2}(\tilde{\theta}_n - \theta_0) = D_n^{-1/2} \boldsymbol{S}_n(\tilde{\theta}_0) + o_p(1),$$

and so we obtain

$$D_n^{1/2}(\tilde{\theta}_n - \dot{\theta}_n) = D_n^{1/2}(\tilde{\theta}_n - \hat{\theta}_0) - D_n^{1/2}(\tilde{\theta}_n - \hat{\theta}_0) = P D_n^{-1/2} \boldsymbol{S}_n(\theta_0) + o_p(1). \quad (8.85)$$

It is easy to check the P is idempotent of rank $K - 1$, so there is an orthogonal matrix A such that

$$P = A^T \begin{pmatrix} I_{K-1} & \boldsymbol{0} \\ \boldsymbol{0} & \boldsymbol{0} \end{pmatrix} A.$$

Let $\boldsymbol{Z}_n = A D_n^{-1/2} \boldsymbol{S}_n(\theta_0)$. Then by the central limit theorem, as used in Theorem 8.2, we have

$$\boldsymbol{Z}_n \xrightarrow{D} \boldsymbol{Z} \sim N(0, I).$$

Thus by (8.85)

$$(\tilde{\theta}_n - \dot{\theta}_n)^T D_n (\tilde{\theta}_n - \dot{\theta}_n) = (A^T \boldsymbol{Z}_n)^T P A^T \boldsymbol{Z}_n + o_p(1)$$
$$= \boldsymbol{Z}_n^T \begin{pmatrix} I_{K-1} & \boldsymbol{0} \\ \boldsymbol{0} & \boldsymbol{0} \end{pmatrix} \boldsymbol{Z}_n + o_p(1) \xrightarrow{D} \boldsymbol{Z}^T \begin{pmatrix} I_{K-1} & \boldsymbol{0} \\ \boldsymbol{0} & \boldsymbol{0} \end{pmatrix} \boldsymbol{Z} \sim \chi^2_{K-1}. \quad (8.86)$$

Finally, we have by Taylor expansion

$$d_n = -2[l_n(\dot{\theta}_n) - l_n(\tilde{\theta}_n)]$$
$$= -2(\dot{\theta}_n - \tilde{\theta}_n)^T \boldsymbol{S}_n(\tilde{\theta}_n) + (\dot{\theta}_n - \tilde{\theta}_n)^T F_n(\tilde{\theta})(\dot{\theta}_n - \tilde{\theta}_n)$$

where $\tilde{\theta}$ lies between $\tilde{\theta}_n$ and $\dot{\theta}_n$. Thus $\tilde{\theta}$ is in $N_n(A)$ WPA1 for each $A > 0$. We can replace $F_n(\tilde{\theta})$ by D_n by an argument like that in (8.61). Since $\boldsymbol{S}_n(\tilde{\theta}_n) = \boldsymbol{0}$ WPA1, it follows then from (8.86) that

$$d_n = (\tilde{\theta}_n - \dot{\theta}_n)^T D_n (\tilde{\theta}_n - \dot{\theta}_n) + o_p(1) \xrightarrow{D} \chi^2_{K-1}, \text{ as } n \to \infty.$$

This completes the proof of Theorem 8.4. ■

CHAPTER 9

Further Topics

In this chapter we cover some further topics of interest. In particular, we review ideas associated with the modelling of hazard functions, since these are common in survival analysis. We look briefly at the idea of competing risks, concerned with dividing a population at risk into components in which various causes 'compete' to bring about the death of failure of an individual; it is also closely connected with the idea of one component of a population being immune to death or failure. We look at how to estimate the *censoring* distribution from a set of survival data. And we discuss a topic which is of special interest in the analysis of data with immunes: how to estimate the probability that an individual is immune. We begin with the idea of modelling the hazard function.

9.1 HAZARDS AND HAZARD FUNCTION MODELLING

Hazard functions

Let T be a nonnegative random variable representing the survival time of an individual, with $F(t)$ the cumulative distribution function (c.d.f.) of T. The *hazard function* associated with F, evaluated at time $t \geq 0$, is the instantaneous probability of death at time t, given survival of the individual up to time t. If F is absolutely continuous with density $f(t) = \mathrm{d}F(t)/\mathrm{d}t$, then its hazard function $\lambda(t)$ is

$$\lambda(t) = \frac{f(t)}{1 - F(t)}, \tag{9.1}$$

and is defined at points $0 \leq t < \tau_F$ (for which $F(t) < 1$). (There is a similar formula for the case when T is a discrete random variable but we will restrict our discussion to the continuous case). Clearly $\lambda(t) \geq 0$ for all $t \geq 0$.

Associated with $\lambda(t)$ is the *cumulative hazard function*

$$\Lambda(t) = \int_0^t \lambda(y)\,\mathrm{d}y. \tag{9.2}$$

By integrating (9.1), we see that $\Lambda(t)$ satisfies

$$\Lambda(t) = -\log(1 - F(t)), \tag{9.3}$$

or equivalently, we have

$$F(t) = 1 - e^{-\Lambda(t)}. \tag{9.4}$$

Thus F is uniquely determined by its hazard function, and conversely. If F is proper, i.e. $F(\tau_F) = 1$, then $\Lambda(\tau_F) = \infty$, or equivalently,

$$\int_0^{\tau_F} \lambda(t)\,dt = \infty. \tag{9.5}$$

In other words, $\lambda(t)$ is not integrable on $[0, \tau_F)$. Conversely, *an improper F corresponds to the case when $\lambda(t)$ is integrable on* $[0, \tau_F)$, i.e. to the condition

$$\int_0^{\tau_F} \lambda(t)\,dt < \infty.$$

In the latter case, we have

$$F(\tau_F) = 1 - \exp\left(-\int_0^{\tau_F} \lambda(t)\,dt\right) \tag{9.6}$$

and so if we represent F as

$$F(t) = pF_0(t),$$

where $1 - p$ is the probability that an individual is immune, and $F_0(t)$ is the proper c.d.f. of the susceptibles, then

$$p = 1 - \exp\left(-\int_0^{\tau_F} \lambda(t)\,dt\right). \tag{9.7}$$

Suppose now that $t_1, \ldots, t_n$ is a sample of survival times with censor indicators $c_1, \ldots, c_n$, according to our usual notation. An alternative way of formulating models for the distribution of the survival time is to think in terms of hazard functions, rather than in terms of distribution functions. The approaches are essentially equivalent, of course, but there are various reasons why the first is often found to be more convenient in the analysis of (ordinary) survival data. One reason perhaps is the following. When $F(t) = 1 - e^{-\lambda t}$ is the exponential c.d.f., we see from (9.1) that

$$\lambda(t) = \lambda$$

for all $t \geq 0$, i.e. the hazard of the exponential distribution is the constant λ. For hazard functions of some other standard survival distributions see Kalbfleisch and Prentice (1980, Ch. 2).

We can think of the hazard function of an arbitrary c.d.f. as being approximated by a step function which is constant on subintervals of $[0, \infty)$, and in this way a general c.d.f. can be approximated arbitrarily closely, in principle, by a 'piecewise exponential' distribution, whose hazard is constant on those subintervals. In a given data set we are limited by the discreteness of the data as to the accuracy with which we can make this approximation, but if we pursue the subdivision to the extent that each uncensored survival time determines its own subinterval, we arrive at the KME as the ultimate 'piecewise constant hazard' estimator of an arbitrary c.d.f., constructed from the sample. See Aitkin *et al.* (1989, p. 272) for an exposition of this. By contrast, the above development does not seem so natural if constructed in terms of distribution function approximations rather than hazard function approximations.

Another reason for thinking in terms of hazards is that the likelihood function L of a sample, for a parametrically specified c.d.f. $F(t)$, is easily written in terms of its hazard function. Indeed we saw in Section 5.1 that L has, in general, the form

$$\begin{aligned} L(t_1, \ldots, t_n) &= \prod_{i=1}^{n} (f(t_i))^{c_i} (1 - F(t_i))^{1-c_i} \\ &= \prod_{i=1}^{n} \left(\frac{f(t_i)}{1 - F(t_i)} \right)^{c_i} (1 - F(t_i)) \\ &= \prod_{i=1}^{n} (\lambda(t_i))^{c_i} \exp\left(- \int_0^{t_i} \lambda(y)\, \mathrm{d}y \right). \end{aligned} \tag{9.8}$$

Especially when it comes to incorporating covariates in the model, equation (9.8) lends itself to convenient formulations. For example, covariates are often assumed to affect the hazard 'proportionally'—see below.

A third reason for using hazards is that much of the modern theory of survival analysis is best phrased in terms of an application of the martingale theory to the counting process which represents the occurrence of the survival times, and this can be conveniently formulated in terms of hazard functions. We promised in the preface to the book not to go into this theory, so we will merely mention that much of the initial development of the counting process approach is attributed to Aalen, especially Aalen (1978), and refer the reader to those authors who treat the subject in this way, particularly Fleming and Harrington (1993, Ch. 1) and Andersen *et al.* (1993, Ch. III).

Estimation of the cumulative hazard function is usually done via the Nelson (1969, 1972) estimator. This is defined by

$$\tilde{\Lambda}_n(t) = \sum_{j=1}^{i} \frac{d_j}{n_j} \qquad \text{when } t_{(i)}^{(d)} \leq t < t_{(i+1)}^{(d)}, \tag{9.9}$$

in the notation which we used for the KME in Section 1.2. (Thus the $t_{(i)}^{(d)}$ denote the distinct ordered survival or censored times, and the d_i and n_i are the number failing and the number at risk just prior to time $t_{(i)}^{(d)}$, respectively.)

Despite the simple form of (9.9), $\tilde{\Lambda}_n(t)$ is a biased estimator of $\Lambda(t)$ in general, just as the KME, $\hat{F}_n(t)$, is biased in general. Using (9.4), we can construct from

$\tilde{\Lambda}_n(t)$ an estimator

$$\tilde{F}_n(t) = 1 - \mathrm{e}^{-\tilde{\Lambda}_n(t)} = 1 - \prod_{j=1}^{i} \mathrm{e}^{-d_j/n_j} \tag{9.10}$$

(when $t_{(i)}^{(d)} \leq t < t_{(i+1)}^{(d)}$) of $F(t)$, which is usually close in value to the KME $\hat{F}_n(t)$. The estimator $\tilde{F}_n(t)$ is also biased for $F(t)$; in fact, since $1 - \mathrm{e}^{-x} \leq x$ for $x > 0$, we have

$$\tilde{F}_n(t) \leq 1 - \prod_{j=1}^{i} \left(1 - \frac{d_j}{n_j}\right) = \hat{F}_n(t). \tag{9.11}$$

Since, as we saw in Section 3.3, the KME is biased *downwards* for $F(t)$, $\tilde{F}_n(t)$ is further biased downwards for $F(t)$. As we know from Section 3.3, the bias of $\hat{F}_n(t)$ disappears in large samples *provided follow-up is sufficient.* Fleming and Harrington (1991, pp. 100–101) give an example to suggest that the biases in $\tilde{F}_n(t)$ and $\hat{F}_n(t)$ are typically small. For examples of hazard function estimation and modelling of 'ordinary' survival data, see Andersen *et al.* (1993, pp. 446–447) and Fleming and Harrington (1991, pp. 5–7). Bie *et al.* (1987) discuss the accuracy of estimation of $\Lambda(t)$ via $\tilde{\Lambda}_n(t)$.

Despite the above ideas, we have promulgated in this book the idea of looking at the KME as an estimator of the c.d.f. $F(t)$ of the survival times, rather than looking at a sample estimator of the hazard, or of the cumulative hazard. The reason is simple. Our interest is focused on the question of whether or not immunes exist in the population we are studying, and it is much easier to infer their presence from a levelling off of the KME below 1, rather than (what is essentially equivalent) trying to decide whether the estimated hazard is integrable on $[0, \infty)$ or not (see (9.5)). Of course we need not necessarily use the KME in this context; we could consider the use of the distribution function estimator given by (9.10) in this regard, and we have not investigated which (or which other estimator) may be the best choice.

Proportional hazard models

Suppose we have survival times, assumed exponentially distributed, with associated covariates. We suggested (in Section 6.1) modelling the c.d.f. of the survival time of individual i by

$$1 - \mathrm{e}^{-\lambda_i t}, \qquad t \geq 0,$$

where $\lambda_i = \mathrm{e}^{\boldsymbol{\beta}^T \boldsymbol{x}_i}$. This represents a 'log-linear link' to the covariate $\boldsymbol{x}_i$, $\boldsymbol{\beta}$ being as usual a vector of parameters to be estimated. The hazard function associated with individual i is in this case λ_i, which does not depend on t at all.

More generally, the hazard function for i may satisfy

$$\lambda_i(t) = \lambda(t)\mathrm{e}^{\boldsymbol{\beta}^T \boldsymbol{x}_i} \tag{9.12}$$

for some positive function $\lambda(t)$. If so, we say that we have a 'proportional hazards model' for the influence of the covariates on survival, in that the effect of the covariates is simply to scale the 'baseline hazard' $\lambda(t)$ by the factor $e^{\beta^T x_i}$. (More generally, we could replace $e^{\beta^T x_i}$ by a positive function $g(\boldsymbol{\beta}^T \boldsymbol{x}_i)$, but we take the special case for simplicity of exposition.) When $\lambda(t) = 1$ for all $t \geq 0$, (9.12) reduces to the exponential model with covariates.

If we think, as we did in the previous section, of approximating an arbitrary individual hazard $h_i(t)$ by a step function corresponding to a piecewise exponential distribution whose hazard is constant on subintervals, but still keeping the hazards proportional with respect to the covariates, we obtain a piecewise exponential model with covariates which is often used in practice; see for example Aitkin *et al.* (1989, pp. 288–291), Larson and Dinse (1985) and Section 9.2. Taking the subintervals small enough, so that only one (distinct) death time occurs per subinterval, leads to the famous *Cox proportional hazards regression model* (Cox, 1972), which appears in this way as a very natural generalisation of the KME, if we recall the similar heuristic derivation of the KME given in the previous section.

In generalising these ideas to survival models with immunes, we can consider both proportional and nonproportional hazards models. One compelling reason for the use of the proportional hazards model in ordinary survival analysis is its analytic tractability. From this point of view, proportional hazards models do not seem particularly relevant for data with immunes. Consider the exponential mixture model of Section 5.1, which has c.d.f. $p(1 - e^{-\lambda t})$, for $t \geq 0$, and hazard function

$$\lambda(t) = \frac{d(p(1 - e^{-\lambda t}))/dt}{1 - p(1 - e^{-\lambda t})} = \frac{p\lambda e^{-\lambda t}}{1 - p + pe^{-\lambda t}} = \frac{p\lambda}{(1 - p)e^{\lambda t} + p}. \tag{9.13}$$

When $p < 1$, the hazard is not constant in t, and forcing a proportional hazards relationship like (9.12) seems rather unnatural here. What does seem reasonable, for data with immunes, is to suppose that the distribution of the survival times of the *susceptibles* follows a proportional hazards model with respect to the covariates, while also supposing that the probability that individual i is immune is related to the covariates in some canonical manner such as is given by

$$\frac{e^{\delta^T x_i}}{1 + e^{\delta^T x_i}}.$$

With $\lambda = \lambda_i = e^{\beta^T x_i}$, this is of course precisely the model we proposed and analysed in Section 8.1.

An alternative class of models to the above, which *does* have proportional hazards and also allows for immune individuals, can be constructed as follows. To obtain the survival function of the survival time of individual i, take any survival c.d.f. $F(t)$ and raise its survival function $S(t)$ to the power $e^{\beta^T x_i}$. Letting

$e_i = e^{\beta^T x_i}$ for brevity, this means that the hazard for individual i is then

$$-\frac{1}{(S(t))^{e_i}}\frac{d(S(t))^{e_i}}{dt} = -\frac{e_i}{S(t)}\frac{dS(t)}{dt} = \lambda(t)e^{\beta^T x_i}$$

where $\lambda(t)$ is the hazard function associated with F. Consequently the proportional hazards formulation holds.

If F is improper, so will be the resulting survival distribution for individual i. For example, suppose F is the exponential mixture distribution $F(t) = p(1 - e^{-\lambda t})$. This has survivor function $S(t) = 1 - p + pe^{-\lambda t}$, so the c.d.f. of survival time for individual i under the above model would be

$$1 - (1 - p + pe^{-\lambda t})^{e_i}. \tag{9.14}$$

This is not a mixture model, at least in the sense we have used the term, but it is improper, with a mass at infinity of size $1 - (1 - p)^{e_i}$, and therefore an immune component of size $(1 - p)^{e_i}$. As far as we know, the properties of this model have not been investigated further.

We conclude this section by mentioning that a Cox proportional hazard regression, or some modification of it, seems to be one of the most common methods currently used to test for differences between groups, or more generally for the effects of covariates, in survival data. This method is sometimes referred to as 'semiparametric' because it combines a nonparametric approach to the modelling of the underling survival distribution with a parametric representation of the covariates in the hazard function. Such an approach to the analysis of survival data with immunes possibly present is perfectly appropriate, but we should realise that the usual Cox model (or its equivalent in the two-group special case, the log-rank test) can only test for effects of the covariates in general, and does not take account of the immune component, if present. If, for example, two groups are found by the log-rank test to be significantly different, we are uncertain whether to ascribe this to a difference between the groups in survival rates, to a difference in immune proportions or to both. See Kalbfleisch and Prentice (1980, pp. 16–19) for a discussion of the log-rank test.

Kuk and Chen (1992) suggest a method of incorporating mixture components of a population into a nonparametric proportional hazards model which we have not investigated. Not all models are proportional hazards models, of course; Arbutski (1985) gives a family of 'multiplicative' models based on the Weibull distribution which allows for long-term survivors, and fits them to some of the data of Boag and Farewell. Taylor (1995) suggests a mixed parametric/nonparametric model and an iterative EM-type procedure, similar in spirit to that of Kuk and Chen, in which censored observations are assigned a probability of corresponding to immunes. When this probability is zero, Taylor's estimator reduces to the KME.

9.2 COMPETING RISKS

When discussing the concept of 'curability' (of breast cancer, say) in Chapter 1, we mentioned that, when analysing survival times of patients, a number of authors have taken into account the possibility of the patient's death occurring from a cause other than the one under study. In medical data, one should perhaps always consider this eventuality when attempting to estimate an immune proportion from a data set, since all patients will ultimately die from some cause, if not from the disease under study. We are only justified in overlooking this if patients' lifetimes are relatively long, compared to the amount of follow-up in the experiment.

Thus, as we argued in Section 1.4, when the data is such that the short-term risk of death from another cause is small, we may be prepared to neglect that risk. But in a study in which the follow-up time is long, an appreciable proportion of an individual's life span, say, we may observe a sizeable proportion of deaths from one or more 'competing' causes (or from a cause which is just designated as 'natural causes' or 'other'), and we may wish to account for or to model such data in a more formal way. These kinds of studies come into the class of 'competing risks' analyses, and as we will demonstrate, they can be closely connected with or considered as extensions of the mixture models we have been fitting so far in the book.

The idea of competing risks applies in disciplines other than biostatistics and epidemiology. The basic idea is that we can classify death or failure into one or more mutually exclusive *types* or *causes*. In criminology an individual's return to prison after his/her first release, if it ever occurs, may be for an offence of a certain type — homicide, sexual assault, minor larceny, etc. — and we might simply classify it as being of lesser, equal or greater severity than the first, according to some offence severity scale supplied by criminologists. We can observe the type of failure for those who return to prison, but for those releasees who have not recidivated by the cutoff data — the time of analysis — no second imprisonment has occurred, so no 'type of failure' can be ascribed to them.

In a medical survival analysis, the variable of interest may be the time to recurrence of a particular kind of cancer; deaths from causes other than this cancer are from the competing causes, whereas patients withdrawn from the study or alive and cancer-free at the end of the study cannot be ascribed a failure type at all. For a final example, we could think of the failure time of a car engine, which may be due to the failure of a single component. These would constitute the 'types' of failure.

Various models have been proposed for the analysis of competing risks data, and the literature contains much discussion about their appropriateness, the assumptions under which they are thought to be applicable and the value of the conclusions which can be drawn. Here we intend only to sketch a couple, comparing them with the kind of mixture model analyses we have been advocating for survival data with immunes. We will show that some of the ideas we have developed in the process can also be applied to some models for competing risks data.

First we need some notation. As usual, we have n survival times $t_1, t_2, \ldots, t_n$ with associated censor indicators $c_1, c_2, \ldots, c_n$; as usual, $c_i = 1$ if t_i is uncensored, and $c_i = 0$ if t_i is censored. In addition, we have associated with those individuals i for which $c_i = 1$, an indicator $j(i)$ of the cause of death. Suppose $j(i)$ may take values in $\{1, 2, \ldots, J\}$, so there are J 'causes' or 'types' of death or failure. We can define (and observe) another indicator c_{ij} for individual i as

$$c_{ij} = \begin{cases} 1 & \text{if } i \text{ dies from cause } j \\ 0 & \text{otherwise.} \end{cases} \tag{9.15}$$

We assume that the causes are mutually exclusive, so

$$c_i = \sum_{j=1}^{J} c_{ij}, \tag{9.16}$$

and, as usual, we assume that observations on random variables representing different individuals are independent. In what follows we will ignore covariate information, if any, so we are in a 'single-sample' situation.

The Larson–Dinse formulation

Larson and Dinse (1985) suggest a mixture model approach which is based on a semi-Markov formulation of Lagakos *et al.* (1978) (see also Lagakos, 1979). Larson and Dinse assume that the cause of death of individual i is chosen at the outset by a stochastic mechanism from the list of possible causes $\{1, 2, \ldots, J\}$. We can thus define

$$p_j = P\{\text{individual } i \text{ dies from cause } j\}, \qquad 1 \le j \le J \tag{9.17}$$

(which is assumed not to depend on i). The survival time is an observation on a random variable T, and the model is based on the 'cause-specific' conditional distributions

$$F_j(t) = P(T \le t | \text{ death is from cause } j). \tag{9.18}$$

These are assumed to be *proper* distributions with densities $f_j(t)$, say.

At this stage we do not assume any individuals are immune to death, so we can write, for $t \ge 0$,

$$\begin{aligned} P(T \le t) &= \sum_{j=1}^{J} P(T \le t, \text{ death is from cause } j) \\ &= \sum_{j=1}^{J} p_j F_j(t) = F(t), \text{ say.} \end{aligned} \tag{9.19}$$

Larson and Dinse then suggest basing inference on a likelihood constructed from the (improper) *cause-specific joint densities* $p_j f_j(t)$ and the survival function from (9.19). Thus they define

$$L_n = \prod_{i=1}^{n} (p_{j(i)} f_{j(i)}(t_i))^{c_i} \left(1 - \sum_{j=1}^{J} p_j F_j(t_i)\right)^{1-c_i}, \tag{9.20}$$

and consider maximising this for some parametrisation of the c.d.f.'s $F_j(t)$. If we index the survival times according to the cause of death, i.e. we let $t_{ij} = t_i$ if $j(i) = j$, then

$$L_n = \prod_{i=1}^{n} \left(\prod_{j=1}^{J} (p_j f_j(t_{ij}))^{c_{ij}}\right) \left(1 - \sum_{j=1}^{J} p_j F_j(t_i)\right)^{1-c_i} \tag{9.21}$$

where the c_{ij} are defined in (9.15).

Larson and Dinse assume a piecewise exponential distribution for the $F_j(t)$ and fit this model to the famous Stanford transplant data set. To illustrate some of the theoretical properties of this model, we will simply assume an exponential distribution for the $F_j(t)$. Thus

$$F_j(t) = 1 - \mathrm{e}^{-\lambda_j t} \text{ and } f_j(t) = \lambda_j \mathrm{e}^{-\lambda_j t}, \qquad t \geq 0,$$

where we assume that no two of the λ_j are the same. So by (9.21) we can write

$$\log L_n = \sum_{i=1}^{n} \left\{ \sum_{j=1}^{J} c_{ij} (\log p_j + \log \lambda_j - \lambda_j t_{ij}) + (1 - c_i) \log \left(1 - \sum_{j=1}^{J} p_j (1 - \mathrm{e}^{-\lambda_j t_i})\right)\right\}. \tag{9.22}$$

Differentiating with respect to λ_j and p_j gives

$$\frac{\partial \log L_n}{\partial \lambda_j} = \sum_{i=1}^{n} \left\{ c_{ij} \left(\frac{1}{\lambda_j} - t_{ij}\right) - \frac{(1 - c_i) p_j t_i \mathrm{e}^{-\lambda_j t_i}}{1 - \sum_{j=1}^{J} p_j (1 - \mathrm{e}^{-\lambda_j t_i})} \right\} \tag{9.23a}$$

and

$$\frac{\partial \log L_n}{\partial p_j} = \sum_{i=1}^{n} \left\{ \frac{c_{ij}}{p_j} - \frac{(1 - c_i)(1 - \mathrm{e}^{-\lambda_j t_i})}{1 - \sum_{j=1}^{J} p_j (1 - \mathrm{e}^{-\lambda_j t_i})} \right\}, \tag{9.23b}$$

and we can attempt to solve the equations (9.23) to obtain estimators of the $2J$ parameters $\boldsymbol{\theta} = (\lambda_1, \lambda_2, \ldots, \lambda_J, p_1, p_2, \ldots, p_J)$.

We do not wish to go much more deeply into this analysis here, interesting though it would be, but we can at least indicate that the methods of Chapter 7 can be applied to suggest that consistent estimation of $\boldsymbol{\theta}$ will result from the solution of equations (9.23) *under certain conditions*. We saw in the proofs of the theorems in Chapter 7 that a crucial requirement for this is (at least) the unbiasedness of the estimating equation

$$\frac{\partial \log L_n}{\partial \boldsymbol{\theta}} = 0,$$

i.e. we require its expectation (with respect to the 'true' distribution(s) governing the survival times) to be 0. This holds here as we can show by, essentially, the method of Theorem 7.2, provided certain independence assumptions are made.

Consider the expectation of the summands in (9.23a). We can write $c_{ij} = I(t^*_{ij} \leq u_i)$, where t^*_{ij} is a random variable with (sub)distribution $p_j F_j(t)$, and u_i is a censoring random variable *independent* of t^*_{ij}. Then we calculate

$$\begin{aligned}
E\left(c_{ij}\left(\frac{1}{\lambda_j} - t_{ij}\right)\right) &= E\left(\int_0^u \left(\frac{1}{\lambda_j} - t\right) p_j \,\mathrm{d}F_j(t)\right) \\
&= p_j E\left(\int_0^u \left(\frac{1}{\lambda_j} - t\right) \lambda_j \mathrm{e}^{-\lambda_j t}\,\mathrm{d}t\right) \\
&= \frac{p_j}{\lambda_j} E(1 - \mathrm{e}^{-\lambda_j u}) - p_j \lambda_j E\left(\int_0^u t\mathrm{e}^{-\lambda_j t}\,\mathrm{d}t\right) \\
&= \frac{p_j}{\lambda_j} E(1 - \mathrm{e}^{-\lambda_j u}) + p_j E(u\mathrm{e}^{-\lambda_j u}) - \frac{p_j}{\lambda_j} E(1 - \mathrm{e}^{-\lambda_j u}) \\
&= p_j E(u\mathrm{e}^{-\lambda_j u}). \qquad (9.24)
\end{aligned}$$

Here u is any random variable having the same distribution as the u_i. Also, we can write $c_i = I(T_i > u_i)$, where T_i has c.d.f. given by $F(t)$ in (9.19). If we assume that each T_i is *independent* of each u_i, then

$$\begin{aligned}
E\left(\frac{(1 - c_i) p_j t_i \mathrm{e}^{-\lambda_j t_i}}{1 - \sum_{j=1}^{J} p_j (1 - \mathrm{e}^{-\lambda_j t_i})}\right) &= E\left(\frac{p_j (1 - F(u)) u\mathrm{e}^{-\lambda_j u}}{1 - \sum_{j=1}^{J} p_j (1 - \mathrm{e}^{-\lambda_j u})}\right) \\
&= p_j E(u\mathrm{e}^{-\lambda_j u}). \qquad (9.25)
\end{aligned}$$

Since (9.24) and (9.25) are equal, we see that

$$E\left(\frac{\partial \log L_n}{\partial \lambda_j}\right) = 0.$$

Similarly, under the same conditions,

$$\frac{1}{n} E\left(\frac{\partial \log L_n}{\partial p_j}\right) = \sum_{i=1}^{n} \left\{ E\left(\frac{I(t^*_{ij} \leq u_i)}{p_j}\right) - E\left(\frac{(1 - F(u))(1 - \mathrm{e}^{-\lambda_j u})}{1 - \sum_{j=1}^{J} p_j (1 - \mathrm{e}^{-\lambda_j u})}\right) \right\}$$

$$= E\left(\frac{p_j F_j(u)}{p_j}\right) - E(F_j(u)) = 0.$$

In a similar way we can calculate the expected value of $-\partial^2 \log L_n/\partial\boldsymbol{\theta}^2$ under the true distribution, obtaining an expected information matrix D_n. If the information 'grows' in the sense that $\lambda_{\min}(D_n) \to \infty$ as $n \to \infty$, we conjecture that the estimator $\tilde{\boldsymbol{\theta}}_n$ obtained by solving (9.23) will be consistent for the true value of $\boldsymbol{\theta}$ and asymptotically normally distributed around it.

Note that as sufficient conditions for the above calculations, we assumed the independence of the t_{ij}^* and the u_i, for each $j \in \{1, 2, \ldots, J\}$ and of the T_i and the u_i. Much discussion can take place on whether these and the other assumptions are realistic, and how the results are to be interpreted; for some such discussion refer to Kalbfleisch and Prentice (1980 Ch. 7), Prentice *et al.* (1978) and Gail (1975). It is not our task here to add to this discussion other than perhaps to point out that we suggested that consistency of the Larson–Dinse estimates will obtain under certain *sufficient* independence assumptions which we have nowhere proved to be *necessary*. The same is true for all of our analyses of the mixture models in previous chapters.

As another point, notice that the calculations of the unbiasedness of the Larson–Dinse estimating equations did not in fact require the restriction that $\sum_{j=1}^{J} p_j = 1$, i.e. that all individuals fail eventually. It is easy to extend the Larson–Dinse mixture model by adding in a component allowing for an immune proportion, who are not subject to failure at all. This of course provides a generalisation of our mixture model for immunes to the competing risks situation. Since the details appear straightforward we do not develop them further.

Supposing a parametric model such as the exponential is assumed for the cause-specific (sub)distribution function (for failure from cause j, say) and fitted to data, how do we assess goodness of fit, and generally, how can we obtain some visual assessment of the data? Kalbfleisch and Prentice (1980), and before them Hoel (1972), suggested calculating a cause-specific Kaplan–Meier Estimator for cause j by treating deaths from other causes as censored observations (censored at their death times), as well as including actual censored observations. Kalbfleisch and Prentice (1980 Ch. 7) show that this is optimal (maximal likelihood in a certain sense) in the same way that the ordinary KME is optimal for estimating the underlying failure distribution in ordinary (single-cause) censored survival data.

The David and Moeschberger formulation

The approaches of Larson–Dinse and Kalbfleisch–Prentice outlined above avoid the assumption that the competing causes of death act *independently*, and as we noted above, such an assumption is controversial anyway, perhaps seldom applicable except as an approximation. Think of the automobile engine example, when the approach of some component near to its breakdown point is likely to increase the stress on other components.

However, if we are prepared to assume independence of risks, we obtain a number of useful simplifications. In outlining this approach we follow the development of Moeschberger and David (1971) and David and Moeschberger (1978, Chs. 1 and 2). Let T_j be a random variable representing the lifetime of individuals exposed only to risk of death from cause j, $1 \le j \le J$. Also let U be a censoring random variable. The independent risks model postulates that we observe

$$\begin{cases} (T_j, j), & \text{if } \min(T_1, \ldots, T_J) \le U \text{ and } j \in \{k : T_k = \min(T_1, \ldots, T_J)\} \\ U, & \text{if } \min(T_1, \ldots, T_J) > U. \end{cases} \tag{9.26}$$

The data for individual i, consisting of survival time t_i and cause indicator $j(i)$ if t_i is uncensored, and censored survival time t_i, otherwise, are then thought of as being independent realisations of the above random variables.

Now define the *cause-specific* survival c.d.f.'s

$$F_j(t) = P(T_j \le t), \qquad 1 \le j \le J$$

and assume they are *proper*, so that even in the absence of all other risks, an individual will die of cause j eventually. Suppose that the $F_j(t)$ are absolutely continuous with densities $f_j(t)$ (as usual, the discrete case is easily handled too). At this stage assume that *the T_j are independent of each other and of the censoring random variable U*. Then we can calculate

$$P(T_j \le t, \text{ death is from cause } j) = \int_0^t \prod_{k \ne j} (1 - F_k(y))\, \mathrm{d}F_j(y) \tag{9.27}$$

and so

$$\frac{\mathrm{d}}{\mathrm{d}t} P(T_j \le t, \text{ death is from cause } j) = f_j(t) \prod_{k \ne j} (1 - F_k(t)). \tag{9.28}$$

Define an overall survival time random variable T by

$$T = \min(T_1, T_2, \ldots, T_J)$$

so that

$$P(T > t) = \prod_{k=1}^{J} (1 - F_k(t)). \tag{9.29}$$

According to our usual procedure, we take as the likelihood contribution from individual i the quantity

$$\left(f_{j(i)}(t_i) \prod_{k \ne j(i)} (1 - F_k(t_i)) \right)^{c_i} \left(\prod_{k=1}^{J} (1 - F_k(t_i)) \right)^{1-c_i} \tag{9.30}$$

$$= \left(\frac{f_{j(i)}(t_i)}{1 - F_{j(i)}(t_i)} \right)^{c_i} \prod_{k=1}^{J} (1 - F_k(t_i)). \tag{9.31}$$

Thus the likelihood based on n individuals is (apart from a constant factor)

$$L = \prod_{i=1}^{n} \left(\frac{f_{j(i)}(t_i)}{1 - F_{j(i)}(t_i)} \right)^{c_i} \prod_{k=1}^{J} (1 - F_k(t_i)). \tag{9.32}$$

Notice that this is quite different from the corresponding quantity for the Larson–Dinse mixture model (9.20). Nevertheless, we will show that it also leads to consistent estimation of the parameters involved, under reasonable conditions.

To see this, suppose for simplicity that $J = 2$. Go back to (9.30) to write the likelihood as

$$L = \prod_{i=1}^{n} \prod_{j=1}^{2} \left(f_j(t_{ij}) \prod_{k \neq j} (1 - F_k(t_{ij})) \right)^{c_{ij}} \prod_{k=1}^{2} (1 - F_k(t_i))^{1-c_i} \tag{9.33}$$

where $t_{ij} = t_i$ when $j(i) = j$, and c_{ij} is defined as in (9.15). Collecting together terms involving the same subscripts for f and F gives

$$L = \prod_{i=1}^{n} (f_1(t_{i1}))^{c_{i1}} (1 - F_1(t_{i2}))^{c_{i2}} (1 - F_1(t_i))^{1-c_i}$$

$$\times (f_2(t_{i2}))^{c_{i2}} (1 - F_2(t_{i1}))^{c_{i1}} (1 - F_2(t_i))^{1-c_i}.$$

Thus the likelihood factors into two components, the first of which,

$$L_1 = \prod_{i=1}^{n} (f_1(t_{i1}))^{c_{i1}} (1 - F_1(t_{i2}))^{c_{i2}} (1 - F_1(t_i))^{1-c_i}, \tag{9.34}$$

is precisely the likelihood (apart from a constant factor) that would be obtained by regarding individuals dying from cause 2 to be censored at their observed death times. Similarly for L_2. It follows that estimation of the parameters of $F_1(t)$ can be done completely separately from those of $F_2(t)$, provided there are no parameters in common. Suppose further that each F_j is exponential, so that

$$1 - F_j(t) = e^{-\lambda_j t}, \qquad t \geq 0,$$

with $\lambda_1 \neq \lambda_2$. Then we can apply the results of Section 7.2 for the ordinary exponential model to deduce that consistent estimation of each λ_j will take place provided the censoring distribution (the distribution of U) does not degenerate at 0.

So we see that the competing risk formulation of David and Moeschberger can be analysed by the methods we have developed for ordinary survival models, when the risks are independent. There is also a natural connection with models with immunes in this setup, if we suppose that some individuals are immune to death from cause j. Such a model was put forward by Hoel (1972); see also Sampford (1954) and David and Moeschberger (1978, Section 2.5). We expect that these models could be analysed by the methods we have given in previous chapters for the single-cause mixture model with immunes. As we mentioned at the beginning of this section, there are many important applications of such models.

We conclude this section by mentioning that David and Moeschberger (1978, Ch. 4) generalise their method to cases when the risks are not independent, but with a corresponding increase in complexity. Finally, they (and most of the other authors we have mentioned), also consider the use of competing risks models with covariates. Such applications suggest a fertile area for further analysis of the kind we gave in Chapter 8.

9.3 ESTIMATING THE PROBABILITY OF BEING IMMUNE

Suppose an individual has been followed up for a time $t > 0$ and his/her death, relapse, recidivism or failure has not been observed. What is the probability that the individual is immune to or cured of the disease? This is obviously important for prediction or assessment of the *individual's* status as to his/her risk of ultimate death from the disease under study. We can calculate this probability under the censoring model of Section 2.1, and estimate it either nonparametrically by the methods of Chapter 4, or parametrically by the methods of Chapter 5.

Under the model of Section 2.1, susceptibles are associated with the value 1 of an unobservable Bernoulli random variable B and, for them,

$$P\{t^* \le t|B = 1\} = F_0(t). \tag{9.35}$$

Here t^* is the survival time and $F_0(t)$ is the c.d.f. of survival time for susceptibles. For immunes, we have $B = 0$ and, for each $t > 0$,

$$P\{t^* \le t|B = 0\} = 0. \tag{9.36}$$

Given that an individual has survived for a time $t > 0$, denote the probability that he/she is immune by

$$p(t) = P\{B = 0|t^* > t\}. \tag{9.37}$$

Since $t^* \le \tau_G$ a.s., where τ_G is the right extreme of the censoring distribution G, it only makes sense to calculate (9.37) for values of $t \le \tau_G$. So take $t \le \tau_G$. Then we have by Bayes' theorem

$$p(t) = \frac{P\{B = 0, t^* > t\}}{P\{t^* > t\}} = \frac{P\{t^* > t|B = 0\}P\{B = 0\}}{P\{t^* > t\}}.$$

From (9.36) we have $P\{t^* > t|B = 0\} = 1$, and our model hypothesis is $P\{B = 0\} = 1 - p$. Consequently

$$\begin{aligned} P\{t^* > t\} &= P\{t^* > t|B = 0\}P\{B = 0\} + P\{t^* > t|B = 1\}P\{B = 1\} \\ &= 1 - p + p(1 - F_0(t)) = 1 - pF_0(t), \end{aligned} \tag{9.38}$$

and so we deduce that

$$p(t) = \frac{1-p}{1-pF_0(t)}. \tag{9.39}$$

The function on the right-hand side of (9.39) is increasing in t, from a value of $1-p$ at $t=0$ (corresponding to no information regarding the immunity of our individual, other than the overall probability of being immune), to a value of 1 when $t=\infty$ (corresponding to certainty of immunity if the individual's lifetime is very large). Figure 9.1 illustrates the function for the case when $F_0(t) = 1-e^{-\lambda_0 t}$, $\lambda_0 > 0$.

Let us consider nonparametric estimation of $p(t)$ based on an independent sample of n individuals, from which we have constructed the KME $\hat{F}_n(t)$. As we know from Chapter 3, $\hat{F}_n(t)$ consistently estimates $F(t)$, the distribution of t^*, under some circumstances. Thus it is natural to propose as an estimator of $p(t)$ the statistic

$$\hat{p}_n(t) = \frac{1-\hat{p}_n}{1-\hat{F}_n(t)} \tag{9.40}$$

where $\hat{p}_n = \hat{F}_n(t_{(n)})$, the maximum value of the KME. We only define (9.40) if $\hat{p}_n < 1$, so there is some evidence of immunes in the population.

The analyses of Chapter 4 lead us to conjecture that $\hat{p}_n(t)$ is a consistent estimator of $p(t)$ if follow-up in the sample is sufficient, and we now prove that to be the case. Suppose then that $\tau_{F_0} \le \tau_G$ and assume that F is continuous at $\tau_H = \tau_{F_0}$ in case $\tau_H < \infty$. Then by Theorem 4.1, we know that $\hat{p}_n \xrightarrow{P} p$. Also by Theorem 3.8, $\hat{F}_n(t) \to F(t) = pF_0(t)$ for each $t \ge 0$. It follows that,

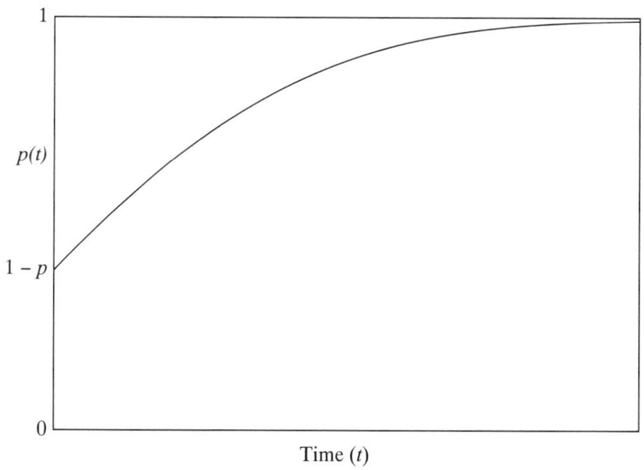

Figure 9.1 Schematic diagram of the probability of being immune as a function of elapsed survival time.

as $n \to \infty$,

$$\hat{p}_n(t) \xrightarrow{P} \frac{1-p}{1-pF_0(t)} = p(t),$$

showing the consistency of $\hat{p}_n(t)$ for $p(t)$. We expect, but have not proven, that $\hat{p}_n(t)$ is asymptotically normally distributed around $p(t)$ in the interior case—i.e. when there are actually immunes in the population ($p < 1$)—but we currently have no conjecture for the boundary case. (See also the discussion in Section 4.1 concerning the case $p = 1$ in Theorem 4.3).

The function in (9.40) gives the estimated probability that an individual selected independently from the same population, but independent of the sample from which (9.40) is calculated, is immune. But we can also apply (9.40) in practice to the individuals in the sample from which (9.40) was calculated, if the sample size is large. To do so, we plot the function described by (9.40) on a graph such as Figure 9.1, and from this we can read off a selected individual's probability of being immune, given his/her currently elapsed survival time. We do this only for individuals that are censored; the others have zero probability of being immune!

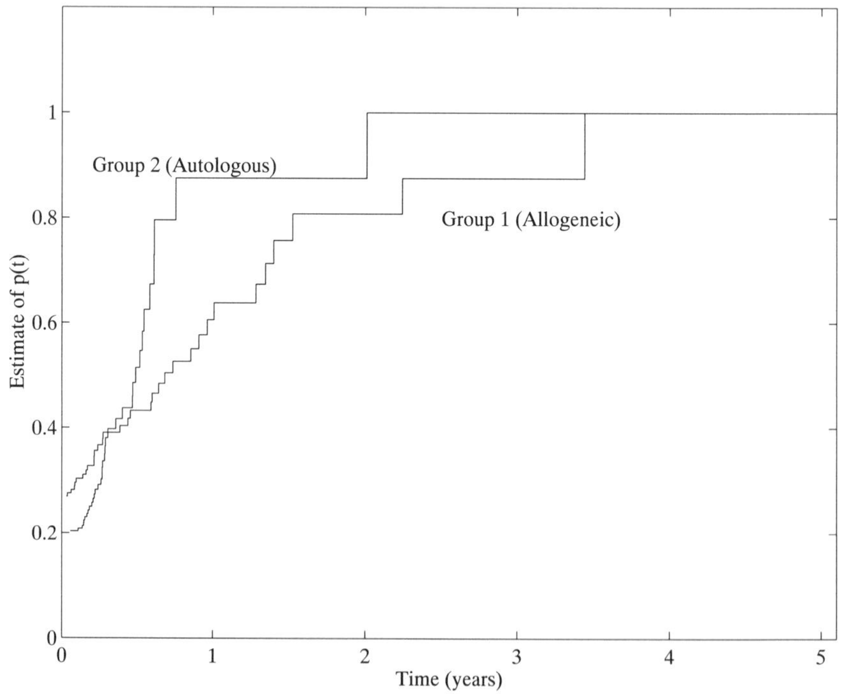

Figure 9.2 Estimate of the probability of being immune for leukaemia data (Kersey *et al.* 1987).

Figure 9.2 shows the estimates of $\hat{p}_n(t)$ for the autologous and allogeneic groups in the leukaemia data of Table 4.1 (Kersey *et al.* 1987). For a person in the population from where the data was selected, who has survived around 2 years from the beginning of the treatments, we estimate the probability of indefinite survival (as far as the disease is concerned) to be around 0.8. Precision of estimation of this figure is low, due to the small sample sizes involved.

9.4 ESTIMATING THE CENSORING DISTRIBUTION

As we mentioned earlier, the censoring distribution can also be estimated from the data, although we do not often do so; it is usually viewed as a nuisance while the main interest is on the survival distribution. But, as we have seen in Chapters 4 and 5, the censoring distribution can have a significant impact on the distributions of the various estimates we are interested in, so sometimes we may wish to have some idea of the kind of censoring distribution we are dealing with. Here we give a brief discussion on how to estimate the censoring distribution from a sample of censored survival data.

A natural approach to estimating the censoring distribution function $G(t)$ of the u_i is to reverse the roles between censored and uncensored observations. Thus, using the notation introduced in Section 2.1, we view the u_i's as the lifetimes and the t_i^*'s become the censoring random variables. We then apply the KME to estimate the distribution of u_i's. We should note a slight asymmetry in the procedures, in that when $t_i^* = u_i$, we defined observation i as being uncensored in Section 2.1. This means that, unless there are no ties between t_i^* and u_i, the indicator random variable $I_{\{u_i \leq t_i^*\}}$ is not exactly the counterpart of the censoring indicator $c_i = I_{\{t_i^* \leq u_i\}}$ as we have used it so far. So instead we take $c_i' = 1 - c_i$ as the appropriate indicator and define our *first* estimator of $G(t)$ by

$$\hat{G}_n(t) = 1 - \prod_{i:t_{(i)} \leq t} \left(1 - \frac{c'_{(i)}}{n - i + 1}\right), \tag{9.41}$$

in parallel with the KME $\hat{F}_n(t)$ defined in (1.7). We will call the $\hat{G}_n(t)$ in (9.41) the *KME of the censoring distribution.*

Another obvious approach to estimating $G(t)$ can be based on the KME $\hat{F}_n(t)$ of $F(t)$, the distribution function of t_i^*. Our next estimator of the censoring distribution exploits an analytical relationship between $G(t)$ and $F(t)$, namely that

$$H(t) = 1 - [1 - F(t)][1 - G(t)]$$

is the distribution function of the observed, censored survival times $t_i = \min\{t_i^*, u_i\}$. Since the proper c.d.f. $H(t)$ can be consistently estimated by the empirical distribution function $\hat{H}_n(t)$ of t_i, our *second* estimator of $G(t)$ will be defined by the sample function

$$1 - \frac{1 - \hat{H}_n(t)}{1 - \hat{F}_n(t)}, \tag{9.42}$$

with the convention that $[1 - \hat{H}_n(t)]/[1 - \hat{F}_n(t)] = 0$ if $\hat{F}_n(t) = 1$.

Which of the two estimators should we use? It turns out that they are equal, in general, as we will show directly. Note that when the t_i^*'s and the u_i's are continuous, so that the probability of $t_i^* = u_i$ occurring is zero, we have $c_i' = I_{\{u_i \le t_i^*\}}$ almost surely, so the theory concerning the KME discussed in Section 3.1 (see Theorem 3.3) can be applied to show that the $\hat{G}_n(t)$ defined in (9.41) is a consistent estimator of $G(t)$. But in the general case, when ties between t_i^* and u_i are possible, we can no longer apply our theory directly. On the other hand, as $\hat{F}_n(t)$ is a (uniformly) consistent estimator of $F(t)$ by Theorem 3.3, we would expect that the estimator in (9.42) should be consistent for $G(t)$ in all cases. We will show in the next theorem that this is indeed so, and moreover, that the estimators in (9.41) and (9.42) are actually identical.

Theorem 9.1 *Assume the i.i.d. censoring model. Let $F(t)$ and $G(t)$ be the distribution functions of t_i^* and u_i respectively, and $H(t) = 1 - [1 - F(t)][1 - G(t)]$ denote the distribution function of $t_i = \min\{t_i^*, u_i\}$. Then*

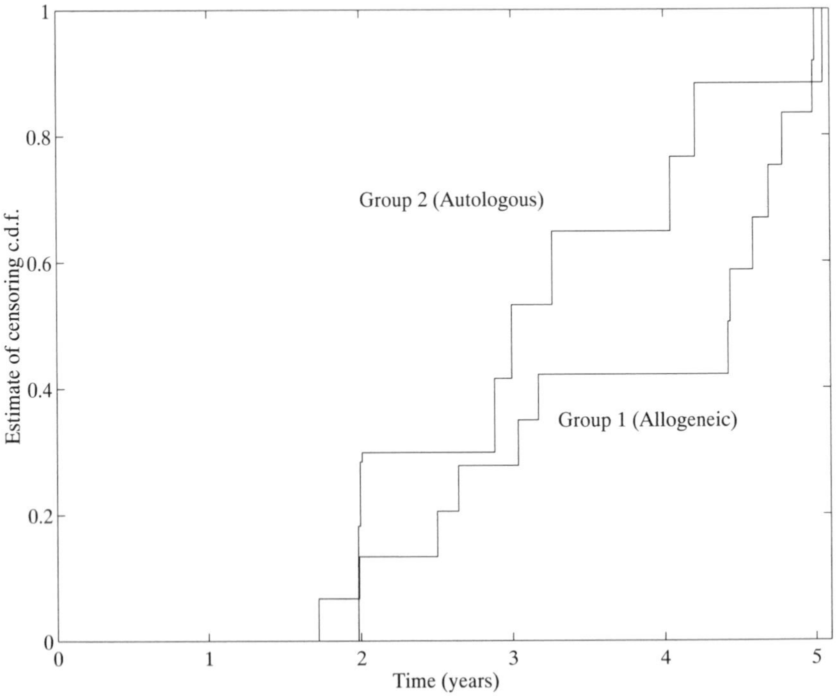

Figure 9.3 KME of censoring distribution for leukaemia data.

(a) *The estimator $\hat{G}_n(t)$ defined in (9.41) satisfies*

$$[1-\hat{F}_n(t)][1-\hat{G}_n(t)] = 1-\hat{H}_n(t). \tag{9.43}$$

(b) *If the susceptible proportion $p < 1$, then*

$$\sup_{0\le t\le t_{(n)}} |\hat{G}_n(t)-G(t)| \xrightarrow{P} 0 \qquad \text{as } n\to\infty, \tag{9.44}$$

(c) *but if $p = 1$, then*

$$\sup_{0\le t\le \tau} |\hat{G}_n(t)-G(t)| \xrightarrow{P} 0 \qquad \text{as } n\to\infty \text{ for any } \tau<\tau_H. \tag{9.45}$$

Proof Suppose $t_{(k)} \le t < t_{(k+1)}$. Then by (9.41),

$$1-\hat{G}_n(t) = \prod_{i:t_{(i)}\le t}\left(\frac{n-i+1-c'_{(i)}}{n-i+1}\right) = \prod_{i=1}^{k}\left(\frac{n-i+1-(1-c_{(i)})}{n-i+1}\right)$$

$$= \prod_{i=1}^{k}\left(\frac{n-i+c_{(i)}}{n-i+1}\right). \tag{9.46}$$

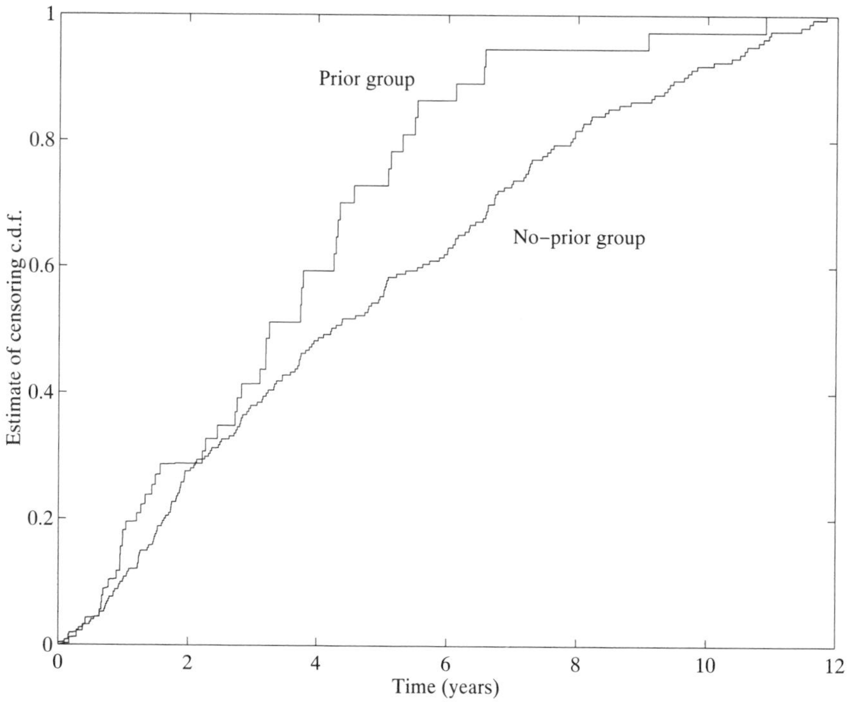

Figure 9.4 KME of censoring distribution for recidivism data of Section 4.3.

Note that

$$(n-i+1-c_{(i)})(n-i+c_{(i)}) = (n-i+1)(n-i).$$

Hence by (9.46) and (1.7),

$$\begin{aligned}
[1-\hat{F}_n(t)][1-\hat{G}_n(t)] &= \prod_{i=1}^{k} \frac{n-i+1-c_{(i)}}{n-i+1} \prod_{i=1}^{k} \frac{n-i+c_{(i)}}{n-i+1} = \prod_{i=1}^{k} \frac{n-i}{n-i+1} \\
&= \prod_{i=1}^{k}(n-i)\prod_{i=1}^{k}\frac{1}{n-i+1} = \prod_{i=1}^{k}(n-i)\prod_{i=0}^{k-1}\frac{1}{n-i} \\
&= (n-k)\frac{1}{n} = 1-\frac{k}{n} = 1-\hat{H}_n(t).
\end{aligned}$$

This proves (a); (b) and (c) follow from (a) together with the results that $\hat{F}_n(t)$ and $\hat{H}_n(t)$ are uniformly consistent for $F(t)$ and $H(t)$ respectively over $[0, t_{(n)}]$, and the fact that $1-F(t)$ is bounded away from zero either when $p<1$ or over $[0,\tau]$ for $\tau < \tau_H$. ■

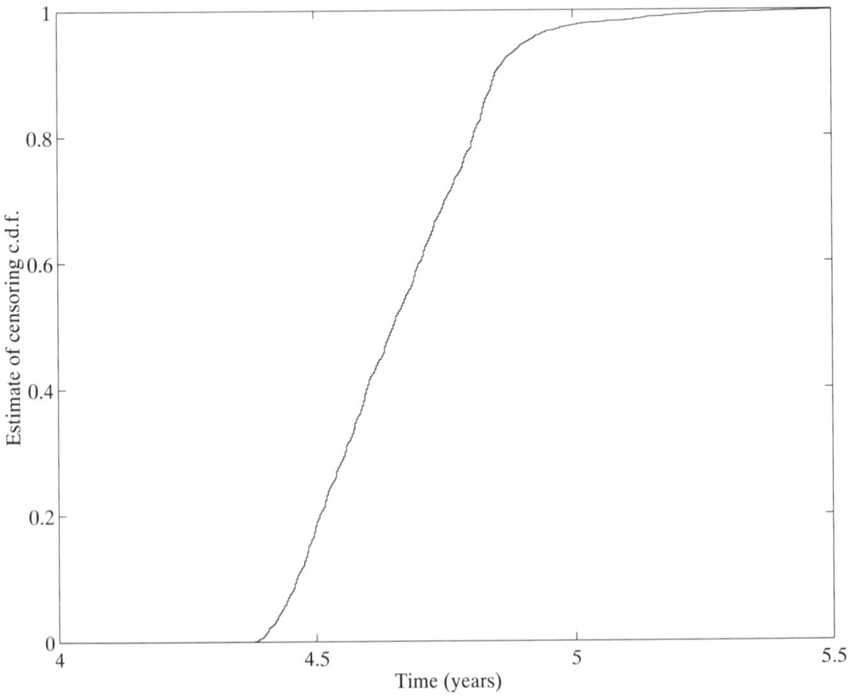

Figure 9.5 KME of censoring distribution for recidivism arrest data of Section 5.1.

Figures 9.3 and 9.4 show the KMEs of the censoring distributions for the leukaemia data introduced in Figure 4.3 and the recidivism data in Figure 4.6.

Figure 9.5 shows the censoring distribution estimated from a subsample of 5000 male non-Aborigines from the large arrest data set described in Section 5.1. It was taken from near the middle of the data. Figure 9.5 shows clearly that there is follow-up of at least 4 years and at most 5.5 years available for most individuals in this cohort, and that the censoring is closely approximated by a uniform distribution within 4.4 and 4.9 years, which is the range containing the main body of $\hat{G}_n(t)$.

A uniform censoring distribution is precisely what we would expect from data such as this: the arrestees are released at random times, probably approximating fairly well to a Poisson process, times to rearrest (for susceptibles) are approximately exponential (as we demonstrated in Figure 5.1) and individuals are followed up to a fixed 'cutoff' censoring time.

We would expect something similar for the censoring distributions shown in Figure 9.4, and in fact they are reasonably close to uniform, except possibly for some departure from the uniform near the right extreme of the distribution for the prior group. Such a departure from uniform is interesting for criminologists to ponder, once the statistician brings it to their attention.

The occurrence of a uniform censoring distribution in the above data sets justifies to some extent our calculating the percentiles of the various statistics we consider in the appended statistical tables based on uniform and/or exponential censoring distributions. Many published papers concerned with simulating values of test statistics in survival data use these distributions too.

References

Aalen, O. O. (1978) Nonparametric inference for a family of counting processes, *Ann. Statist.*, **6**, 701–726.

Aitkin, M. (1996) A general maximum likelihood analysis of overdispersion in generalized linear models, *Statistics and Computing* (to appear).

Aitkin, M., Anderson, D., Francis, B. and Hinde, J. (1989) *Statistical Modelling in GLIM*, Oxford Science Publications, New York.

Andersen, P. K., Borgan, O., Gill, R. D. and Keiding, N. (1993) *Statistical Models Based on Counting Processes*, Springer-Verlag, New York.

Anscombe, F. J. (1961) Estimating a mixed-exponential response law, *J. Amer. Statist. Assoc.*, **56**, 493–502.

Ansell, J. I. and Phillips, M. J. (1994) *Practical Methods for Reliability Data Analysis*, Oxford Science Publications, New York.

Arbutski, T. (1985) A family of multiplicative survival models incorporating a long-term survivorship parameter *C* as a function of covariates, *Comm. Statist.—Theory and Methods*, **14**, 1627–1642.

Bailey, N. T. J. and Thomas, A. S. (1971) The estimation of parameters from population data on the general stochastic epidemic, *Theor. Pop. Biol.*, **2**, 253–270.

Baker, R. J. (1985) GLIM 3.77 Reference guide. In: *Glim 3.77 Reference Manual*, Numerical Algorithms Group, Oxford.

Baker, R. J. and Nelder, J. A. (1978) *The GLIM System, Generalised Linear Interactive Modelling*, Numerical Algorithms Group, Oxford.

Barnett, V. D. and Lewis, T. (1984) *Outliers in Statistical Data*, 2nd Ed, Wiley, Chichester.

Becker, N. G. (1983) Analysis of data from a single epidemic, *Austral. J. Statist.*, **25**, 191–197.

Berkson, J. and Gage, R. P. (1952) Survival curve for cancer patients following treatment, *J. Amer. Statist. Assoc.*, **47**, 501–515.

Bie, O., Borgan, Ø. and Liestøl, K. (1987) Confidence intervals and confidence bands for the cumulative hazard rate function and their small sample properties, *Scand. J. Statist.*, **14**, 221–233.

Bingham, N. H., Goldie, C. M. and Teugels, J. L. (1987) *Regular Variation*, Cambridge University Press, London.

Bithel, J. F. and Upton, R. G. (1977) A mixed model for survival applied to British children with neuroblastoma, In: *Recent Developments in Statistics*, Barra, J. R. *et al.* (Eds), North-Holland, Amsterdam, pp. 635–646.

Bloom, H. S. (1979) Evaluating human service and correctional programs by modelling the timing of recidivism, *Sociological Methods and Research*, **8**, 179–208.

Boag, J. W. (1949) Maximum likelihood estimates of the proportion of patients cured by cancer therapy, *J. Royal Statist. Soc.*, **11B**, 15–44.

Borgan, Ø. and Liestøl, K. (1990) A note on confidence intervals and bands for the survival function based on transformations, *Scand. J. Statist.*, **17**, 35–41.

Breslow, N. E. (1992) Introduction to: Kaplan and Meier (1958) 'Nonparametric Estimation from Incomplete Observations', In: *Breakthroughs in Statistics, Vol. II*, Kotz, S. and Johnson, N. L. (Eds), Springer-Verlag, New York.

Breslow, N. E. and Crowley, J. J. (1974) A large sample study of the life table and product limit estimates under random censorship, *Ann. Statist.*, **2**, 437–453.

Brinkley, D. and Haybittle, J. L. (1975) The curability of breast cancer, *Lancet* **2**, 95–97.

Broadhurst, R. G. and Loh, N. S. (1995) Rearrest probabilities for the 1984–1993 Western Australian population: a survival analysis, *J. Quant. Crim.*, **11**, 289–313.

Broadhurst, R. G. and Maller, R. A. (1990) The recidivism of prisoners released for the first time: reconsidering the effectiveness question, *Aust. N.Z. J. Crim.* **23** 88–103.

Broadhurst, R. G. and Maller, R. A. (1991) Estimating the numbers of terms in criminal careers from one-step probabilities of recidivism, *J. Quant. Crim.*, **7**, 275–290.

Broadhurst, R. G. and Maller, R. A. (1992) An analysis of the recidivism of sex offenders in Western Australian prisons, *Brit. J. Crim.*, **32**, 54–80.

Broadhurst, R. G., Maller, R. A., Maller, M. G. and Duffecy, J. (1988) Aboriginal and non-Aboriginal recidivism in Western Australia: a failure rate analysis, *J. Research in Crime and Delinquency*, **25**, 83–108.

Cantor, A. B. and Shuster, J. J. (1992) Parametric versus non-parametric methods for estimating cure rates based on censored survival data, *Statistics in Medicine*, **11**, 931–937.

Cantor, A. B., and Shuster, J. J. (1994) Reply to A. Yu. Yakovlev, *Statistics in Medicine*, **13**, 986.

Chen, W. C., Hill, B. M., Greenhouse, J. B. and Fayos, J. V. (1985) Bayesian analysis of survival curves for cancer patients following treatments, In: *Bayesian Statistics*, Elsevier, Amsterdam.

Chernoff, H. (1954) On the distribution of the likelihood ratio, *Ann. Math. Statist.*, **25**, 573–578.

Chow, Y. S. and Teicher, H. (1988) *Probability Theory: Independence, Interchangeability, Martingales*, 2nd Ed., Springer-Verlag, New York.

Collett, D. (1994) *Modelling Survival Data in Medical Research*, Chapman and Hall, London.

Cox, D. R. (1972) Regression models and life tables (with discussion), *J. Royal Statist. Soc., Series B*, **34**, 187–220.

Cox, D. R. and Hinkley, D. V. (1974) *Theoretical Statistics*, Chapman and Hall, London.

Cox, D. R. and Oakes, D. (1984) *Analysis of Survival Data*, Chapman and Hall, London.

Crowder, M. J., Kimber, A. C., Smith, R. L. and Sweeting, T. J. (1991) *Statistical Analysis of Reliability Data*, Chapman and Hall, London.

Crowley, J. J and Hu, M. (1977) Covariance analysis of heart transplant survival data, *J. Amer. Statist. Assoc.*, **72**, 27–36.

David, H. A. and Moeschberger, M. L. (1978) *The Theory of Competing Risks*, Griffin, London.

Dinse, G. E. (1985) An alternative to Efron's redistribution-of-mass construction of the Kaplan–Meier Estimator, *Amer. Statistician*, **39**, 299–300.

Drygas, H. (1971) Consistency of the least squares and Gauss–Markov estimators in regression models, *Z. Wahrscheinlichkeitstheorie und verw. Gebiete*, **17**, 309–326.

Drygas, H. (1976) Weak and strong consistency of the least squares estimators in regression models, *Z. Wahrscheinlichkeitstheorie und verw. Gebiete*, **34**, 119–127.

Dunsmuir, W., Tweedie, R., Flack, L. and Mengersen, K. (1989) Modelling of transitions between employment states for young Australians, *Austral. J. Statist.*, **31A**, 165–196.

Edmunson, J. H., Fleming, T. R., Decker, D. G., Malkasian, G .D., Jorgenson, E. O., Jeffries, J. A., Webb, M. J., and Kvols, L. K. (1979) Different chemotherapeutic sensitivities and host factors affecting prognosis in advanced ovarian carcinoma versus minimal residual disease. *Cancer Treatment Reports*, **63**, 241–247.

Efron, B. (1967) The two-sample problem with censored data, In: *Proceedings of the Fifth Berkeley Symposium*, University of California Press, Berkeley, CA, Vol. 4, pp. 831–853.

Eicker, F. (1963) Asymptotic normality and consistency of the least squares estimators for families of linear regressions, *Ann. Math. Statist.*, **34**, 447–456.
Elandt-Johnson, R. C. and Johnson, N. L. (1980) *Survival Models and Data Analysis*, Wiley, New York.
Fahrmeir, L. and Kaufmann, H. (1985) Consistency and asymptotic normality of the maximum likelihood estimator in generalized linear models, *Ann. Statist.*, **13**, 342–368.
Fallen, D. L. (1996) A comparison of four parametric models of estimating cure rates with censored data, Preprint.
Farewell, V. T. (1977a) A model for a binary variable with time censored observations, *Biometrika* **64** 43–46.
Farewell, V. T. (1977b) The combined effect of breast cancer risk factors, *Cancer* **40** 931–936.
Farewell, V. T. (1982) The use of mixture models for the analysis of survival data with long-term survivors, *Biometrics* **38**, 1041–1046.
Farewell, V. T. (1986) Mixture models in survival analysis: are they worth the risk? *Canad. J. Statist.* **14**, 257–262.
Feder, P. I. (1968) On the distribution of the log likelihood ratio test statistic when the true parameter is 'near' the boundaries of the hypothesis regions, *Ann. Math. Statist.*, **39**, 2044–2055.
Filliben, J. J. (1975) The probability plot correlation coefficient test for normality, *Technometrics* **17**, 111–117.
Fleming, T. R. and Harrington, D. P. (1991) *Counting Processes and Survival Analysis*, Wiley, New York.
Freireich, E. J., Gehan, E., Frei, E., Schroeder, L. R., Wolman, I. J., Anbari, R., Burgert, E. O., Mills, S. D., Pinkel, D., Selawry, O. S., Moon, J. H., Gendel, B. R., Spurr, C. L., Storrs, R., Haurani, F., Hoogstraten, B., and Lee, S. (1963) The effect of 6-Mercaptopurine on the duration of steroid-induced remissions in acute leukaemia; a model for evaluation of other potentially useful therapy, *Blood*, **21**, 699–716.
Gail, M. (1975) A review and critique of some models used in competing risk analysis, *Biometrics*, **31**, 209–222.
Gamel, J. W., McLean, I. W., and Rosenberg, S. H. (1990) Proportion cured and mean log survival time as functions of tumour size, *Statistics in Medicine*, **9**, 999–1006.
Gehan, E. A. (1965) A generalized Wilcoxon test for comparing arbitrarily singly censored samples, *Biometrika*, **52**, 203–223.
Ghitany, M. E. (1993) On the information matrix of exponential mixture models with long-term survivors, *Biom. J.*, **35**, 15–27.
Ghitany, M. E. and Maller, R. A. (1992) Asymptotic results for exponential mixture models with long-term survivors, *Statistics*, **23**, 321–336.
Ghitany, M. E., Maller, R. A. and Zhou, S. (1994) Exponential mixture models with long-term survivors and covariates, *J. Mult. Analysis*, **49**, 218–241.
Ghitany, M. E., Maller, R. A. and Zhou, S. (1995) Estimating the proportion of immunes in censored samples: a simulation study, *Statistics in Medicine*, **14**, 39–49.
Gill, R. D. (1980) *Censoring and Stochastic Integrals*, Math Centre Tracts, **124** (Math Centrum, Amsterdam).
Gill, R. D. (1983) Large sample behaviour of the product limit estimator on the whole line, *Ann. Statist.*, **11**, 49–58.
Gill, R. D. (1994) Lectures on Survival Analysis. In: *Lectures on Probability Theory*, D. Bakry, R. D. Gill, S. A. Molchanov. Eds. Springer-Verlag, No.1581.
Goldman, A. I. (1984) Survivorship analysis when cure is a possibility: a Monte Carlo study, *Statistics in Medicine*, **3**, 153–163.
Goldman, A. I. (1991) The cure model and time confounded risk in the analysis of survival and other timed events, *J. Clin. Epidemiology*, **44**, 1327–1340.

Gordon, N. H. (1990) Application of the theory of finite mixtures for the estimation of 'cure' rates of treated cancer patients, *Statistics in Medicine*, **9**, 397–407.

Gray, R. J., and Tsiatis, A. A. (1989) A linear rank test for use when the main interest is in differences in cure rates, *Biometrics*, **45**, 899–904.

Greenhouse, J. B. and Wolfe, R. A. (1984) A competing risks derivation of a mixture model for the analysis of survival data, *Comm. Statist.—Theor. Meth.*, **13**, 3133–3154.

Greenwood, M. (1926) A report on the natural duration of cancer, *Reports on Public Health and Medical Subjects*, **33**, 1–26.

Grimmett, G. R. and Stirzaker, D. R. (1992) *Probability and Random Processes*, 2nd Ed., Clarendon Press, Oxford.

Guilbaud, O. (1988) Exact Kolmogorov-type tests for left-truncated and/or right-censored data, *J. Amer. Statist. Assoc.*, **83**, 213–221.

Gulland, J. A. (1955) On the estimation of population parameters from marked members, *Biometrika*, **42**, 269–270.

Hall, W. J. and Wellner, J. A. (1980) Confidence bands for a survival curve from censored data, *Biometrika*, **67**, 133–143.

Halpern, J. and Brown, B. W. Jr. (1987) Cure rate models: power of the log-rank and generalized Wilcoxon tests, *Statistics in Medicine*, **6**, 483–489.

Harris, E. K. and Albert, A. (1991) *Survivorship Analysis for Clinical Studies*, Marcel Dekker, New York.

Haybittle, J. L. (1959) The estimation of the proportion of patients cured after treatment for cancer of the breast, *Brit. J. Radiology*, **32**, 725–733.

Haybittle, L. (1965) A two-parameter model for the survival curve of treated cancer patients, *J. Amer. Statist. Assoc.*, **60**, 16–26.

Hoel, D. G. (1972) A representation of mortality data by competing risks, *Biometrics*, **28**, 475–488.

Kalbfleisch, J. D. and Prentice, R. L. (1980) *The Statistical Analysis of Failure Time Data*, Wiley, New York.

Kaplan, E. L. and Meier, P. (1958) Nonparametric estimation from incomplete observations, *J. Amer. Statist. Assoc.*, **53**, 457–481.

Kersey, J. H., Weisdorf, D., Nesbit, M. E., LeBien, T. W., Woods, W. G., McGlave, P. B., Kim, T., Vallera, D. A., Goldman, A. I., Bostrom, B., Hurd, D. and Ramsay, N. K. C. (1987) Comparison of autologous and allogeneic bone marrow transplantation for treatment of high-risk refractory acute lymphoblastic leukaemia, *New England Journal of Medicine*, **317**, 461–467.

Kimber, H. and Crowder, M. (1984) An analysis of resistance times to infection under treatment, *Statistics in Medicine*, **3**, 165–171.

Kuk, A. Y. C. and Chen, C. H. (1992) A mixture model combining logistic regression with proportional hazards regression, *Biometrika*, **79**, 531–541.

Lai T. L., Robbins, H. and Wei, C. Z. (1979) Strong consistency of least squares estimates in multiple regression II, *J. Mult. Analysis*, **9**, 343–361.

Lagakos, S. W. (1979) General right censoring and its impact on the analysis of survival data, *Biometrics*, **35**, 139–156.

Lagakos, S. W., Sommer, C. J. and Zelen, M. (1978) Semi-Markov models for partially censored data, *Biometrika*, **65**, 311–317.

Langlands, A. O., Pocock, S. J., Kerr, G. R. and Gore S. M. (1979) Long-term survival of patients with breast cancer: a study of the curability of the disease, *Brit. Med. J.*, **2**, 1247–1251.

Larson, M. G. and Dinse, G. E. (1985) A mixture model for the regression analysis of competing risk data, *Appl. Statist.*, **34**, 201–211.

Laska, E. M., and Meisner, M. J. (1992) Nonparametric estimation and testing in a cure model, *Biometrics*, **48**, 1223–1234.

Lawless, J. F. (1982) *Statistical Models and Methods for Lifetime Data*, Wiley, New York.

Leathem, A. J. and Brooks, S. A. (1987) Predictive value of lectin binding on breast cancer recurrence and survival, *The Lancet*, **1**, 1054–1056.
Maller, R. A. (1988) On the exponential model for survival, Biometrika, **75**, 582–586.
Maller, R. A. (1993) Factorial analysis of recidivist data, *Austral. J. Statist.*, **35**, 5–18.
Maller, R. A. and Zhou, S. (1992) Estimating the proportion of immunes in a censored sample, *Biometrika*, **79**, 731–739.
Maller, R. A. and Zhou, S. (1993) The probability that the largest observation is censored, *J. Appl. Prob.*, **30**, 602–615.
Maller, R. A. and Zhou, S. (1994) Testing for sufficient followup and outliers in survival data *J. Amer. Stat. Assoc.*, **89**, 1499–1506.
Maltz, M. D. (1984) *Recidivism*, Academic Press, New York.
Maltz, M. D. and MacCleary, R. (1977) The mathematics of behavioral change: recidivism and construct validity, *Evaluation Quarterly*, **1** 421–438.
MathWorks (1994) *Matlab, The MathWorks, Inc.*, 24 Prime Park Way, Natick, MA 01760.
Matthews, D. E. (1984) Some observations on semi-Markov models for partially censored data, *Canad. J. Statist.*, **12**, 201–205.
McCullagh, P. and Nelder, J. A. (1989) *Generalised Linear Models.* 2nd Ed., Chapman - Hall, London.
Meeker, W. Q. (1987) Limited failure population life tests: application to integrated circuit reliability, *Technometrics*, **29**, 51–65.
Meeker, W. Q. and LuValle, M. J. (1995) An accelerated life test model based on reliability kinetics, *Technometrics*, **37**, 133–146.
Meier, P. (1975) Estimation of a distribution function from incomplete observations, In: *Perspectives in Probability and Statistics*, Gani, J. (Ed.), Academic Press, New York, pp. 67–87.
Miller, R. G. Jr. (1981) *Survival Analysis*, Wiley, New York.
Milner, P. C., and Watts, M. (1987) Effect of socioeconomic status on survival from cervical cancer in Sheffield, *J. Epidemiology and Community Health*, **41**, 200–203.
Moeschberger, M. C. and David, H. A. (1971) Life tests under competing causes of failure and the theory of competing risks, *Biometrics*, **27**, 909–933.
Moran, P. A. P. (1971) Maximum-likelihood estimation in non-standard conditions, *Proc. Camb. Phil. Soc.*, **70**, 441–450.
Mosteller, F. and Tukey, J. W. (1977) *Data Analysis and Regression: A Second course in Statistics*, Addison-Wesley, Reading, Ma.
Mould, R. F. and Boag, J. W. (1975) A test of several parametric statistical models for estimating success rate in the treatment of carcinoma cervix uteri, *Brit. J. Cancer*, **32**, 529–550.
Nelder, J. A. and Wedderburn, R. W. M. (1972) Generalized linear models, *J. Royal Statist. Soc. Series A*, **135**, 370–384.
Nelson, W. (1969) Hazard plotting for incomplete failure data, *J. Qual. Tech.*, **1**, 27–52.
Nelson, W. (1972) Theory and applications of hazard plotting for censored failure data, *Technometrics*, **14**, 945–965.
Nelson, W. (1982) *Applied Life Data Analysis*, Wiley, New York.
Partanen, J. (1969) On waiting time distributions, *Acta Sociologica*, **12**, 132–143.
Pocock, S. J., Gore, S. M. and Kerr, G. (1982) Long-term survival analysis: the curability of breast cancer, *Statistics in Medicine*, **1**, 93–104.
Prentice, R. L., Kalbfleisch, J. D., Peterson, A. V., Jr., Flournoy, N., Farewell, V. T. and Breslow, N. E. (1978) The analysis of failure times in the presence of competing risks, *Biometrics*, **34**, 541–554.
Regal, R. R. and Larntz, K. (1978) Likelihood methods for testing group problem solving models with censored data, *Psychometrika*, **45**, 353–366.
Reiss, R.-D. (1989) *Approximate Distributions of Order Statistics: With Applications to Nonparametric Statistics.* Springer-Verlag, New York.

Rutqvist, L. E. (1985) On the utility of the lognormal model for analysis of breast cancer survival in Sweden 1961–1973, *Brit. J. Cancer*, **52**, 875–883.

Rutqvist, L. E. and Wallgren, A. (1985) Long-term survival of 458 young breast cancer patients, *Cancer*, **55**, 658–665.

Rutqvist, L. E., Wallgren, A. and Nilsson, B. (1984) Is breast cancer a curable disease? *Cancer*, **53**, 1793–1800.

Sampford, M. R. (1954) The estimation of response-time distributions, III. Truncation and survival, *Biometrics*, **10**, 13–32.

Schmidt, P. and Witte, A. D. (1988) *Predicting Recidivism using Survival Models*, Springer-Verlag, New York.

Self, S. G. and Liang, K. (1987) Asymptotic properties of maximum likelihood estimators and likelihood ratio tests under nonstandard conditions, *J. Amer. Statist. Assoc.*, **82**, 605–610.

Shorack, G. R. and Wellner, J. A. (1986) *Empirical Processes with Applications to Statistics*, Wiley, New York.

Sposto, R., Sather, H. N. and Baker, S. A. (1992) A comparison of tests of the difference in the proportion of patients who are cured, *Biometrics*, **48**, 87–99.

Struthers, C. A. and Farewell, V. T. (1989) A mixture model for time to AIDS data with left truncation and an uncertain origin, *Biometrika*, **76**, 814–817.

Stute, W. (1992a) Strong consistency of the MLE under random censoring, *Metrika*, **39**, 257–267.

Stute, W. (1992b) Strong consistency under the Koziol–Green model, *Statist. Prob. Letters*, **14**, 313–320.

Stute, W. (1993) Consistent estimation under random censorship when covariables are present, *J. Mult. Analysis*, **45**, 89–103.

Stute, W. (1994) The bias of Kaplan–Meier integrals, *Scand. J. Statist.*, **21**, 475–484.

Stute, W. and Wang, J. L. (1993) The strong law under random censorship, *Ann. Statist.*, **21**, 1591–1607.

Stute, W. and Wang, J. L. (1994) The jackknife estimate of a Kaplan–Meier integral, *Biometrika*, **81**, 602–606.

Taylor, J. M. G. (1995) Semi-parametric estimation in failure time mixture distributions, *Biometrics*, **51**, 814–817.

Therneau, T. M. (1986) The COXREGR procedure, In: *SAS SUGI Supplemental Library User's Guide*, Version 5, SAS Institute Inc., Gary, IN.

Thomas, G. M. (1960) *Calculus and Analytic Geometry*, 3rd Ed., Addison-Wesley, Reading, MA.

Tukey, J. W. (1977) *Exploratory Data Analysis*, Addison-Wesley, Reading, MA.

Wang, J. (1987) A note on the uniform consistency of the Kaplan–Meier estimator, *Ann. Statist.*, **15**, 1313–1316.

Vu, H. T. V. (1995) The asymptotic theory of maximum estimators with applications: a unified approach for both interior and boundary cases, PhD Thesis, University of Western Australia.

Vu, H. T. V, Maller, R. A. and Klass, M. J. (1996) On the Studentisation of random vectors, *J. Mult. Analysis*, **57**, 142–155.

Vu, H. T. V., and Zhou, X. (1997) Generalisation of likelihood ratio tests under nonstandard conditions. *Ann. Statist.*, to appear.

Weibull, W. (1939) A statistical theory of the strength of material, *Ingeniors Vetenscaps Akademiens Handligar*, no. 151, Stockholm.

Weibull, W. (1951) A statistical distribution function of wide applicability, *J. Appl. Mech.*, **18**, 293–297.

Wellner, J. A. (1985) A heavy censoring limit theorem for the product limit estimator, *Ann. Statist.*, **13**, 150–162.

Winter, B. B. (1987) Nonparametric estimation with censored data from a distribution with nonincreasing density, *Comm. Statist.—Theor. Meth.*, **16**, 93–120.

Woods, W. G., Ramsay, N. K. C., Weisdorf, D. J., Haake, R., Vallera, D. A., Kim, T., H., Lasky, L., Nesbit, M. E., Bostrom, B., Uckun, F., Goldman, A. I. and Kersey, J. H. (1990) Bone marrow transplantation for acute lymphocytic leukaemia utilizing total body irradiation followed by high doses of cytosine arabinoside: lack of superiority over cyclophosphamide-containing conditioning regimens, *Bone Marrow Transplantation*, **6**, 9–16.

Yakovlev, A. Yu. (1994) Letter to the Editor, *Statistics in Medicine*, **13**, 983–985.

Yamaguchi, K. (1992) Accelerated failure time regression models with a regression model of surviving fraction: An application to the analysis of 'permanent employment', *J. Amer. Statist. Assoc.*, **87**, 284–292.

Zhou, S. and Maller, R. A. (1995) The likelihood ratio test for the presence of immunes in a censored sample, *Statistics*, **27**, 181–201.

Statistical Tables

Table A.1 Percentiles of $\hat{p}_n$ with $\text{Exp}(1/\mu)$ censoring.

%	$\mu = 1$	$\mu = 2$	$\mu = 3$	$\mu = 4$	$\mu = 5$	$\mu = 1$	$\mu = 2$	$\mu = 3$	$\mu = 4$	$\mu = 5$	%
			$n = 20$					$n = 150$			
1	0.4034	0.6607	0.7565	0.8075	0.8329	0.7872	0.9095	0.9476	0.9610	0.9700	1
5	0.5699	0.7734	0.8510	0.8830	0.9052	0.8424	0.9396	0.9664	0.9760	0.9824	5
10	0.6500	0.8273	0.8890	0.9167	0.9297	0.8712	0.9539	0.9753	0.9829	0.9871	10
20	0.7412	0.8871	0.9299	1.0000	1.0000	0.9048	0.9687	0.9845	0.9900	1.0000	20
			$n = 30$					$n = 200$			
1	0.5306	0.7431	0.8118	0.8616	0.8876	0.8160	0.9281	0.9588	0.9696	0.9763	1
5	0.6500	0.8297	0.8871	0.9163	0.9302	0.8652	0.9525	0.9734	0.9813	0.9860	5
10	0.7139	0.8689	0.9174	0.9377	0.9495	0.8891	0.9634	0.9800	0.9865	0.9898	10
20	0.7881	0.9123	0.9464	0.9622	1.0000	0.9180	0.9751	0.9874	0.9923	1.0000	20
			$n = 40$					$n = 250$			
1	0.6005	0.7913	0.8600	0.8926	0.9131	0.8306	0.9365	0.9621	0.9752	0.9804	1
5	0.6998	0.8569	0.9095	0.9361	0.9469	0.8776	0.9571	0.9772	0.9843	0.9882	5
10	0.7557	0.8921	0.9324	0.9532	0.9610	0.9013	0.9674	0.9832	0.9887	0.9915	10
20	0.8212	0.9292	0.9568	1.0000	1.0000	0.9281	0.9786	0.9892	1.0000	1.0000	20
			$n = 50$					$n = 300$			
1	0.6413	0.8123	0.8750	0.9087	0.9266	0.8495	0.9430	0.9680	0.9781	0.9835	1
5	0.7340	0.8746	0.9228	0.9427	0.9556	0.8895	0.9626	0.9795	0.9865	0.9902	5
10	0.7786	0.9042	0.9433	0.9594	0.9671	0.9095	0.9718	0.9852	0.9904	0.9927	10
20	0.8376	0.9370	0.9644	1.0000	1.0000	0.9333	0.9809	0.9905	0.9951	1.0000	20
			$n = 60$					$n = 350$			
1	0.6668	0.8335	0.8970	0.9206	0.9359	0.8591	0.9496	0.9716	0.9805	0.9851	1
5	0.7515	0.8925	0.9325	0.9509	0.9614	0.8958	0.9663	0.9823	0.9882	0.9911	5
10	0.7976	0.9177	0.9503	0.9645	0.9721	0.9159	0.9744	0.9869	0.9916	0.9935	10
20	0.8500	0.9449	0.9684	0.9799	1.0000	0.9384	0.9830	0.9916	1.0000	1.0000	20
			$n = 70$					$n = 400$			
1	0.6890	0.8531	0.9055	0.9315	0.9418	0.8725	0.9531	0.9750	0.9821	0.9870	1
5	0.7736	0.8999	0.9409	0.9570	0.9652	0.9039	0.9697	0.9845	0.9893	0.9923	5
10	0.8147	0.9250	0.9560	0.9695	0.9750	0.9218	0.9768	0.9884	0.9923	0.9943	10
20	0.8637	0.9494	0.9728	1.0000	1.0000	0.9429	0.9844	0.9925	0.9954	1.0000	20
			$n = 80$					$n = 450$			
1	0.7026	0.8665	0.9169	0.9382	0.9485	0.8744	0.9560	0.9757	0.9840	0.9880	1
5	0.7807	0.9102	0.9460	0.9619	0.9704	0.9074	0.9715	0.9848	0.9902	0.9926	5
10	0.8226	0.9323	0.9610	0.9720	0.9780	0.9247	0.9787	0.9891	0.9930	0.9947	10
20	0.8672	0.9541	0.9754	0.9846	1.0000	0.9447	0.9856	0.9931	1.0000	1.0000	20
			$n = 90$					$n = 500$			
1	0.7230	0.8715	0.9195	0.9419	0.9546	0.8817	0.9601	0.9781	0.9852	0.9892	1
5	0.7945	0.9152	0.9516	0.9650	0.9729	0.9137	0.9732	0.9862	0.9910	0.9935	5
10	0.8332	0.9362	0.9651	0.9746	0.9800	0.9294	0.9795	0.9901	0.9935	0.9952	10
20	0.8785	0.9577	0.9773	0.9849	1.0000	0.9479	0.9864	0.9938	1.0000	1.0000	20
			$n = 100$					$n = 1000$			
1	0.7427	0.8856	0.9286	0.9476	0.9575	0.9170	0.9746	0.9866	0.9913	0.9939	1
5	0.8087	0.9227	0.9560	0.9677	0.9753	0.9386	0.9829	0.9917	0.9949	0.9965	5
10	0.8430	0.9405	0.9667	0.9768	0.9821	0.9504	0.9873	0.9940	0.9963	0.9974	10
20	0.8841	0.9604	0.9791	1.0000	1.0000	0.9635	0.9914	0.9962	1.0000	1.0000	20

Table A.2 Percentiles for $\hat{p}_n$ with $U[0, B]$ censoring.

%	$B=2$	$B=4$	$B=6$	$B=8$	$B=10$	$B=2$	$B=4$	$B=6$	$B=8$	$B=10$	%
			$n=20$					$n=150$			
1	0.7204	0.7763	0.8150	0.8391	0.8607	0.8966	0.9296	0.9509	0.9615	0.9680	1
5	0.8000	0.8487	0.8788	0.8980	0.9164	0.9211	0.9494	0.9659	0.9744	0.9802	5
10	0.8391	0.8831	0.9105	0.9258	0.9363	0.9338	0.9596	0.9734	0.9808	0.9851	10
20	0.8872	0.9202	0.9381	0.9500	1.0000	0.9489	0.9702	0.9813	0.9871	0.9898	20
			$n=30$					$n=200$			
1	0.7665	0.8230	0.8541	0.8795	0.8945	0.9048	0.9399	0.9586	0.9678	0.9738	1
5	0.8372	0.8819	0.9056	0.9238	0.9368	0.9287	0.9564	0.9699	0.9789	0.9837	5
10	0.8706	0.9070	0.9300	0.9444	0.9535	0.9403	0.9644	0.9763	0.9842	0.9878	10
20	0.9054	0.9372	0.9520	0.9616	1.0000	0.9529	0.9734	0.9836	0.9891	0.9917	20
			$n=40$					$n=250$			
1	0.8040	0.8529	0.8859	0.9001	0.9167	0.9152	0.9463	0.9634	0.9726	0.9786	1
5	0.8571	0.8980	0.9226	0.9369	0.9499	0.9355	0.9609	0.9741	0.9813	0.9861	5
10	0.8859	0.9215	0.9406	0.9537	0.9632	0.9452	0.9675	0.9798	0.9855	0.9896	10
20	0.9169	0.9460	0.9598	0.9676	1.0000	0.9560	0.9755	0.9856	0.9903	0.9929	20
			$n=50$					$n=300$			
1	0.8222	0.8718	0.8994	0.9146	0.9288	0.9219	0.9515	0.9664	0.9750	0.9813	1
5	0.8742	0.9113	0.9340	0.9452	0.9558	0.9391	0.9638	0.9768	0.9832	0.9875	5
10	0.8973	0.9308	0.9504	0.9604	0.9677	0.9478	0.9699	0.9813	0.9871	0.9907	10
20	0.9238	0.9522	0.9664	0.9727	0.9785	0.9578	0.9773	0.9866	0.9913	0.9936	20
			$n=60$					$n=350$			
1	0.8399	0.8875	0.9091	0.9273	0.9387	0.9276	0.9549	0.9700	0.9778	0.9829	1
5	0.8851	0.9207	0.9391	0.9539	0.9627	0.9429	0.9662	0.9783	0.9848	0.9888	5
10	0.9059	0.9370	0.9542	0.9654	0.9725	0.9510	0.9721	0.9828	0.9882	0.9916	10
20	0.9296	0.9556	0.9689	0.9758	0.9807	0.9605	0.9788	0.9877	0.9920	0.9944	20
			$n=70$					$n=400$			
1	0.8509	0.8941	0.9181	0.9354	0.9435	0.9309	0.9573	0.9718	0.9790	0.9845	1
5	0.8915	0.9271	0.9465	0.9579	0.9657	0.9462	0.9681	0.9798	0.9858	0.9900	5
10	0.9113	0.9423	0.9583	0.9687	0.9751	0.9533	0.9735	0.9838	0.9890	0.9925	10
20	0.9329	0.9595	0.9722	0.9788	0.9833	0.9614	0.9798	0.9882	0.9924	0.9948	20
			$n=80$					$n=450$			
1	0.8625	0.8995	0.9229	0.9408	0.9495	0.9338	0.9589	0.9731	0.9809	0.9859	1
5	0.8998	0.9311	0.9509	0.9621	0.9689	0.9472	0.9693	0.9809	0.9871	0.9906	5
10	0.9179	0.9455	0.9623	0.9717	0.9771	0.9540	0.9744	0.9846	0.9899	0.9928	10
20	0.9382	0.9615	0.9744	0.9807	0.9841	0.9623	0.9801	0.9888	0.9930	0.9952	20
			$n=90$					$n=500$			
1	0.8695	0.9054	0.9295	0.9467	0.9545	0.9357	0.9612	0.9741	0.9818	0.9866	1
5	0.9033	0.9352	0.9525	0.9659	0.9710	0.9490	0.9707	0.9817	0.9876	0.9910	5
10	0.9204	0.9491	0.9637	0.9743	0.9786	0.9555	0.9757	0.9853	0.9905	0.9932	10
20	0.9396	0.9633	0.9757	0.9822	0.9853	0.9636	0.9811	0.9892	0.9934	0.9954	20
			$n=100$					$n=1000$			
1	0.8755	0.9160	0.9357	0.9498	0.9585	0.9501	0.9712	0.9822	0.9880	0.9912	1
5	0.9074	0.9405	0.9563	0.9664	0.9736	0.9585	0.9777	0.9869	0.9917	0.9941	5
10	0.9222	0.9526	0.9663	0.9747	0.9806	0.9632	0.9809	0.9891	0.9934	0.9955	10
20	0.9410	0.9654	0.9770	0.9832	0.9868	0.9689	0.9845	0.9918	0.9953	0.9969	20

Table B.1 Percentiles of q_n with $p = 0.2$.

%	$B=2$	$B=4$	$B=6$	$B=8$	$B=10$	$B=2$	$B=4$	$B=6$	$B=8$	$B=10$	%
			$n=20$					$n=200$			
5	0.0000	0.0500	0.0500	0.0500	0.0500	0.0050	0.0050	0.0050	0.0150	0.0500	5
10	0.0500	0.0500	0.0500	0.0500	0.0500	0.0050	0.0050	0.0100	0.0350	0.1250	10
90	0.2000	0.2500	0.2500	0.3000	0.3000	0.0800	0.1500	0.1850	0.2050	0.2150	90
95	0.2000	0.2500	0.3000	0.3000	0.3500	0.1000	0.1650	0.2000	0.2150	0.2250	95
			$n=30$					$n=300$			
5	0.0333	0.0333	0.0333	0.0667	0.0667	0.0033	0.0033	0.0033	0.0067	0.0233	5
10	0.0333	0.0333	0.0667	0.0667	0.1000	0.0033	0.0033	0.0067	0.0200	0.0800	10
90	0.1667	0.2333	0.2333	0.2667	0.2667	0.0533	0.1267	0.1767	0.1967	0.2067	90
95	0.2000	0.2333	0.2667	0.3000	0.3000	0.0700	0.1467	0.1867	0.2067	0.2133	95
			$n=40$					$n=400$			
5	0.0250	0.0250	0.0500	0.0750	0.0750	0.0025	0.0025	0.0025	0.0050	0.0150	5
10	0.0250	0.0250	0.0750	0.1000	0.1000	0.0025	0.0025	0.0050	0.0125	0.0500	10
90	0.1500	0.2000	0.2250	0.2500	0.2500	0.0350	0.0950	0.1700	0.1925	0.2025	90
95	0.1750	0.2250	0.2500	0.2750	0.2750	0.0500	0.1300	0.1800	0.2000	0.2100	95
			$n=50$					$n=500$			
5	0.0200	0.0200	0.0400	0.0600	0.0800	0.0020	0.0020	0.0020	0.0040	0.0120	5
10	0.0200	0.0400	0.0600	0.1000	0.1000	0.0020	0.0020	0.0040	0.0080	0.0340	10
90	0.1400	0.2000	0.2200	0.2400	0.2400	0.0280	0.0680	0.1640	0.1880	0.1980	90
95	0.1600	0.2200	0.2400	0.2600	0.2800	0.0380	0.1000	0.1740	0.1960	0.2060	95
			$n=60$					$n=600$			
5	0.0167	0.0167	0.0333	0.0667	0.0833	0.0017	0.0017	0.0017	0.0033	0.0083	5
10	0.0167	0.0333	0.0500	0.1000	0.1167	0.0017	0.0017	0.0033	0.0067	0.0250	10
90	0.1333	0.1833	0.2167	0.2333	0.2500	0.0217	0.0517	0.1567	0.1850	0.1967	90
95	0.1500	0.2167	0.2333	0.2500	0.2667	0.0300	0.0750	0.1700	0.1933	0.2033	95
			$n=70$					$n=700$			
5	0.0143	0.0143	0.0286	0.0571	0.1000	0.0014	0.0014	0.0014	0.0029	0.0071	5
10	0.0143	0.0143	0.0429	0.1000	0.1143	0.0014	0.0014	0.0029	0.0057	0.0200	10
90	0.1286	0.1857	0.2143	0.2286	0.2429	0.0186	0.0414	0.1457	0.1829	0.1943	90
95	0.1429	0.2000	0.2286	0.2429	0.2571	0.0257	0.0614	0.1643	0.1900	0.2014	95
			$n=80$					$n=800$			
5	0.0125	0.0125	0.0250	0.0500	0.1000	0.0012	0.0012	0.0012	0.0025	0.0050	5
10	0.0125	0.0125	0.0375	0.1000	0.1125	0.0012	0.0012	0.0025	0.0037	0.0150	10
90	0.1250	0.1750	0.2125	0.2250	0.2375	0.0162	0.0350	0.1225	0.1813	0.1937	90
95	0.1375	0.2000	0.2250	0.2375	0.2500	0.0213	0.0500	0.1575	0.1875	0.1988	95
			$n=90$					$n=900$			
5	0.0111	0.0111	0.0111	0.0444	0.1000	0.0011	0.0011	0.0011	0.0022	0.0044	5
10	0.0111	0.0111	0.0333	0.0889	0.1222	0.0011	0.0011	0.0011	0.0022	0.0044	10
90	0.1222	0.1778	0.2111	0.2222	0.2333	0.0133	0.0289	0.1033	0.1789	0.1922	90
95	0.1333	0.1889	0.2222	0.2333	0.2444	0.0189	0.0422	0.1489	0.1856	0.1978	95
			$n=100$					$n=1000$			
5	0.0100	0.0100	0.0100	0.0400	0.1000	0.0010	0.0010	0.0010	0.0010	0.0040	5
10	0.0100	0.0100	0.0300	0.0900	0.1200	0.0010	0.0010	0.0020	0.0030	0.0110	10
90	0.1200	0.1700	0.2000	0.2200	0.2300	0.0120	0.0250	0.0870	0.1770	0.1910	90
95	0.1300	0.1900	0.2200	0.2300	0.2400	0.0170	0.0370	0.1340	0.1840	0.1960	95

Table B.2 Percentiles of q_n with $p = 0.3$.

%	$B=2$	$B=4$	$B=6$	$B=8$	$B=10$	$B=2$	$B=4$	$B=6$	$B=8$	$B=10$	%
	$n = 20$					$n = 200$					
5	0.0500	0.0500	0.0500	0.1000	0.1000	0.0050	0.0050	0.0050	0.0100	0.0350	5
10	0.0500	0.0500	0.1000	0.1000	0.1500	0.0050	0.0050	0.0100	0.0250	0.1150	10
90	0.2500	0.3000	0.3500	0.4000	0.4000	0.0750	0.1900	0.2650	0.2950	0.3050	90
95	0.2500	0.3500	0.4000	0.4000	0.4500	0.1050	0.2200	0.2800	0.3050	0.3200	95
	$n = 30$					$n = 300$					
5	0.0333	0.0333	0.0333	0.1000	0.1333	0.0033	0.0033	0.0033	0.0067	0.0200	5
10	0.0333	0.0333	0.1000	0.1333	0.1667	0.0033	0.0033	0.0067	0.0133	0.0600	10
90	0.2000	0.3000	0.3333	0.3667	0.3667	0.0467	0.1200	0.2500	0.2833	0.3000	90
95	0.2333	0.3333	0.3667	0.4000	0.4000	0.0633	0.1733	0.2663	0.2933	0.3100	95
	$n = 40$					$n = 400$					
5	0.0250	0.0250	0.0250	0.1000	0.1500	0.0025	0.0025	0.0025	0.0050	0.0125	5
10	0.0250	0.0250	0.0750	0.1500	0.1750	0.0025	0.0025	0.0050	0.0100	0.0375	10
90	0.2000	0.2750	0.3250	0.3500	0.3500	0.0325	0.0775	0.2350	0.2775	0.2925	90
95	0.2250	0.3000	0.3500	0.3750	0.3750	0.0450	0.1125	0.2525	0.2875	0.3025	95
	$n = 50$					$n = 500$					
5	0.0200	0.0200	0.0200	0.0800	0.1400	0.0020	0.0020	0.0020	0.0040	0.0080	5
10	0.0200	0.0200	0.0600	0.1400	0.1800	0.0020	0.0020	0.0040	0.0080	0.0240	10
90	0.1800	0.2800	0.3200	0.3400	0.3400	0.0260	0.0560	0.2040	0.2720	0.2900	90
95	0.2200	0.3000	0.3400	0.3600	0.3800	0.0340	0.0820	0.2420	0.2820	0.2980	95
	$n = 60$					$n = 600$					
5	0.0167	0.0167	0.0167	0.0500	0.1500	0.0017	0.0017	0.0017	0.0033	0.0067	5
10	0.0167	0.0167	0.0500	0.1333	0.1833	0.0017	0.0017	0.0033	0.0050	0.0183	10
90	0.1833	0.2667	0.3000	0.3333	0.3333	0.0200	0.0433	0.1517	0.2683	0.2867	90
95	0.2000	0.2833	0.3333	0.3500	0.3667	0.0283	0.0633	0.2233	0.2783	0.2950	95
	$n = 70$					$n = 700$					
5	0.0143	0.0143	0.0143	0.0429	0.1429	0.0014	0.0014	0.0014	0.0014	0.0043	5
10	0.0143	0.0143	0.0429	0.1286	0.1857	0.0014	0.0014	0.0014	0.0043	0.0143	10
90	0.1714	0.2571	0.3000	0.3286	0.3429	0.0171	0.0357	0.1186	0.2643	0.2843	90
95	0.1857	0.2714	0.3286	0.3429	0.3571	0.0229	0.0514	0.1814	0.2743	0.2929	95
	$n = 80$					$n = 800$					
5	0.0125	0.0125	0.0125	0.0375	0.1375	0.0012	0.0012	0.0012	0.0012	0.0037	5
10	0.0125	0.0125	0.0375	0.1125	0.1875	0.0012	0.0012	0.0012	0.0037	0.0125	10
90	0.1625	0.2500	0.3000	0.3250	0.3375	0.0150	0.0300	0.0962	0.2612	0.2825	90
95	0.1875	0.2750	0.3125	0.3375	0.3500	0.0200	0.0425	0.1462	0.2713	0.2900	95
	$n = 90$					$n = 900$					
5	0.0111	0.0111	0.0111	0.0333	0.1222	0.0011	0.0011	0.0011	0.0011	0.0033	5
10	0.0111	0.0111	0.0222	0.0889	0.1889	0.0011	0.0011	0.0011	0.0033	0.0100	10
90	0.1556	0.2444	0.2889	0.3111	0.3333	0.0133	0.0256	0.0800	0.2567	0.2811	90
95	0.1778	0.2667	0.3111	0.3333	0.3444	0.0178	0.0367	0.1211	0.2678	0.2878	95
	$n = 100$					$n = 1000$					
5	0.0100	0.0100	0.0100	0.0300	0.1100	0.0010	0.0010	0.0010	0.0010	0.0030	5
10	0.0100	0.0100	0.0200	0.0800	0.1900	0.0010	0.0010	0.0010	0.0030	0.0080	10
90	0.1500	0.2400	0.2900	0.3100	0.3200	0.0110	0.0220	0.0680	0.2520	0.2800	90
95	0.1700	0.2600	0.3100	0.3300	0.3400	0.0150	0.0310	0.1040	0.2650	0.2870	95

Table B.3 Percentiles of q_n with $p = 0.4$.

%	$B=2$	$B=4$	$B=6$	$B=8$	$B=10$	$B=2$	$B=4$	$B=6$	$B=8$	$B=10$	%
			$n=20$					$n=200$			
5	0.0500	0.0500	0.0500	0.1000	0.1500	0.0050	0.0050	0.0050	0.0100	0.0300	5
10	0.0500	0.0500	0.1000	0.2000	0.2000	0.0050	0.0050	0.0100	0.0250	0.0950	10
90	0.3000	0.4000	0.4500	0.5000	0.5000	0.0700	0.1850	0.3400	0.3800	0.4000	90
95	0.3500	0.4500	0.5000	0.5000	0.5500	0.1000	0.2550	0.3550	0.3950	0.4100	95
			$n=30$					$n=300$			
5	0.0333	0.0333	0.0333	0.1000	0.2000	0.0033	0.0033	0.0033	0.0067	0.0167	5
10	0.0333	0.0333	0.1000	0.2000	0.2333	0.0033	0.0033	0.0067	0.0133	0.0467	10
90	0.2667	0.3667	0.4333	0.4667	0.4667	0.0433	0.1033	0.3133	0.3700	0.3900	90
95	0.3000	0.4000	0.4667	0.5000	0.5000	0.0600	0.1500	0.3367	0.3800	0.4000	95
			$n=40$					$n=400$			
5	0.0250	0.0250	0.0250	0.0750	0.2000	0.0025	0.0025	0.0025	0.0050	0.0100	5
10	0.0250	0.0250	0.0750	0.1750	0.2500	0.0025	0.0025	0.0050	0.0075	0.0300	10
90	0.2500	0.3500	0.4000	0.4500	0.4500	0.0300	0.0675	0.2425	0.3600	0.3825	90
95	0.2750	0.3750	0.4500	0.4750	0.4750	0.0425	0.1000	0.3150	0.3725	0.3925	95
			$n=50$					$n=500$			
5	0.0200	0.0200	0.0200	0.0600	0.2000	0.0020	0.0020	0.0020	0.0020	0.0080	5
10	0.0200	0.0200	0.0600	0.1600	0.2600	0.0020	0.0020	0.0020	0.0060	0.0220	10
90	0.2200	0.3400	0.4000	0.4200	0.4400	0.0240	0.0500	0.1700	0.3540	0.3800	90
95	0.2600	0.3600	0.4200	0.4600	0.4800	0.0320	0.0720	0.2580	0.3660	0.3880	95
			$n=60$					$n=600$			
5	0.0167	0.0167	0.0167	0.0500	0.1833	0.0017	0.0017	0.0017	0.0017	0.0050	5
10	0.0167	0.0167	0.0333	0.1333	0.2500	0.0017	0.0017	0.0017	0.0050	0.0167	10
90	0.2167	0.3333	0.3833	0.4167	0.4333	0.0200	0.0400	0.1283	0.3467	0.3767	90
95	0.2500	0.3500	0.4167	0.4500	0.4667	0.0267	0.0567	0.1967	0.3600	0.3850	95
			$n=70$					$n=700$			
5	0.0143	0.0143	0.0143	0.0429	0.1429	0.0014	0.0014	0.0014	0.0014	0.0043	5
10	0.0143	0.0143	0.0286	0.1000	0.2571	0.0014	0.0014	0.0014	0.0043	0.0129	10
90	0.2000	0.3143	0.3857	0.4143	0.4286	0.0171	0.0314	0.1000	0.3400	0.3743	90
95	0.2286	0.3429	0.4000	0.4286	0.4571	0.0214	0.0457	0.1514	0.3557	0.3814	95
			$n=80$					$n=800$			
5	0.0125	0.0125	0.0125	0.0375	0.1250	0.0012	0.0012	0.0012	0.0012	0.0037	5
10	0.0125	0.0125	0.0250	0.0875	0.2500	0.0012	0.0012	0.0012	0.0037	0.0100	10
90	0.1875	0.3125	0.3750	0.4125	0.4250	0.0137	0.0262	0.0812	0.3313	0.3713	90
95	0.2250	0.3375	0.4000	0.4250	0.4500	0.0188	0.0375	0.1225	0.3500	0.3800	95
			$n=90$					$n=900$			
5	0.0111	0.0111	0.0111	0.0333	0.1000	0.0011	0.0011	0.0011	0.0011	0.0033	5
10	0.0111	0.0111	0.0222	0.0778	0.2556	0.0011	0.0011	0.0011	0.0022	0.0089	10
90	0.1778	0.3000	0.3778	0.4000	0.4222	0.0122	0.0222	0.0678	0.3022	0.3700	90
95	0.2111	0.3333	0.3889	0.4222	0.4444	0.0167	0.0322	0.1022	0.3444	0.3778	95
			$n=100$					$n=1000$			
5	0.0100	0.0100	0.0100	0.0200	0.0900	0.0010	0.0010	0.0010	0.0010	0.0030	5
10	0.0100	0.0100	0.0200	0.0600	0.2500	0.0010	0.0010	0.0010	0.0020	0.0070	10
90	0.1600	0.3000	0.3700	0.4000	0.4200	0.0110	0.0200	0.0570	0.2510	0.3680	90
95	0.2000	0.3200	0.3900	0.4200	0.4400	0.0150	0.0280	0.0870	0.3380	0.3750	95

Table B.4 Percentiles of q_n with $p = 0.5$.

%	$B = 2$	$B = 4$	$B = 6$	$B = 8$	$B = 10$	$B = 2$	$B = 4$	$B = 6$	$B = 8$	$B = 10$	%
			$n = 20$					$n = 200$			
5	0.0500	0.0500	0.0500	0.1000	0.2000	0.0050	0.0050	0.0050	0.0100	0.0250	5
10	0.0500	0.0500	0.1000	0.2000	0.3000	0.0050	0.0050	0.0100	0.0200	0.0850	10
90	0.3500	0.4500	0.5500	0.5500	0.6000	0.0650	0.1650	0.4050	0.4650	0.4900	90
95	0.4000	0.5000	0.5500	0.6000	0.6500	0.0900	0.2400	0.4300	0.4800	0.5000	95
			$n = 30$					$n = 300$			
5	0.0333	0.0333	0.0333	0.1000	0.2333	0.0033	0.0033	0.0033	0.0067	0.0133	5
10	0.0333	0.0333	0.0667	0.2333	0.3000	0.0033	0.0033	0.0067	0.0100	0.0400	10
90	0.3000	0.4333	0.5000	0.5333	0.5667	0.0433	0.0933	0.3333	0.4533	0.4800	90
95	0.3333	0.4667	0.5333	0.5667	0.6000	0.0567	0.1367	0.4033	0.4667	0.4900	95
			$n = 40$					$n = 400$			
5	0.0250	0.0250	0.0250	0.0750	0.2250	0.0025	0.0025	0.0025	0.0025	0.0100	5
10	0.0250	0.0250	0.0500	0.1750	0.3250	0.0025	0.0025	0.0025	0.0075	0.0250	10
90	0.2750	0.4250	0.5000	0.5250	0.5500	0.0300	0.0625	0.2125	0.4425	0.4725	90
95	0.3250	0.4500	0.5250	0.5500	0.5750	0.0400	0.0900	0.3250	0.4550	0.4825	95
			$n = 50$					$n = 500$			
5	0.0200	0.0200	0.0200	0.0600	0.1800	0.0020	0.0020	0.0020	0.0020	0.0060	5
10	0.0200	0.0200	0.0400	0.1400	0.3200	0.0020	0.0020	0.0020	0.0060	0.0180	10
90	0.2600	0.4000	0.4800	0.5200	0.5400	0.0240	0.0460	0.1480	0.4320	0.4680	90
95	0.3000	0.4400	0.5000	0.5400	0.5600	0.0320	0.0660	0.2260	0.4480	0.4780	95
			$n = 60$					$n = 600$			
5	0.0167	0.0167	0.0167	0.0500	0.1333	0.0017	0.0017	0.0017	0.0017	0.0050	5
10	0.0167	0.0167	0.0333	0.1167	0.3167	0.0017	0.0017	0.0017	0.0050	0.0133	10
90	0.2333	0.3833	0.4667	0.5000	0.5333	0.0183	0.0350	0.1117	0.4200	0.4650	90
95	0.2833	0.4167	0.5000	0.5333	0.5500	0.0250	0.0517	0.1683	0.4417	0.4733	95
			$n = 70$					$n = 700$			
5	0.0143	0.0143	0.0143	0.0286	0.1143	0.0014	0.0014	0.0014	0.0014	0.0043	5
10	0.0143	0.0143	0.0286	0.0857	0.3143	0.0014	0.0014	0.0014	0.0029	0.0100	10
90	0.2143	0.3857	0.4571	0.5000	0.5286	0.0157	0.0286	0.0886	0.3886	0.4614	90
95	0.2571	0.4143	0.4857	0.5286	0.5429	0.0214	0.0414	0.1314	0.4329	0.4700	95
			$n = 80$					$n = 800$			
5	0.0125	0.0125	0.0125	0.0250	0.1000	0.0012	0.0012	0.0012	0.0012	0.0037	5
10	0.0125	0.0125	0.0250	0.0750	0.3000	0.0012	0.0012	0.0012	0.0025	0.0088	10
90	0.2000	0.3625	0.4625	0.5000	0.5125	0.0137	0.0250	0.0700	0.3100	0.4588	90
95	0.2500	0.4000	0.4750	0.5125	0.5375	0.0188	0.0350	0.1063	0.4225	0.4675	95
			$n = 90$					$n = 900$			
5	0.0111	0.0111	0.0111	0.0222	0.0889	0.0011	0.0011	0.0011	0.0011	0.0022	5
10	0.0111	0.0111	0.0222	0.0667	0.2778	0.0011	0.0011	0.0011	0.0022	0.0078	10
90	0.1778	0.3556	0.4556	0.4889	0.5111	0.0122	0.0211	0.0589	0.2533	0.4567	90
95	0.2333	0.3889	0.4778	0.5111	0.5333	0.0156	0.0300	0.0900	0.3944	0.4656	95
			$n = 100$					$n = 1000$			
5	0.0100	0.0100	0.0100	0.0200	0.0700	0.0010	0.0010	0.0010	0.0010	0.0020	5
10	0.0100	0.0100	0.0200	0.0500	0.2300	0.0010	0.0010	0.0010	0.0020	0.0060	10
90	0.1500	0.3500	0.4500	0.4900	0.5100	0.0110	0.0180	0.0500	0.2130	0.4540	90
95	0.2100	0.3800	0.4700	0.5100	0.5300	0.0140	0.0260	0.0750	0.3360	0.4630	95

Table B.5 Percentiles of q_n with $p = 0.6$.

%	$B=2$	$B=4$	$B=6$	$B=8$	$B=10$	$B=2$	$B=4$	$B=6$	$B=8$	$B=10$	%
	$n=20$					$n=200$					
5	0.0500	0.0500	0.0500	0.1000	0.2000	0.0050	0.0050	0.0050	0.0100	0.0200	5
10	0.0500	0.0500	0.1000	0.2000	0.3500	0.0050	0.0050	0.0100	0.0200	0.0700	10
90	0.4000	0.5500	0.6000	0.6500	0.6500	0.0650	0.1500	0.4700	0.5500	0.5750	90
95	0.4000	0.5500	0.6500	0.7000	0.7000	0.0900	0.2200	0.5050	0.5650	0.5900	95
	$n=30$					$n=300$					
5	0.0333	0.0333	0.0333	0.0667	0.2000	0.0033	0.0033	0.0033	0.0033	0.0133	5
10	0.0333	0.0333	0.0667	0.1667	0.3667	0.0033	0.0033	0.0033	0.0100	0.0367	10
90	0.3333	0.5000	0.5667	0.6333	0.6333	0.0400	0.0833	0.2900	0.5333	0.5667	90
95	0.4000	0.5333	0.6333	0.6667	0.6667	0.0533	0.1233	0.4400	0.5500	0.5800	95
	$n=40$					$n=400$					
5	0.0250	0.0250	0.0250	0.0500	0.1750	0.0025	0.0025	0.0025	0.0025	0.0075	5
10	0.0250	0.0250	0.0500	0.1500	0.3750	0.0025	0.0025	0.0025	0.0075	0.0225	10
90	0.3250	0.4750	0.5750	0.6000	0.6250	0.0300	0.0575	0.1850	0.5200	0.5600	90
95	0.3500	0.5250	0.6000	0.6500	0.6750	0.0400	0.0825	0.2850	0.5375	0.5725	95
	$n=50$					$n=500$					
5	0.0200	0.0200	0.0200	0.0400	0.1400	0.0020	0.0020	0.0020	0.0020	0.0060	5
10	0.0200	0.0200	0.0400	0.1200	0.3800	0.0020	0.0020	0.0020	0.0060	0.0160	10
90	0.2800	0.4600	0.5600	0.6000	0.6200	0.0220	0.0420	0.1340	0.5060	0.5560	90
95	0.3400	0.5000	0.5800	0.6200	0.6400	0.0300	0.0600	0.2040	0.5280	0.5660	95
	$n=60$					$n=600$					
5	0.0167	0.0167	0.0167	0.0333	0.1167	0.0017	0.0017	0.0017	0.0017	0.0050	5
10	0.0167	0.0167	0.0333	0.0833	0.3667	0.0017	0.0017	0.0017	0.0033	0.0117	10
90	0.2500	0.4500	0.5500	0.6000	0.6167	0.0183	0.0333	0.1000	0.4400	0.5517	90
95	0.3167	0.4833	0.5833	0.6167	0.6333	0.0250	0.0467	0.1500	0.5167	0.5617	95
	$n=70$					$n=700$					
5	0.0143	0.0143	0.0143	0.0286	0.1000	0.0014	0.0014	0.0014	0.0014	0.0043	5
10	0.0143	0.0143	0.0286	0.0714	0.3143	0.0014	0.0014	0.0014	0.0029	0.0100	10
90	0.2143	0.4286	0.5429	0.5857	0.6143	0.0157	0.0271	0.0786	0.3386	0.5486	90
95	0.2857	0.4714	0.5714	0.6143	0.6286	0.0200	0.0386	0.1186	0.5000	0.5586	95
	$n=80$					$n=800$					
5	0.0125	0.0125	0.0125	0.0250	0.0750	0.0012	0.0012	0.0012	0.0012	0.0025	5
10	0.0125	0.0125	0.0250	0.0625	0.2625	0.0012	0.0012	0.0012	0.0025	0.0075	10
90	0.1875	0.4125	0.5375	0.5750	0.6000	0.0137	0.0225	0.0650	0.2688	0.5450	90
95	0.2500	0.4625	0.5625	0.6000	0.6250	0.0175	0.0325	0.0975	0.4238	0.5550	95
	$n=90$					$n=900$					
5	0.0111	0.0111	0.0111	0.0222	0.0667	0.0011	0.0011	0.0011	0.0011	0.0022	5
10	0.0111	0.0111	0.0222	0.0556	0.2222	0.0011	0.0011	0.0011	0.0022	0.0067	10
90	0.1667	0.4000	0.5222	0.5778	0.6000	0.0122	0.0200	0.0533	0.2233	0.5422	90
95	0.2333	0.4444	0.5556	0.6000	0.6222	0.0156	0.0278	0.0800	0.3467	0.5522	95
	$n=100$					$n=1000$					
5	0.0100	0.0100	0.0100	0.0200	0.0600	0.0010	0.0010	0.0010	0.0010	0.0020	5
10	0.0100	0.0100	0.0200	0.0400	0.2000	0.0010	0.0010	0.0010	0.0020	0.0050	10
90	0.1500	0.3700	0.5200	0.5700	0.6000	0.0100	0.0170	0.0460	0.1860	0.5390	90
95	0.2000	0.4400	0.5500	0.5900	0.6200	0.0140	0.0240	0.0690	0.2920	0.5500	95

Table B.6 Percentiles of q_n with $p = 0.7$.

%	$B=2$	$B=4$	$B=6$	$B=8$	$B=10$	$B=2$	$B=4$	$B=6$	$B=8$	$B=10$	%
	$n=20$					$n=200$					
5	0.0500	0.0500	0.0500	0.0500	0.1000	0.0050	0.0050	0.0050	0.0050	0.0200	5
10	0.0500	0.0500	0.0500	0.1000	0.3500	0.0050	0.0050	0.0050	0.0150	0.0600	10
90	0.4000	0.6000	0.7000	0.7000	0.7500	0.0600	0.1350	0.4900	0.6300	0.6650	90
95	0.4500	0.6500	0.7000	0.7500	0.8000	0.0850	0.2000	0.5700	0.6450	0.6800	95
	$n=30$					$n=300$					
5	0.0333	0.0333	0.0333	0.0333	0.1333	0.0033	0.0033	0.0033	0.0033	0.0100	5
10	0.0333	0.0333	0.0333	0.1000	0.4000	0.0033	0.0033	0.0033	0.0100	0.0300	10
90	0.3667	0.5667	0.6667	0.7000	0.7333	0.0400	0.0767	0.2633	0.6133	0.6533	90
95	0.4333	0.6000	0.7000	0.7333	0.7667	0.0533	0.1100	0.4033	0.6333	0.6667	95
	$n=40$					$n=400$					
5	0.0250	0.0250	0.0250	0.0500	0.1250	0.0025	0.0025	0.0025	0.0025	0.0075	5
10	0.0250	0.0250	0.0500	0.1000	0.3750	0.0025	0.0025	0.0025	0.0075	0.0200	10
90	0.3250	0.5250	0.6500	0.7000	0.7250	0.0275	0.0525	0.1675	0.5975	0.6475	90
95	0.4000	0.5750	0.6750	0.7250	0.7500	0.0375	0.0775	0.2575	0.6200	0.6600	95
	$n=50$					$n=500$					
5	0.0200	0.0200	0.0200	0.0400	0.1000	0.0020	0.0020	0.0020	0.0020	0.0060	5
10	0.0200	0.0200	0.0400	0.0800	0.3400	0.0020	0.0020	0.0020	0.0040	0.0140	10
90	0.3000	0.5200	0.6400	0.6800	0.7000	0.0220	0.0400	0.1180	0.5260	0.6440	90
95	0.3600	0.5600	0.6600	0.7000	0.7400	0.0300	0.0560	0.1780	0.6060	0.6540	95
	$n=60$					$n=600$					
5	0.0167	0.0167	0.0167	0.0333	0.0833	0.0017	0.0017	0.0017	0.0017	0.0033	5
10	0.0167	0.0167	0.0333	0.0667	0.2667	0.0017	0.0017	0.0017	0.0033	0.0100	10
90	0.2500	0.5000	0.6167	0.6667	0.7000	0.0183	0.0317	0.0900	0.3967	0.6383	90
95	0.3333	0.5500	0.6500	0.7000	0.7167	0.0233	0.0433	0.1367	0.5867	0.6483	95
	$n=70$					$n=700$					
5	0.0143	0.0143	0.0143	0.0286	0.0714	0.0014	0.0014	0.0014	0.0014	0.0029	5
10	0.0143	0.0143	0.0286	0.0571	0.2286	0.0014	0.0014	0.0014	0.0029	0.0086	10
90	0.2143	0.4714	0.6143	0.6714	0.7000	0.0143	0.0257	0.0714	0.3057	0.6343	90
95	0.2857	0.5286	0.6429	0.6857	0.7143	0.0200	0.0357	0.1071	0.4757	0.6457	95
	$n=80$					$n=800$					
5	0.0125	0.0125	0.0125	0.0250	0.0625	0.0012	0.0012	0.0012	0.0012	0.0025	5
10	0.0125	0.0125	0.0250	0.0500	0.2000	0.0012	0.0012	0.0012	0.0025	0.0063	10
90	0.1750	0.4500	0.6125	0.6625	0.6875	0.0125	0.0225	0.0587	0.2412	0.6313	90
95	0.2500	0.5125	0.6375	0.6875	0.7125	0.0175	0.0312	0.0875	0.3787	0.6413	95
	$n=90$					$n=900$					
5	0.0111	0.0111	0.0111	0.0222	0.0556	0.0011	0.0011	0.0011	0.0011	0.0022	5
10	0.0111	0.0111	0.0222	0.0444	0.1778	0.0011	0.0011	0.0011	0.0022	0.0056	10
90	0.1556	0.4000	0.6000	0.6556	0.6889	0.0111	0.0189	0.0489	0.1989	0.6267	90
95	0.2222	0.5000	0.6222	0.6778	0.7111	0.0156	0.0267	0.0733	0.3133	0.6378	95
	$n=100$					$n=1000$					
5	0.0100	0.0100	0.0100	0.0200	0.0500	0.0010	0.0010	0.0010	0.0010	0.0020	5
10	0.0100	0.0100	0.0200	0.0400	0.1600	0.0010	0.0010	0.0010	0.0020	0.0050	10
90	0.1400	0.3500	0.5900	0.6500	0.6800	0.0100	0.0160	0.0430	0.1680	0.6230	90
95	0.1900	0.4800	0.6200	0.6800	0.7000	0.0140	0.0230	0.0630	0.2610	0.6360	95

Table B.7 Percentiles of q_n with $p = 0.8$.

%	$B=2$	$B=4$	$B=6$	$B=8$	$B=10$	$B=2$	$B=4$	$B=6$	$B=8$	$B=10$	%
	$n=20$					$n=200$					
5	0.0500	0.0500	0.0500	0.0500	0.0500	0.0050	0.0050	0.0050	0.0050	0.0150	5
10	0.0500	0.0500	0.0500	0.0500	0.1000	0.0050	0.0050	0.0050	0.0150	0.0500	10
90	0.4500	0.6500	0.7500	0.8000	0.8000	0.0600	0.1200	0.4300	0.7100	0.7500	90
95	0.5000	0.7000	0.7500	0.8000	0.8500	0.0800	0.1800	0.6200	0.7300	0.7650	95
	$n=30$					$n=300$					
5	0.0333	0.0333	0.0333	0.0333	0.0333	0.0033	0.0033	0.0033	0.0033	0.0100	5
10	0.0333	0.0333	0.0333	0.0667	0.1333	0.0033	0.0033	0.0033	0.0100	0.0267	10
90	0.4000	0.6000	0.7333	0.7667	0.8000	0.0367	0.0733	0.2300	0.6933	0.7400	90
95	0.4667	0.6667	0.7667	0.8000	0.8333	0.0500	0.1033	0.3533	0.7133	0.7533	95
	$n=40$					$n=400$					
5	0.0250	0.0250	0.0250	0.0250	0.0500	0.0025	0.0025	0.0025	0.0025	0.0075	5
10	0.0250	0.0250	0.0250	0.0500	0.1750	0.0025	0.0025	0.0025	0.0050	0.0175	10
90	0.3500	0.5750	0.7000	0.7750	0.8000	0.0275	0.0500	0.1525	0.6575	0.7350	90
95	0.4250	0.6250	0.7500	0.8000	0.8250	0.0350	0.0700	0.2275	0.6975	0.7450	95
	$n=50$					$n=500$					
5	0.0200	0.0200	0.0200	0.0200	0.0600	0.0020	0.0020	0.0020	0.0020	0.0040	5
10	0.0200	0.0200	0.0200	0.0600	0.1800	0.0020	0.0020	0.0020	0.0040	0.0120	10
90	0.2800	0.5600	0.7000	0.7600	0.7800	0.0220	0.0360	0.1060	0.4720	0.7300	90
95	0.3800	0.6200	0.7400	0.7800	0.8200	0.0280	0.0520	0.1620	0.6780	0.7400	95
	$n=60$					$n=600$					
5	0.0167	0.0167	0.0167	0.0167	0.0500	0.0017	0.0017	0.0017	0.0017	0.0033	5
10	0.0167	0.0167	0.0167	0.0500	0.1667	0.0017	0.0017	0.0017	0.0033	0.0100	10
90	0.2333	0.5333	0.6833	0.7500	0.7833	0.0167	0.0300	0.0817	0.3483	0.7250	90
95	0.3167	0.6000	0.7167	0.7833	0.8000	0.0233	0.0417	0.1217	0.5483	0.7350	95
	$n=70$					$n=700$					
5	0.0143	0.0143	0.0143	0.0143	0.0429	0.0014	0.0014	0.0014	0.0014	0.0029	5
10	0.0143	0.0143	0.0143	0.0429	0.1571	0.0014	0.0014	0.0014	0.0029	0.0071	10
90	0.2000	0.4857	0.6857	0.7429	0.7714	0.0143	0.0243	0.0657	0.2700	0.7200	90
95	0.2714	0.5714	0.7143	0.7714	0.8000	0.0200	0.0343	0.0986	0.4229	0.7314	95
	$n=80$					$n=800$					
5	0.0125	0.0125	0.0125	0.0125	0.0500	0.0012	0.0012	0.0012	0.0012	0.0025	5
10	0.0125	0.0125	0.0125	0.0375	0.1375	0.0012	0.0012	0.0012	0.0025	0.0063	10
90	0.1625	0.4000	0.6750	0.7375	0.7750	0.0125	0.0213	0.0538	0.2150	0.7163	90
95	0.2375	0.5500	0.7000	0.7625	0.7875	0.0162	0.0288	0.0800	0.3313	0.7275	95
	$n=90$					$n=900$					
5	0.0111	0.0111	0.0111	0.0111	0.0444	0.0011	0.0011	0.0011	0.0011	0.0022	5
10	0.0111	0.0111	0.0111	0.0333	0.1222	0.0011	0.0011	0.0011	0.0022	0.0056	10
90	0.1444	0.3556	0.6667	0.7333	0.7667	0.0111	0.0178	0.0456	0.1789	0.7100	90
95	0.2000	0.5111	0.7000	0.7556	0.7889	0.0144	0.0244	0.0678	0.2788	0.7244	95
	$n=100$					$n=1000$					
5	0.0100	0.0100	0.0100	0.0100	0.0400	0.0010	0.0010	0.0010	0.0010	0.0020	5
10	0.0100	0.0100	0.0100	0.0300	0.1100	0.0010	0.0010	0.0010	0.0020	0.0040	10
90	0.1300	0.3000	0.6600	0.7300	0.7700	0.0100	0.0160	0.0390	0.1500	0.7010	90
95	0.1800	0.4500	0.6900	0.7500	0.7800	0.0130	0.0220	0.0570	0.2360	0.7210	95

Table B.8 Percentiles of q_n with $p = 0.9$.

%	$B = 2$	$B = 4$	$B = 6$	$B = 8$	$B = 10$	$B = 2$	$B = 4$	$B = 6$	$B = 8$	$B = 10$	%
	$n = 20$					$n = 200$					
5	0.0500	0.0500	0.0500	0.0500	0.0500	0.0050	0.0050	0.0050	0.0050	0.0100	5
10	0.0500	0.0500	0.0500	0.0500	0.0500	0.0050	0.0050	0.0050	0.0100	0.0300	10
90	0.4500	0.6500	0.7500	0.8500	0.8500	0.0550	0.1050	0.3450	0.7850	0.8350	90
95	0.5500	0.7000	0.8000	0.8500	0.9000	0.0750	0.1550	0.5400	0.8050	0.8450	95
	$n = 30$					$n = 300$					
5	0.0333	0.0333	0.0333	0.0333	0.0333	0.0033	0.0033	0.0033	0.0033	0.0067	5
10	0.0333	0.0333	0.0333	0.0333	0.0333	0.0033	0.0033	0.0033	0.0067	0.0200	10
90	0.4000	0.6333	0.7667	0.8333	0.8667	0.0367	0.0633	0.1967	0.7600	0.8233	90
95	0.4667	0.7000	0.8000	0.8667	0.9000	0.0467	0.0933	0.3033	0.7900	0.8367	95
	$n = 40$					$n = 400$					
5	0.0250	0.0250	0.0250	0.0250	0.0250	0.0025	0.0025	0.0025	0.0025	0.0050	5
10	0.0250	0.0250	0.0250	0.0250	0.0250	0.0025	0.0025	0.0025	0.0050	0.0150	10
90	0.3250	0.6000	0.7500	0.8250	0.8500	0.0275	0.0450	0.1300	0.5600	0.8200	90
95	0.4250	0.6750	0.8000	0.8500	0.8750	0.0350	0.0650	0.2000	0.7700	0.8300	95
	$n = 50$					$n = 500$					
5	0.0200	0.0200	0.0200	0.0200	0.0200	0.0020	0.0020	0.0020	0.0020	0.0040	5
10	0.0200	0.0200	0.0200	0.0200	0.0400	0.0020	0.0020	0.0020	0.0040	0.0100	10
90	0.2600	0.5400	0.7600	0.8200	0.8600	0.0200	0.0340	0.0940	0.3960	0.8140	90
95	0.3600	0.6400	0.7800	0.8400	0.8800	0.0280	0.0500	0.1440	0.6280	0.8240	95
	$n = 60$					$n = 600$					
5	0.0167	0.0167	0.0167	0.0167	0.0167	0.0017	0.0017	0.0017	0.0017	0.0033	5
10	0.0167	0.0167	0.0167	0.0167	0.0333	0.0017	0.0017	0.0017	0.0033	0.0083	10
90	0.2167	0.4500	0.7500	0.8167	0.8500	0.0167	0.0267	0.0733	0.2950	0.8100	90
95	0.3000	0.6167	0.7833	0.8333	0.8667	0.0217	0.0383	0.1100	0.4617	0.8200	95
	$n = 70$					$n = 700$					
5	0.0143	0.0143	0.0143	0.0143	0.0143	0.0014	0.0014	0.0014	0.0014	0.0029	5
10	0.0143	0.0143	0.0143	0.0143	0.0429	0.0014	0.0014	0.0014	0.0029	0.0071	10
90	0.1714	0.3857	0.7429	0.8143	0.8571	0.0143	0.0229	0.0586	0.2314	0.8043	90
95	0.2571	0.5714	0.7714	0.8429	0.8714	0.0186	0.0314	0.0871	0.3643	0.8157	95
	$n = 80$					$n = 800$					
5	0.0125	0.0125	0.0125	0.0125	0.0125	0.0012	0.0012	0.0012	0.0012	0.0025	5
10	0.0125	0.0125	0.0125	0.0125	0.0500	0.0012	0.0012	0.0012	0.0025	0.0050	10
90	0.1500	0.3250	0.7250	0.8125	0.8500	0.0125	0.0188	0.0487	0.1875	0.7987	90
95	0.2125	0.5000	0.7625	0.8375	0.8625	0.0162	0.0262	0.0725	0.2925	0.8125	95
	$n = 90$					$n = 900$					
5	0.0111	0.0111	0.0111	0.0111	0.0222	0.0011	0.0011	0.0011	0.0011	0.0022	5
10	0.0111	0.0111	0.0111	0.0222	0.0444	0.0011	0.0011	0.0011	0.0022	0.0044	10
90	0.1333	0.2889	0.7222	0.8111	0.8444	0.0111	0.0167	0.0411	0.1556	0.7811	90
95	0.1889	0.4333	0.7556	0.8333	0.8667	0.0144	0.0233	0.0611	0.2456	0.8089	95
	$n = 100$					$n = 1000$					
5	0.0100	0.0100	0.0100	0.0100	0.0200	0.0010	0.0010	0.0010	0.0010	0.0020	5
10	0.0100	0.0100	0.0100	0.0200	0.0500	0.0010	0.0010	0.0010	0.0020	0.0040	10
90	0.1200	0.2500	0.7100	0.8100	0.8500	0.0100	0.0140	0.0350	0.1320	0.6520	90
95	0.1600	0.3800	0.7500	0.8300	0.8600	0.0130	0.0200	0.0520	0.2050	0.8040	95

Table C.1 Percentiles of correlation coefficient r with $p = 0.2$.

%	$B=2$	$B=4$	$B=6$	$B=8$	$B=10$	$B=2$	$B=4$	$B=6$	$B=8$	$B=10$	%
			$n=20$					$n=200$			
1	0.3964	0.4598	0.5210	0.5364	0.5546	0.8866	0.9357	0.9558	0.9631	0.9668	1
5	0.5262	0.6192	0.6542	0.6837	0.6965	0.9362	0.9612	0.9709	0.9760	0.9780	5
10	0.6299	0.7065	0.7345	0.7632	0.7728	0.9515	0.9710	0.9774	0.9808	0.9827	10
20	0.7392	0.7949	0.8197	0.8368	0.8461	0.9649	0.9782	0.9831	0.9861	0.9877	20
			$n=30$					$n=300$			
1	0.4036	0.5015	0.5870	0.6168	0.6625	0.9278	0.9523	0.9689	0.9753	0.9784	1
5	0.6050	0.6885	0.7629	0.7816	0.8058	0.9561	0.9740	0.9810	0.9844	0.9866	5
10	0.6983	0.7782	0.8223	0.8375	0.8517	0.9670	0.9798	0.9853	0.9876	0.9892	10
20	0.7942	0.8442	0.8750	0.8889	0.8963	0.9761	0.9855	0.9889	0.9908	0.9918	20
			$n=40$					$n=400$			
1	0.4905	0.6246	0.6722	0.7230	0.7670	0.9466	0.9703	0.9787	0.9827	0.9840	1
5	0.6889	0.7767	0.8192	0.8413	0.8613	0.9677	0.9812	0.9860	0.9884	0.9896	5
10	0.7664	0.8375	0.8690	0.8854	0.8975	0.9748	0.9854	0.9890	0.9909	0.9918	10
20	0.8362	0.8852	0.9089	0.9191	0.9276	0.9813	0.9891	0.9918	0.9932	0.9939	20
			$n=50$					$n=500$			
1	0.5515	0.6983	0.7420	0.7967	0.8188	0.9583	0.9761	0.9830	0.9858	0.9876	1
5	0.7340	0.8321	0.8649	0.8883	0.8932	0.9752	0.9849	0.9886	0.9907	0.9919	5
10	0.8022	0.8736	0.8995	0.9136	0.9201	0.9807	0.9882	0.9911	0.9926	0.9935	10
20	0.8625	0.9098	0.9262	0.9388	0.9421	0.9858	0.9912	0.9934	0.9945	0.9952	20
			$n=60$					$n=600$			
1	0.6132	0.7510	0.8104	0.8333	0.8477	0.9659	0.9803	0.9858	0.9888	0.9893	1
5	0.7766	0.8554	0.8880	0.9054	0.9167	0.9781	0.9874	0.9908	0.9924	0.9931	5
10	0.8319	0.8932	0.9163	0.9283	0.9371	0.9833	0.9901	0.9929	0.9939	0.9946	10
20	0.8845	0.9255	0.9400	0.9492	0.9551	0.9878	0.9926	0.9947	0.9955	0.9960	20
			$n=70$					$n=700$			
1	0.6817	0.7858	0.8373	0.8715	0.8937	0.9683	0.9832	0.9884	0.9897	0.9917	1
5	0.8081	0.8738	0.9065	0.9201	0.9305	0.9815	0.9892	0.9922	0.9935	0.9943	5
10	0.8574	0.9082	0.9288	0.9402	0.9477	0.9858	0.9914	0.9938	0.9950	0.9955	10
20	0.9025	0.9356	0.9494	0.9568	0.9619	0.9894	0.9936	0.9953	0.9961	0.9966	20
			$n=80$					$n=800$			
1	0.7184	0.8212	0.8614	0.8977	0.8999	0.9738	0.9852	0.9898	0.9915	0.9922	1
5	0.8349	0.8941	0.9227	0.9332	0.9391	0.9839	0.9906	0.9932	0.9943	0.9950	5
10	0.8779	0.9206	0.9394	0.9491	0.9527	0.9874	0.9926	0.9946	0.9955	0.9961	10
20	0.9141	0.9454	0.9562	0.9625	0.9663	0.9907	0.9945	0.9960	0.9966	0.9970	20
			$n=90$					$n=900$			
1	0.7309	0.8383	0.8826	0.9115	0.9161	0.9773	0.9864	0.9901	0.9922	0.9928	1
5	0.8535	0.9095	0.9313	0.9404	0.9471	0.9857	0.9916	0.9938	0.9949	0.9956	5
10	0.8909	0.9321	0.9464	0.9557	0.9606	0.9887	0.9932	0.9951	0.9960	0.9964	10
20	0.9220	0.9504	0.9615	0.9682	0.9710	0.9918	0.9951	0.9963	0.9970	0.9973	20
			$n=100$					$n=1000$			
1	0.7691	0.8517	0.8877	0.9214	0.9269	0.9787	0.9879	0.9915	0.9934	0.9938	1
5	0.8698	0.9193	0.9353	0.9490	0.9537	0.9872	0.9923	0.9945	0.9954	0.9960	5
10	0.9033	0.9390	0.9514	0.9602	0.9644	0.9901	0.9939	0.9956	0.9965	0.9969	10
20	0.9314	0.9562	0.9654	0.9707	0.9744	0.9927	0.9955	0.9967	0.9973	0.9976	20

Table C.2 Percentiles of correlation coefficient r with $p = 0.3$.

%	$B=2$	$B=4$	$B=6$	$B=8$	$B=10$	$B=2$	$B=4$	$B=6$	$B=8$	$B=10$	%
	$n=20$					$n=200$					
1	0.4337	0.5259	0.5973	0.6471	0.6621	0.9330	0.9620	0.9738	0.9778	0.9801	1
5	0.6111	0.7110	0.7644	0.7901	0.8084	0.9582	0.9757	0.9826	0.9849	0.9872	5
10	0.7099	0.7869	0.8240	0.8453	0.8554	0.9685	0.9810	0.9861	0.9882	0.9898	10
20	0.7930	0.8509	0.8787	0.8923	0.9027	0.9770	0.9861	0.9897	0.9912	0.9923	20
	$n=30$					$n=300$					
1	0.5187	0.6661	0.7516	0.7633	0.8074	0.9514	0.9775	0.9826	0.9852	0.9871	1
5	0.7129	0.8099	0.8505	0.8685	0.8874	0.9720	0.9846	0.9888	0.9906	0.9911	5
10	0.7851	0.8604	0.8868	0.9022	0.9158	0.9787	0.9879	0.9910	0.9925	0.9931	10
20	0.8511	0.9024	0.9211	0.9327	0.9393	0.9843	0.9908	0.9932	0.9943	0.9949	20
	$n=40$					$n=400$					
1	0.6270	0.7612	0.8212	0.8441	0.8730	0.9656	0.9832	0.9866	0.9890	0.9900	1
5	0.7883	0.8674	0.8973	0.9104	0.9200	0.9801	0.9888	0.9912	0.9927	0.9936	5
10	0.8424	0.8995	0.9228	0.9320	0.9392	0.9845	0.9910	0.9932	0.9941	0.9948	10
20	0.8867	0.9288	0.9447	0.9508	0.9568	0.9884	0.9932	0.9949	0.9956	0.9961	20
	$n=50$					$n=500$					
1	0.7346	0.8101	0.8636	0.8919	0.9043	0.9751	0.9860	0.9890	0.9915	0.9924	1
5	0.8365	0.8958	0.9184	0.9328	0.9424	0.9839	0.9905	0.9931	0.9942	0.9951	5
10	0.8777	0.9199	0.9388	0.9490	0.9545	0.9875	0.9927	0.9945	0.9954	0.9960	10
20	0.9118	0.9423	0.9559	0.9624	0.9661	0.9908	0.9945	0.9959	0.9965	0.9970	20
	$n=60$					$n=600$					
1	0.7351	0.8453	0.8908	0.9148	0.9237	0.9789	0.9883	0.9915	0.9929	0.9940	1
5	0.8530	0.9129	0.9343	0.9461	0.9536	0.9868	0.9924	0.9944	0.9953	0.9958	5
10	0.8931	0.9341	0.9505	0.9584	0.9633	0.9898	0.9940	0.9955	0.9962	0.9966	10
20	0.9229	0.9524	0.9639	0.9697	0.9728	0.9924	0.9955	0.9966	0.9972	0.9975	20
	$n=70$					$n=700$					
1	0.7980	0.8756	0.8999	0.9291	0.9379	0.9827	0.9901	0.9928	0.9938	0.9945	1
5	0.8844	0.9280	0.9436	0.9536	0.9607	0.9887	0.9935	0.9952	0.9959	0.9964	5
10	0.9116	0.9459	0.9571	0.9644	0.9696	0.9912	0.9949	0.9961	0.9967	0.9971	10
20	0.9365	0.9601	0.9687	0.9736	0.9769	0.9934	0.9962	0.9971	0.9975	0.9979	20
	$n=80$					$n=800$					
1	0.8172	0.8973	0.9270	0.9365	0.9482	0.9834	0.9912	0.9939	0.9947	0.9950	1
5	0.8975	0.9375	0.9540	0.9596	0.9659	0.9898	0.9943	0.9958	0.9964	0.9968	5
10	0.9225	0.9521	0.9633	0.9686	0.9726	0.9921	0.9954	0.9966	0.9972	0.9975	10
20	0.9432	0.9651	0.9731	0.9773	0.9793	0.9941	0.9965	0.9975	0.9979	0.9981	20
	$n=90$					$n=900$					
1	0.8400	0.9087	0.9366	0.9460	0.9519	0.9869	0.9922	0.9941	0.9953	0.9958	1
5	0.9077	0.9451	0.9600	0.9651	0.9681	0.9915	0.9949	0.9963	0.9970	0.9972	5
10	0.9312	0.9571	0.9684	0.9727	0.9749	0.9931	0.9958	0.9971	0.9975	0.9978	10
20	0.9497	0.9692	0.9765	0.9801	0.9817	0.9949	0.9968	0.9978	0.9981	0.9984	20
	$n=100$					$n=1000$					
1	0.8466	0.9149	0.9363	0.9511	0.9586	0.9884	0.9932	0.9948	0.9957	0.9963	1
5	0.9165	0.9508	0.9626	0.9700	0.9723	0.9921	0.9954	0.9966	0.9971	0.9975	5
10	0.9373	0.9626	0.9712	0.9763	0.9781	0.9938	0.9964	0.9973	0.9977	0.9980	10
20	0.9552	0.9722	0.9784	0.9825	0.9841	0.9953	0.9972	0.9980	0.9983	0.9985	20

Table C.3 Percentiles of correlation coefficient r with $p = 0.4$.

%	$B=2$	$B=4$	$B=6$	$B=8$	$B=10$	$B=2$	$B=4$	$B=6$	$B=8$	$B=10$	%
	$n = 20$					$n = 200$					
1	0.5168	0.6571	0.7436	0.7686	0.7833	0.9494	0.9740	0.9807	0.9834	0.9849	1
5	0.6936	0.7938	0.8448	0.8600	0.8693	0.9700	0.9830	0.9875	0.9895	0.9903	5
10	0.7644	0.8478	0.8801	0.8929	0.9012	0.9767	0.9864	0.9901	0.9916	0.9922	10
20	0.8357	0.8935	0.9124	0.9219	0.9309	0.9833	0.9901	0.9925	0.9937	0.9942	20
	$n = 30$					$n = 300$					
1	0.6468	0.7820	0.8297	0.8634	0.8738	0.9713	0.9836	0.9880	0.9898	0.9904	1
5	0.7933	0.8721	0.9009	0.9143	0.9215	0.9813	0.9890	0.9916	0.9930	0.9936	5
10	0.8450	0.9021	0.9244	0.9334	0.9390	0.9851	0.9913	0.9934	0.9945	0.9950	10
20	0.8906	0.9286	0.9451	0.9533	0.9561	0.9889	0.9935	0.9950	0.9958	0.9963	20
	$n = 40$					$n = 400$					
1	0.7596	0.8493	0.8843	0.9036	0.9158	0.9777	0.9878	0.9908	0.9919	0.9930	1
5	0.8502	0.9051	0.9306	0.9394	0.9448	0.9851	0.9920	0.9938	0.9946	0.9953	5
10	0.8843	0.9270	0.9472	0.9535	0.9585	0.9886	0.9936	0.9951	0.9959	0.9963	10
20	0.9182	0.9482	0.9604	0.9663	0.9693	0.9915	0.9952	0.9963	0.9969	0.9972	20
	$n = 50$					$n = 500$					
1	0.8040	0.8755	0.9106	0.9265	0.9348	0.9828	0.9906	0.9922	0.9935	0.9945	1
5	0.8771	0.9264	0.9459	0.9529	0.9572	0.9887	0.9934	0.9951	0.9957	0.9964	5
10	0.9065	0.9448	0.9576	0.9635	0.9665	0.9911	0.9948	0.9961	0.9967	0.9971	10
20	0.9329	0.9591	0.9687	0.9730	0.9747	0.9934	0.9960	0.9971	0.9975	0.9978	20
	$n = 60$					$n = 600$					
1	0.8162	0.9026	0.9259	0.9396	0.9501	0.9859	0.9918	0.9937	0.9949	0.9956	1
5	0.8962	0.9389	0.9534	0.9613	0.9671	0.9908	0.9946	0.9957	0.9966	0.9969	5
10	0.9224	0.9523	0.9641	0.9700	0.9736	0.9927	0.9956	0.9967	0.9972	0.9975	10
20	0.9444	0.9656	0.9739	0.9777	0.9803	0.9945	0.9966	0.9975	0.9979	0.9982	20
	$n = 70$					$n = 700$					
1	0.8500	0.9155	0.9381	0.9501	0.9560	0.9881	0.9932	0.9946	0.9954	0.9959	1
5	0.9137	0.9481	0.9607	0.9672	0.9712	0.9920	0.9954	0.9965	0.9971	0.9973	5
10	0.9349	0.9605	0.9694	0.9745	0.9776	0.9936	0.9963	0.9972	0.9977	0.9979	10
20	0.9531	0.9715	0.9779	0.9812	0.9832	0.9952	0.9972	0.9979	0.9982	0.9984	20
	$n = 80$					$n = 800$					
1	0.8593	0.9266	0.9467	0.9541	0.9616	0.9893	0.9939	0.9956	0.9961	0.9964	1
5	0.9219	0.9563	0.9669	0.9726	0.9750	0.9931	0.9959	0.9970	0.9974	0.9976	5
10	0.9415	0.9655	0.9743	0.9779	0.9803	0.9946	0.9968	0.9976	0.9979	0.9981	10
20	0.9576	0.9750	0.9807	0.9837	0.9854	0.9959	0.9976	0.9982	0.9984	0.9986	20
	$n = 90$					$n = 900$					
1	0.8919	0.9378	0.9541	0.9634	0.9652	0.9911	0.9945	0.9963	0.9966	0.9968	1
5	0.9356	0.9609	0.9710	0.9755	0.9775	0.9938	0.9964	0.9973	0.9977	0.9979	5
10	0.9512	0.9699	0.9768	0.9801	0.9825	0.9950	0.9971	0.9978	0.9981	0.9983	10
20	0.9642	0.9778	0.9829	0.9854	0.9870	0.9963	0.9978	0.9984	0.9986	0.9987	20
	$n = 100$					$n = 1000$					
1	0.8940	0.9463	0.9587	0.9654	0.9692	0.9915	0.9952	0.9963	0.9971	0.9973	1
5	0.9417	0.9654	0.9732	0.9782	0.9795	0.9946	0.9968	0.9976	0.9980	0.9982	5
10	0.9554	0.9731	0.9792	0.9828	0.9840	0.9957	0.9974	0.9981	0.9984	0.9985	10
20	0.9674	0.9800	0.9844	0.9871	0.9879	0.9967	0.9981	0.9985	0.9988	0.9989	20

Table C.4 Percentiles of correlation coefficient r with $p = 0.5$.

%	$B=2$	$B=4$	$B=6$	$B=8$	$B=10$	$B=2$	$B=4$	$B=6$	$B=8$	$B=10$	%
	$n=20$					$n=200$					
1	0.5975	0.7510	0.7975	0.8410	0.8555	0.9648	0.9811	0.9855	0.9868	0.9880	1
5	0.7492	0.8427	0.8787	0.8990	0.9080	0.9780	0.9876	0.9905	0.9915	0.9922	5
10	0.8194	0.8819	0.9067	0.9220	0.9280	0.9832	0.9900	0.9924	0.9933	0.9938	10
20	0.8696	0.9143	0.9330	0.9413	0.9461	0.9875	0.9923	0.9942	0.9949	0.9954	20
	$n=30$					$n=300$					
1	0.7221	0.8361	0.8730	0.9002	0.9112	0.9779	0.9873	0.9904	0.9915	0.9921	1
5	0.8449	0.9003	0.9230	0.9364	0.9445	0.9858	0.9918	0.9936	0.9944	0.9950	5
10	0.8799	0.9261	0.9404	0.9484	0.9551	0.9889	0.9935	0.9948	0.9955	0.9960	10
20	0.9152	0.9466	0.9567	0.9624	0.9666	0.9917	0.9951	0.9961	0.9966	0.9970	20
	$n=40$					$n=400$					
1	0.7955	0.8827	0.9148	0.9286	0.9311	0.9842	0.9908	0.9928	0.9936	0.9942	1
5	0.8823	0.9285	0.9452	0.9551	0.9577	0.9896	0.9940	0.9952	0.9958	0.9962	5
10	0.9106	0.9457	0.9571	0.9634	0.9673	0.9915	0.9952	0.9962	0.9967	0.9970	10
20	0.9364	0.9605	0.9688	0.9724	0.9752	0.9935	0.9963	0.9971	0.9975	0.9977	20
	$n=50$					$n=500$					
1	0.8455	0.9104	0.9307	0.9465	0.9481	0.9878	0.9924	0.9945	0.9948	0.9953	1
5	0.9084	0.9454	0.9563	0.9651	0.9676	0.9917	0.9949	0.9962	0.9966	0.9970	5
10	0.9293	0.9578	0.9665	0.9715	0.9739	0.9935	0.9959	0.9970	0.9973	0.9976	10
20	0.9482	0.9692	0.9753	0.9788	0.9806	0.9950	0.9970	0.9977	0.9980	0.9982	20
	$n=60$					$n=600$					
1	0.8635	0.9270	0.9456	0.9538	0.9601	0.9891	0.9940	0.9954	0.9958	0.9963	1
5	0.9215	0.9534	0.9658	0.9703	0.9730	0.9931	0.9959	0.9969	0.9973	0.9975	5
10	0.9394	0.9631	0.9728	0.9766	0.9787	0.9944	0.9967	0.9975	0.9978	0.9980	10
20	0.9565	0.9732	0.9798	0.9825	0.9843	0.9957	0.9975	0.9981	0.9983	0.9985	20
	$n=70$					$n=700$					
1	0.8887	0.9422	0.9548	0.9637	0.9655	0.9910	0.9946	0.9961	0.9962	0.9968	1
5	0.9370	0.9622	0.9706	0.9745	0.9773	0.9940	0.9965	0.9974	0.9976	0.9978	5
10	0.9509	0.9705	0.9763	0.9798	0.9822	0.9952	0.9972	0.9979	0.9981	0.9983	10
20	0.9634	0.9784	0.9825	0.9852	0.9865	0.9963	0.9978	0.9984	0.9986	0.9987	20
	$n=80$					$n=800$					
1	0.9068	0.9452	0.9592	0.9669	0.9694	0.9920	0.9955	0.9963	0.9969	0.9970	1
5	0.9455	0.9660	0.9746	0.9787	0.9808	0.9948	0.9970	0.9975	0.9979	0.9981	5
10	0.9570	0.9736	0.9797	0.9830	0.9844	0.9958	0.9976	0.9981	0.9983	0.9985	10
20	0.9683	0.9802	0.9849	0.9872	0.9882	0.9968	0.9981	0.9985	0.9987	0.9989	20
	$n=90$					$n=900$					
1	0.9140	0.9548	0.9656	0.9704	0.9722	0.9930	0.9959	0.9967	0.9972	0.9975	1
5	0.9495	0.9709	0.9777	0.9809	0.9821	0.9953	0.9973	0.9979	0.9981	0.9983	5
10	0.9609	0.9769	0.9820	0.9852	0.9857	0.9963	0.9978	0.9983	0.9985	0.9987	10
20	0.9715	0.9825	0.9867	0.9886	0.9893	0.9972	0.9983	0.9987	0.9989	0.9990	20
	$n=100$					$n=1000$					
1	0.9258	0.9569	0.9703	0.9735	0.9772	0.9941	0.9964	0.9972	0.9974	0.9977	1
5	0.9547	0.9729	0.9799	0.9828	0.9843	0.9960	0.9976	0.9981	0.9984	0.9985	5
10	0.9648	0.9788	0.9839	0.9863	0.9877	0.9968	0.9981	0.9985	0.9987	0.9988	10
20	0.9740	0.9845	0.9878	0.9897	0.9908	0.9975	0.9985	0.9988	0.9990	0.9991	20

Table C.5 Percentiles of correlation coefficient r with $p = 0.6$.

%	$B=2$	$B=4$	$B=6$	$B=8$	$B=10$	$B=2$	$B=4$	$B=6$	$B=8$	$B=10$	%
	$n=20$					$n=200$					
1	0.6799	0.8050	0.8480	0.8707	0.8859	0.9735	0.9847	0.9880	0.9897	0.9904	1
5	0.8009	0.8804	0.9042	0.9162	0.9260	0.9827	0.9895	0.9919	0.9928	0.9936	5
10	0.8456	0.9087	0.9253	0.9350	0.9411	0.9864	0.9916	0.9936	0.9943	0.9948	10
20	0.8906	0.9318	0.9444	0.9517	0.9570	0.9897	0.9936	0.9953	0.9957	0.9961	20
	$n=30$					$n=300$					
1	0.7949	0.8783	0.9075	0.9159	0.9247	0.9835	0.9904	0.9920	0.9928	0.9931	1
5	0.8738	0.9208	0.9406	0.9470	0.9509	0.9889	0.9934	0.9946	0.9953	0.9955	5
10	0.9023	0.9400	0.9522	0.9580	0.9610	0.9911	0.9947	0.9957	0.9963	0.9965	10
20	0.9295	0.9555	0.9646	0.9693	0.9708	0.9931	0.9959	0.9968	0.9972	0.9974	20
	$n=40$					$n=400$					
1	0.8514	0.9138	0.9319	0.9419	0.9455	0.9878	0.9923	0.9941	0.9949	0.9951	1
5	0.9081	0.9431	0.9563	0.9620	0.9656	0.9918	0.9950	0.9960	0.9965	0.9967	5
10	0.9287	0.9563	0.9653	0.9705	0.9727	0.9934	0.9959	0.9967	0.9971	0.9974	10
20	0.9476	0.9675	0.9742	0.9782	0.9794	0.9950	0.9969	0.9975	0.9979	0.9981	20
	$n=50$					$n=500$					
1	0.8755	0.9355	0.9467	0.9551	0.9589	0.9896	0.9943	0.9953	0.9956	0.9963	1
5	0.9234	0.9573	0.9664	0.9708	0.9718	0.9934	0.9960	0.9968	0.9971	0.9974	5
10	0.9418	0.9663	0.9735	0.9762	0.9777	0.9946	0.9968	0.9975	0.9977	0.9980	10
20	0.9568	0.9745	0.9796	0.9819	0.9831	0.9959	0.9975	0.9981	0.9983	0.9985	20
	$n=60$					$n=600$					
1	0.9049	0.9457	0.9554	0.9618	0.9666	0.9918	0.9950	0.9962	0.9964	0.9968	1
5	0.9394	0.9643	0.9719	0.9755	0.9781	0.9946	0.9967	0.9974	0.9977	0.9978	5
10	0.9531	0.9719	0.9776	0.9799	0.9825	0.9957	0.9973	0.9979	0.9981	0.9983	10
20	0.9656	0.9787	0.9830	0.9848	0.9866	0.9967	0.9979	0.9984	0.9986	0.9987	20
	$n=70$					$n=700$					
1	0.9217	0.9532	0.9632	0.9685	0.9722	0.9931	0.9958	0.9965	0.9971	0.9972	1
5	0.9497	0.9698	0.9754	0.9784	0.9807	0.9953	0.9972	0.9977	0.9980	0.9981	5
10	0.9606	0.9761	0.9808	0.9833	0.9848	0.9962	0.9977	0.9982	0.9984	0.9985	10
20	0.9706	0.9819	0.9855	0.9874	0.9885	0.9971	0.9982	0.9986	0.9988	0.9989	20
	$n=80$					$n=800$					
1	0.9274	0.9598	0.9684	0.9729	0.9756	0.9939	0.9964	0.9971	0.9972	0.9975	1
5	0.9544	0.9735	0.9789	0.9819	0.9836	0.9958	0.9976	0.9980	0.9982	0.9984	5
10	0.9644	0.9790	0.9835	0.9857	0.9868	0.9967	0.9980	0.9984	0.9986	0.9987	10
20	0.9742	0.9842	0.9875	0.9892	0.9899	0.9975	0.9985	0.9988	0.9989	0.9990	20
	$n=90$					$n=900$					
1	0.9362	0.9626	0.9726	0.9757	0.9774	0.9946	0.9965	0.9975	0.9977	0.9979	1
5	0.9603	0.9761	0.9812	0.9840	0.9848	0.9964	0.9978	0.9983	0.9984	0.9986	5
10	0.9692	0.9811	0.9852	0.9873	0.9881	0.9971	0.9982	0.9986	0.9987	0.9988	10
20	0.9770	0.9858	0.9887	0.9906	0.9910	0.9977	0.9986	0.9989	0.9991	0.9991	20
	$n=100$					$n=1000$					
1	0.9452	0.9678	0.9753	0.9777	0.9798	0.9950	0.9971	0.9977	0.9979	0.9981	1
5	0.9657	0.9793	0.9829	0.9858	0.9864	0.9967	0.9981	0.9984	0.9986	0.9987	5
10	0.9726	0.9833	0.9867	0.9886	0.9893	0.9973	0.9984	0.9987	0.9989	0.9990	10
20	0.9792	0.9873	0.9899	0.9914	0.9919	0.9980	0.9988	0.9990	0.9991	0.9992	20

Table C.6 Percentiles of correlation coefficient r with $p = 0.7$.

%	$B=2$	$B=4$	$B=6$	$B=8$	$B=10$	$B=2$	$B=4$	$B=6$	$B=8$	$B=10$	%
	$n = 20$					$n = 200$					
1	0.7237	0.8509	0.8772	0.8901	0.8987	0.9797	0.9880	0.9890	0.9903	0.9913	1
5	0.8374	0.9008	0.9198	0.9282	0.9317	0.9861	0.9916	0.9929	0.9938	0.9942	5
10	0.8747	0.9223	0.9352	0.9441	0.9463	0.9890	0.9933	0.9945	0.9951	0.9954	10
20	0.9095	0.9418	0.9518	0.9583	0.9607	0.9915	0.9949	0.9958	0.9963	0.9965	20
	$n = 30$					$n = 300$					
1	0.8375	0.9016	0.9220	0.9321	0.9367	0.9868	0.9917	0.9934	0.9935	0.9942	1
5	0.8979	0.9366	0.9496	0.9564	0.9577	0.9906	0.9943	0.9954	0.9958	0.9961	5
10	0.9199	0.9496	0.9598	0.9652	0.9665	0.9926	0.9955	0.9963	0.9968	0.9969	10
20	0.9407	0.9622	0.9699	0.9736	0.9746	0.9943	0.9966	0.9972	0.9975	0.9977	20
	$n = 40$					$n = 400$					
1	0.8837	0.9304	0.9432	0.9503	0.9537	0.9900	0.9940	0.9951	0.9953	0.9956	1
5	0.9244	0.9535	0.9636	0.9673	0.9694	0.9931	0.9958	0.9966	0.9968	0.9971	5
10	0.9401	0.9635	0.9707	0.9738	0.9754	0.9945	0.9965	0.9972	0.9975	0.9977	10
20	0.9564	0.9724	0.9778	0.9799	0.9814	0.9958	0.9974	0.9979	0.9981	0.9983	20
	$n = 50$					$n = 500$					
1	0.9038	0.9479	0.9560	0.9612	0.9631	0.9923	0.9947	0.9958	0.9963	0.9965	1
5	0.9396	0.9644	0.9706	0.9741	0.9754	0.9947	0.9966	0.9973	0.9975	0.9977	5
10	0.9529	0.9715	0.9765	0.9793	0.9808	0.9958	0.9973	0.9978	0.9980	0.9981	10
20	0.9652	0.9781	0.9822	0.9840	0.9856	0.9967	0.9979	0.9983	0.9985	0.9986	20
	$n = 60$					$n = 600$					
1	0.9239	0.9595	0.9641	0.9693	0.9695	0.9932	0.9959	0.9964	0.9969	0.9971	1
5	0.9506	0.9699	0.9757	0.9792	0.9802	0.9955	0.9973	0.9977	0.9979	0.9981	5
10	0.9608	0.9762	0.9808	0.9833	0.9841	0.9964	0.9978	0.9981	0.9983	0.9984	10
20	0.9707	0.9820	0.9854	0.9873	0.9877	0.9972	0.9983	0.9986	0.9987	0.9988	20
	$n = 70$					$n = 700$					
1	0.9376	0.9619	0.9707	0.9732	0.9755	0.9947	0.9966	0.9972	0.9974	0.9974	1
5	0.9586	0.9753	0.9800	0.9818	0.9834	0.9961	0.9976	0.9980	0.9982	0.9984	5
10	0.9672	0.9798	0.9840	0.9852	0.9867	0.9969	0.9981	0.9984	0.9986	0.9987	10
20	0.9759	0.9845	0.9879	0.9889	0.9898	0.9976	0.9985	0.9988	0.9989	0.9990	20
	$n = 80$					$n = 800$					
1	0.9442	0.9665	0.9740	0.9757	0.9777	0.9950	0.9969	0.9975	0.9978	0.9979	1
5	0.9639	0.9780	0.9822	0.9835	0.9849	0.9966	0.9979	0.9983	0.9984	0.9986	5
10	0.9711	0.9823	0.9858	0.9870	0.9879	0.9973	0.9983	0.9987	0.9987	0.9988	10
20	0.9784	0.9865	0.9891	0.9903	0.9910	0.9979	0.9987	0.9990	0.9990	0.9991	20
	$n = 90$					$n = 900$					
1	0.9536	0.9718	0.9772	0.9796	0.9797	0.9958	0.9974	0.9978	0.9979	0.9980	1
5	0.9683	0.9811	0.9842	0.9864	0.9864	0.9971	0.9982	0.9985	0.9986	0.9987	5
10	0.9746	0.9849	0.9872	0.9888	0.9890	0.9976	0.9986	0.9988	0.9989	0.9990	10
20	0.9809	0.9883	0.9904	0.9915	0.9919	0.9982	0.9989	0.9991	0.9991	0.9992	20
	$n = 100$					$n = 1000$					
1	0.9539	0.9732	0.9798	0.9812	0.9828	0.9961	0.9977	0.9980	0.9982	0.9983	1
5	0.9713	0.9829	0.9862	0.9869	0.9883	0.9974	0.9984	0.9986	0.9988	0.9989	5
10	0.9769	0.9862	0.9887	0.9896	0.9905	0.9979	0.9987	0.9989	0.9990	0.9991	10
20	0.9828	0.9894	0.9914	0.9923	0.9928	0.9983	0.9990	0.9992	0.9992	0.9993	20

Table C.7 Percentiles of correlation coefficient r with $p = 0.8$.

%	$B=2$	$B=4$	$B=6$	$B=8$	$B=10$	$B=2$	$B=4$	$B=6$	$B=8$	$B=10$	%
			$n=20$					$n=200$			
1	0.7699	0.8675	0.8953	0.8998	0.9153	0.9827	0.9894	0.9911	0.9916	0.9920	1
5	0.8642	0.9171	0.9320	0.9353	0.9433	0.9880	0.9927	0.9940	0.9945	0.9945	5
10	0.8929	0.9347	0.9449	0.9490	0.9541	0.9904	0.9940	0.9951	0.9956	0.9956	10
20	0.9220	0.9505	0.9591	0.9626	0.9653	0.9927	0.9954	0.9962	0.9966	0.9967	20
			$n=30$					$n=300$			
1	0.8661	0.9224	0.9365	0.9400	0.9415	0.9889	0.9929	0.9939	0.9942	0.9946	1
5	0.9111	0.9476	0.9581	0.9585	0.9608	0.9925	0.9952	0.9958	0.9961	0.9963	5
10	0.9301	0.9585	0.9659	0.9675	0.9690	0.9939	0.9961	0.9967	0.9970	0.9971	10
20	0.9484	0.9682	0.9738	0.9757	0.9767	0.9952	0.9970	0.9974	0.9977	0.9978	20
			$n=40$					$n=400$			
1	0.8971	0.9384	0.9518	0.9553	0.9596	0.9921	0.9948	0.9953	0.9958	0.9960	1
5	0.9382	0.9606	0.9677	0.9713	0.9726	0.9943	0.9964	0.9970	0.9972	0.9972	5
10	0.9514	0.9682	0.9740	0.9768	0.9778	0.9954	0.9971	0.9976	0.9978	0.9978	10
20	0.9635	0.9760	0.9802	0.9822	0.9830	0.9964	0.9977	0.9981	0.9983	0.9984	20
			$n=50$					$n=500$			
1	0.9236	0.9538	0.9608	0.9633	0.9662	0.9933	0.9957	0.9964	0.9966	0.9970	1
5	0.9494	0.9697	0.9751	0.9761	0.9773	0.9955	0.9972	0.9976	0.9977	0.9979	5
10	0.9603	0.9758	0.9802	0.9811	0.9821	0.9963	0.9977	0.9980	0.9982	0.9983	10
20	0.9704	0.9815	0.9846	0.9857	0.9865	0.9971	0.9982	0.9985	0.9986	0.9987	20
			$n=60$					$n=600$			
1	0.9389	0.9617	0.9690	0.9706	0.9717	0.9945	0.9965	0.9971	0.9973	0.9974	1
5	0.9602	0.9743	0.9793	0.9801	0.9817	0.9962	0.9976	0.9980	0.9982	0.9982	5
10	0.9680	0.9793	0.9831	0.9844	0.9857	0.9969	0.9981	0.9984	0.9985	0.9985	10
20	0.9756	0.9842	0.9868	0.9880	0.9890	0.9976	0.9985	0.9987	0.9988	0.9989	20
			$n=70$					$n=700$			
1	0.9479	0.9672	0.9720	0.9760	0.9755	0.9952	0.9971	0.9974	0.9975	0.9977	1
5	0.9655	0.9789	0.9815	0.9837	0.9839	0.9968	0.9980	0.9982	0.9984	0.9985	5
10	0.9724	0.9828	0.9854	0.9867	0.9873	0.9974	0.9984	0.9985	0.9987	0.9988	10
20	0.9790	0.9866	0.9888	0.9899	0.9904	0.9980	0.9987	0.9989	0.9990	0.9991	20
			$n=80$					$n=800$			
1	0.9534	0.9725	0.9763	0.9791	0.9801	0.9960	0.9973	0.9978	0.9980	0.9981	1
5	0.9698	0.9809	0.9844	0.9854	0.9863	0.9972	0.9982	0.9985	0.9986	0.9987	5
10	0.9753	0.9849	0.9872	0.9884	0.9892	0.9977	0.9985	0.9988	0.9989	0.9989	10
20	0.9814	0.9883	0.9902	0.9912	0.9917	0.9982	0.9989	0.9990	0.9991	0.9992	20
			$n=90$					$n=900$			
1	0.9604	0.9747	0.9801	0.9808	0.9818	0.9964	0.9977	0.9981	0.9981	0.9983	1
5	0.9727	0.9829	0.9861	0.9871	0.9878	0.9975	0.9984	0.9987	0.9988	0.9988	5
10	0.9784	0.9864	0.9890	0.9895	0.9901	0.9980	0.9987	0.9989	0.9990	0.9990	10
20	0.9836	0.9897	0.9915	0.9922	0.9924	0.9984	0.9990	0.9992	0.9992	0.9993	20
			$n=100$					$n=1000$			
1	0.9641	0.9774	0.9813	0.9820	0.9835	0.9969	0.9979	0.9983	0.9983	0.9984	1
5	0.9759	0.9853	0.9877	0.9885	0.9892	0.9978	0.9986	0.9988	0.9989	0.9989	5
10	0.9808	0.9881	0.9898	0.9907	0.9914	0.9982	0.9988	0.9990	0.9991	0.9991	10
20	0.9854	0.9908	0.9921	0.9928	0.9934	0.9986	0.9991	0.9992	0.9993	0.9993	20

Table C.8 Percentiles of correlation coefficient r with $p = 0.9$.

%	$B=2$	$B=4$	$B=6$	$B=8$	$B=10$	$B=2$	$B=4$	$B=6$	$B=8$	$B=10$	%
			$n=20$					$n=200$			
1	0.8246	0.8899	0.9101	0.9146	0.9216	0.9865	0.9909	0.9914	0.9921	0.9926	1
5	0.8847	0.9296	0.9400	0.9452	0.9471	0.9903	0.9936	0.9942	0.9947	0.9949	5
10	0.9093	0.9440	0.9514	0.9554	0.9575	0.9920	0.9949	0.9954	0.9958	0.9959	10
20	0.9334	0.9573	0.9634	0.9661	0.9673	0.9939	0.9959	0.9965	0.9967	0.9969	20
			$n=30$					$n=300$			
1	0.8914	0.9341	0.9412	0.9450	0.9476	0.9905	0.9937	0.9945	0.9949	0.9953	1
5	0.9274	0.9547	0.9608	0.9648	0.9646	0.9936	0.9957	0.9963	0.9965	0.9966	5
10	0.9426	0.9635	0.9684	0.9713	0.9723	0.9948	0.9965	0.9970	0.9972	0.9973	10
20	0.9563	0.9717	0.9753	0.9780	0.9787	0.9959	0.9973	0.9977	0.9979	0.9979	20
			$n=40$					$n=400$			
1	0.9176	0.9500	0.9567	0.9588	0.9617	0.9928	0.9954	0.9959	0.9960	0.9961	1
5	0.9481	0.9664	0.9709	0.9724	0.9737	0.9951	0.9968	0.9972	0.9973	0.9975	5
10	0.9575	0.9729	0.9768	0.9781	0.9788	0.9960	0.9975	0.9978	0.9978	0.9980	10
20	0.9678	0.9790	0.9823	0.9830	0.9838	0.9969	0.9980	0.9983	0.9984	0.9984	20
			$n=50$					$n=500$			
1	0.9364	0.9615	0.9672	0.9700	0.9698	0.9944	0.9963	0.9967	0.9969	0.9970	1
5	0.9575	0.9740	0.9769	0.9793	0.9799	0.9963	0.9974	0.9977	0.9978	0.9980	5
10	0.9662	0.9786	0.9816	0.9829	0.9835	0.9970	0.9979	0.9982	0.9983	0.9984	10
20	0.9744	0.9834	0.9858	0.9867	0.9874	0.9976	0.9984	0.9986	0.9987	0.9987	20
			$n=60$					$n=600$			
1	0.9489	0.9697	0.9724	0.9754	0.9756	0.9950	0.9970	0.9973	0.9973	0.9975	1
5	0.9652	0.9781	0.9813	0.9826	0.9828	0.9967	0.9978	0.9981	0.9982	0.9983	5
10	0.9720	0.9822	0.9848	0.9860	0.9861	0.9973	0.9983	0.9985	0.9986	0.9986	10
20	0.9783	0.9862	0.9882	0.9892	0.9891	0.9979	0.9987	0.9988	0.9989	0.9989	20
			$n=70$					$n=700$			
1	0.9587	0.9733	0.9761	0.9771	0.9781	0.9961	0.9974	0.9977	0.9978	0.9979	1
5	0.9712	0.9815	0.9838	0.9850	0.9855	0.9973	0.9982	0.9984	0.9985	0.9986	5
10	0.9768	0.9849	0.9870	0.9881	0.9881	0.9978	0.9985	0.9987	0.9988	0.9988	10
20	0.9821	0.9883	0.9900	0.9907	0.9911	0.9983	0.9989	0.9990	0.9991	0.9991	20
			$n=80$					$n=800$			
1	0.9607	0.9758	0.9792	0.9799	0.9814	0.9965	0.9977	0.9980	0.9982	0.9981	1
5	0.9744	0.9838	0.9856	0.9866	0.9877	0.9976	0.9984	0.9986	0.9987	0.9988	5
10	0.9791	0.9867	0.9886	0.9891	0.9897	0.9981	0.9987	0.9989	0.9989	0.9990	10
20	0.9840	0.9896	0.9912	0.9917	0.9920	0.9985	0.9990	0.9991	0.9992	0.9992	20
			$n=90$					$n=900$			
1	0.9682	0.9793	0.9821	0.9831	0.9825	0.9970	0.9979	0.9981	0.9982	0.9983	1
5	0.9783	0.9857	0.9874	0.9886	0.9883	0.9979	0.9986	0.9987	0.9988	0.9989	5
10	0.9822	0.9882	0.9899	0.9907	0.9906	0.9983	0.9988	0.9990	0.9991	0.9991	10
20	0.9861	0.9910	0.9921	0.9928	0.9929	0.9986	0.9991	0.9992	0.9993	0.9993	20
			$n=100$					$n=1000$			
1	0.9712	0.9812	0.9839	0.9843	0.9846	0.9973	0.9982	0.9984	0.9985	0.9985	1
5	0.9799	0.9872	0.9889	0.9891	0.9897	0.9981	0.9987	0.9989	0.9990	0.9990	5
10	0.9836	0.9894	0.9909	0.9913	0.9918	0.9984	0.9990	0.9991	0.9992	0.9992	10
20	0.9873	0.9919	0.9929	0.9933	0.9937	0.9988	0.9992	0.9993	0.9994	0.9994	20

Table C.9 Percentiles of correlation coefficient r with $p = 1$.

%	$B=2$	$B=4$	$B=6$	$B=8$	$B=10$	$B=2$	$B=4$	$B=6$	$B=8$	$B=10$	%
	$n=20$					$n=200$					
1	0.8445	0.9068	0.9185	0.9259	0.9239	0.9876	0.9918	0.9923	0.9928	0.9930	1
5	0.9041	0.9379	0.9459	0.9510	0.9514	0.9917	0.9942	0.9948	0.9952	0.9952	5
10	0.9247	0.9498	0.9574	0.9600	0.9617	0.9932	0.9952	0.9958	0.9961	0.9961	10
20	0.9418	0.9622	0.9667	0.9697	0.9706	0.9948	0.9964	0.9968	0.9969	0.9970	20
	$n=30$					$n=300$					
1	0.9099	0.9415	0.9458	0.9517	0.9528	0.9916	0.9943	0.9952	0.9954	0.9950	1
5	0.9387	0.9603	0.9643	0.9670	0.9688	0.9943	0.9963	0.9965	0.9967	0.9967	5
10	0.9513	0.9672	0.9716	0.9730	0.9746	0.9954	0.9970	0.9972	0.9974	0.9973	10
20	0.9618	0.9750	0.9781	0.9795	0.9804	0.9965	0.9976	0.9978	0.9979	0.9980	20
	$n=40$					$n=400$					
1	0.9339	0.9565	0.9612	0.9616	0.9651	0.9943	0.9960	0.9963	0.9963	0.9963	1
5	0.9552	0.9705	0.9736	0.9745	0.9766	0.9959	0.9972	0.9975	0.9975	0.9976	5
10	0.9632	0.9755	0.9785	0.9797	0.9810	0.9967	0.9977	0.9979	0.9979	0.9981	10
20	0.9717	0.9812	0.9837	0.9844	0.9854	0.9974	0.9982	0.9984	0.9984	0.9985	20
	$n=50$					$n=500$					
1	0.9466	0.9646	0.9683	0.9706	0.9716	0.9951	0.9966	0.9970	0.9970	0.9970	1
5	0.9643	0.9762	0.9783	0.9805	0.9808	0.9967	0.9977	0.9979	0.9980	0.9980	5
10	0.9712	0.9804	0.9826	0.9840	0.9841	0.9973	0.9981	0.9983	0.9984	0.9984	10
20	0.9776	0.9850	0.9867	0.9878	0.9879	0.9979	0.9986	0.9987	0.9988	0.9988	20
	$n=60$					$n=600$					
1	0.9545	0.9712	0.9755	0.9746	0.9757	0.9960	0.9972	0.9975	0.9976	0.9976	1
5	0.9700	0.9805	0.9831	0.9841	0.9834	0.9972	0.9980	0.9983	0.9984	0.9984	5
10	0.9761	0.9842	0.9859	0.9869	0.9868	0.9977	0.9984	0.9986	0.9987	0.9987	10
20	0.9814	0.9877	0.9891	0.9897	0.9900	0.9982	0.9988	0.9989	0.9990	0.9990	20
	$n=70$					$n=700$					
1	0.9654	0.9760	0.9782	0.9794	0.9798	0.9966	0.9976	0.9979	0.9979	0.9980	1
5	0.9746	0.9834	0.9850	0.9862	0.9864	0.9976	0.9984	0.9985	0.9986	0.9986	5
10	0.9795	0.9864	0.9879	0.9887	0.9892	0.9981	0.9987	0.9988	0.9988	0.9989	10
20	0.9844	0.9895	0.9906	0.9914	0.9917	0.9985	0.9990	0.9991	0.9991	0.9991	20
	$n=80$					$n=800$					
1	0.9692	0.9789	0.9807	0.9826	0.9826	0.9972	0.9980	0.9981	0.9982	0.9982	1
5	0.9783	0.9856	0.9869	0.9880	0.9879	0.9980	0.9986	0.9987	0.9988	0.9988	5
10	0.9822	0.9884	0.9894	0.9901	0.9901	0.9983	0.9989	0.9989	0.9990	0.9990	10
20	0.9864	0.9909	0.9918	0.9923	0.9924	0.9987	0.9991	0.9992	0.9992	0.9993	20
	$n=90$					$n=900$					
1	0.9727	0.9822	0.9837	0.9837	0.9837	0.9975	0.9982	0.9984	0.9983	0.9985	1
5	0.9818	0.9877	0.9886	0.9888	0.9891	0.9982	0.9987	0.9989	0.9989	0.9989	5
10	0.9849	0.9900	0.9909	0.9910	0.9912	0.9985	0.9990	0.9991	0.9991	0.9991	10
20	0.9881	0.9921	0.9929	0.9931	0.9933	0.9989	0.9992	0.9993	0.9993	0.9993	20
	$n=100$					$n=1000$					
1	0.9754	0.9834	0.9837	0.9852	0.9854	0.9976	0.9984	0.9985	0.9986	0.9985	1
5	0.9832	0.9885	0.9895	0.9900	0.9904	0.9983	0.9989	0.9990	0.9990	0.9990	5
10	0.9862	0.9905	0.9916	0.9919	0.9922	0.9986	0.9991	0.9992	0.9992	0.9992	10
20	0.9891	0.9926	0.9934	0.9939	0.9940	0.9989	0.9993	0.9994	0.9994	0.9994	20

Index

WILEY SERIES IN PROBABILITY AND STATISTICS

ESTABLISHED BY WALTER A. SHEWHART AND SAMUEL S. WILKS
Editors
Vic Barnett, Ralph A. Bradley, Nicholas I. Fisher, J. Stuart Hunter, J.B. Kadane, David G. Kendall, David W. Scott, Adrian F. M. Smith, Jozef L. Teugels, Geoffrey S. Watson

Probability and Statistics

ANDERSON • An Introduction to Multivariate Statistical Analysis, *Second Edition*
*ANDERSON • The Statistical Analysis of Time Series
ARNOLD, BALAKRISHNAN and NAGARAJA • A First Course in Order Statistics
BACCELLI, COHEN, OLSDER and QUADRAT • Synchronization and Linearity: An Algebra for Discrete Event Systems
BARTOSZYNSKI and NIEWIADOMSKA-BUGAJ • Probability and Statistical Inference
BASILEVSKY • Statistical Factor Analysis and Related Methods
BERNARDO and SMITH • Bayesian Statistical Concepts and Theory
BHATTACHARYYA and JOHNSON • Statistical Concepts and Methods
BILLINGSLEY • Convergence of Probability Measures
BILLINGSLEY • Probability and Measure, *Third Edition*
BRANDT, FRANKEN and LISEK • Stationary Stochastic Models
CAINES • Linear Stochastic Systems
CAIROLI and DALANG • Sequential Stochastic Optimization
CHEN • Recursive Estimation and Control for Stochastic Systems
CONSTANTINE • Combinatorial Theory and Statistical Design
COOK and WEISBERG • An Introduction to Regression Graphics
COVER and THOMAS • Elements of Information Theory
CSÖRGÓ and HORVATH • Weighted Approximations in Probability and Statistics
*DOOB • Stochastic Processes
DUDEWICZ and MISHRA • Modern Mathematical Statistics
DUPUIS and ELLIS • A Weak Convergence Approach to the Theory of Large Deviations
ENDERS • Applied Econometric Time Series
ETHIER and KURTZ • Markov Processes: Characterization and Convergence
FELLER • An Introduction to Probability Theory and Its Applications, Volume 1, *Third Edition*, Revised; Volume II, *Second Edition*
FREEMAN and SMITH • Aspects of Uncertainty: A Tribute to D.V. Lindley
FULLER • Introduction to Statistical Time Series, *Second Edition*
FULLER • Measurement Error Models
GELFAND and SMITH • Bayesian Computation
GHOSH, MUKHOPADHYAY and SEN • Sequential Estimation
GIFI • Nonlinear Multivariate Analysis
GUTTORP • Statistical Inference for Branching Processes
HALD • A History of Probability and Statistics and Their Applications before 1750
HALL • Introduction to the Theory of Coverage Processes
HAND • Construction and Assessment of Classification Rules
HANNAN and DEISTLER • The Statistical Theory of Linear Systems
*HANSEN, HURWITZ and MADOW • Sample Survey Methods and Theory, 2 Vol Set
HEDAYAT and SINHA • Design and Inference in Finite Population Sampling
HOEL • Introduction to Mathematical Statistics, *Fifth Edition*
HUBER • Robust Statistics
JOHNSON, KOTZ and KEMP • Univariate Discrete Distribution
JUPP and MARDIA • Statistics of Directional Data
JUREK and MASON • Operator-Limit Distributions in Probability Theory
KASS • The Geometrical Foundations of Asymptotic Inference
KAUFMAN and ROUSSEEUW • Finding Groups in Data: An Introduction to Cluster Analysis
KELLY • Probability, Statistics and Optimization

*Now available in a lower priced paperback edition in the Wiley Classics Library

LAMPERTI • Probability: A Survey of the Mathematical Theory, *Second Edition*
LARSON • Introduction to Probability Theory and Statistical Inference, *Third Edition*
LESSLER and KALSBEEK • Nonsampling Error in Surveys
LINDVALL • Lectures on the Coupling Method
McLACHLAN • Discriminant Analysis and Statistical Pattern Recognition
McLACHLAN and KRISHNAN • The EM Algorithm
McNEIL • Epidemiological Research Methods
MANTON, WOODBURY and TOLLEY • Statistical Applications Using Fuzzy Sets
MARDIA • The Art of Statistical Science: A Tribute to G.S. Watson
MARDIA and DRYDEN • Statistical Analysis of Shape
MOLCHANOV • Statistics of the Boolean Model
MORGENTHALER and TUKEY • Configural Polysampling: A Route to Practical Robustness
MUIRHEAD • Aspects of Multivariate Statistical Theory
OLIVER and SMITH • Inference Diagrams, Belief Nets and Decision Analysis
*PARZEN • Modern Probability Theory and Its Applications
PRESS • Bayesian Statistics: Principles, Models, and Applications
PUKELSHEIM • Optimal Experimental Design
PURI and SEN • Nonparametric Methods in General Linear Models
PURI, VILAPLANA and WERTZ • New Perspectives in Theoretical and Applied Statistics
RAO • Asymptotic Theory of Statistical Inference
RAO • Linear Statistical Inference and Its Applications, *Second Edition*
RAO and SHANBHAG • Choquet-Deny Type Functional Equations and Applications to Stochastic Models
RENCHER • Methods of Multivariate Analysis
ROBERTSON, WRIGHT and DYKSTRA • Order Restricted Statistical Inference
ROGERS and WILLIAMS • Diffusions, Markov Processes, and Martingales, Volume I: Foundations, *Second Edition*, Volume II: Itô Calculus
ROHATGI • An Introduction to Probability Theory and Mathematical Statistics
ROSS • Stochastic Processes
RUBINSTEIN • Simulation and the Monte Carlo Method
RUBINSTEIN and SHAPIRO • Discrete Event Systems: Sensitivity Analysis and Stochastic Optimization by the Score Function Method
RUZSA and SZEKELY • Algebraic Probability Theory
SCHEFFE • The Analysis of Variance
SEBER • Linear Regression Analysis
SEBER • Multivariate Observations
SEBER and WILD • Nonlinear Regression
SERFLING • Approximation Theorems of Mathematical Statistics
SHORACK and WELLNER • Empirical Processes with Applications to Statistics
SMALL and McLEISH • Hilbert Space Methods in Probability and Statistical Inference
STAPLETON • Linear Statistical Models
STAUDTE and SHEATHER • Robust Estimation and Testing
STOYANOV • Counterexamples in Probability
STOYANOV • Counterexamples in Probability, *Second Edition*
STYAN • The Collected Papers of T.W. Anderson 1943–1985
TANAKA • Time Series Analysis. Nonstationary and Noninvertible Distribution Theory
THOMPSON and SEBER • Adaptive Sampling
WELSH • Aspects of Statistical Inference
WHITTAKER • Graphic Models in Applied Multivariate Statistics
WILLIAMS • Diffusions, Markov Processes, and Martingales, Volume 1. *Second Edition*
YANG • The Construction Theory of Denumerable Markov Processes

Applied Probability and Statistics
ABRAHAM and LEDOLTER • Statistical Methods for Forecasting
AGRESTI • Analysis of Ordinal Categorical Data
AGRESTI • Categorical Data Analysis
AGRESTI • An Introduction to Categorical Data Analysis

*Now available in a lower priced paperback edition in the Wiley Classics Library

ANDERSON and LOYNES • The Teaching of Practical Statistics
ANDERSON, AUQUIER, HAUCH, OAKES, VANDAELE and WEISBERG • Statistical Methods for Comparative Studies
ARMITAGE and DAVID (editors) • Advances in Biometry
*ARTHANARI and DODGE • Mathematical Programming in Statistics
ASMUSSEN • Applied Probability and Queues
*BAILEY • The Elements of Stochastic Processes with Applications to the Natural Sciences
BARNETT and LEWIS • Outliers in Statistical Data, *Third Edition*
BARTHOLOMEW, FORBES, and McLEAN • Statistical Techniques for Manpower Planning, *Second Edition*
BATES and WATTS • Nonlinear Regression Analysis and Its Applications
BECHOFER, SANTNER and GOLDSMAN • Design and Analysis of Experiments for Statistical Selection, Screening and Multiple Comparisons
BELSLEY • Conditioning Diagnostics: Collinearity and Weak Data in Regression
BELSLEY, KUH and WELSCH • Regression Diagnostics: Identifying Influential Data and Sources of Collinearity •
BERNARDO and SMITH • Bayesian Theory
BERRY, CHALONER and GEWEKE • Bayesian Analysis in Statistics and Econometrics Essays in Honor of Arnold Zellner
BHAT • Elements of Applied Stochastic Processes, *Second Edition*
BHATTACHARYA and WAYMIRE • Stochastic Processes with Applications
BIEMER, GROVES, LYBERG, MATHIOWETZ and SUDMAN • Measurement Errors in Surveys
BIRKES and DODGE • Alternative Methods of Regression
BLOOMFIELD • Fourier Analysis of Time Series: An Introduction
BOLLEN • Structural Equations with Latent Variables
BOULEAU • Numerical Methods for Stochastic Processes
BOX • R.A.Fisher, the Life of a Scientist
BOX and DRAPER • Empirical Model-Building and Response Surfaces
BOX and DRAPER • Evolutionary Operation: A Statistical Method for Process Improvement
BOX, HUNTER and HUNTER • Statistics for Experimenters: An Introduction to Design, Data Analysis, and Model Building
BROWN and HOLLANDER • Statistics: A Biomedical Introduction
BUCKLEW • Large Deviation Techniques in Decision, Simulation, and Estimation
BUNKE and BUNKE • Non-linear Regression, Functional Relations and Robust Methods: Statistical Methods of Model Building
CHATTERJEE and HADI • Sensitivity Analysis in Linear Regression
CHATTERJEE and PRICE • Regression Analysis by Example, *Second Edition*
CLARKE and DISNEY • Probability and Random Processes: A First Course with Applications, *Second Edition*
COCHRAN • Sampling Techniques, *Third Edition*
*COCHRAN and COX • Experimental Designs, *Second Edition*
CONOVER • Practical Nonparametric Statistics, *Second Edition*
CORNELL • Experiments with Mixtures, Designs, Models, and the Analysis of Mixture Data, *Second Edition*
COX • A Handbook of Introductory Statistical Methods
*COX • Planning of Experiments
COX, BINDER, CHINNAPPA, CHRISTIANSON, COLLEDGE, and KOTT • Business Survey Methods
CRESSIE • Statistics for Spatial Data, *Revised Edition*
DANIEL • Applications of Statistics to Industrial Experimentation
DANIEL • Biostatistics: A Foundation for Analysis in the Health Sciences, *Sixth Edition*
DANIEL Fitting Equations into Data: Computer Analysis of Multifactor Data, *Second Edition*
DAVID • Order Statistics, *Second Edition*
*DEGROOT, FIENBERG and KADANE • Statistics and the Law
*DEMING • Sample Design in Business Research
DILLON and GOLDSTEIN • Multivariate Analysis: Methods and Applications
DOWDY and WEARDEN • Statistics for Research, *Second Edition*
DRAPER and SMITH • Applied Regression Analysis, *Second Edition*
DUNN • Basic Statistics: A Primer for the Biomedical Sciences, *Second Edition*

*Now available in a lower priced paperback edition in the Wiley Classics Library

DUNN and CLARK • Applied Statistics: Analysis of Variance and Regression, *Second Edition*
DUPUIS • Large Deviations
ELANDT-JOHNSON and JOHNSON • Survival Models and Data Analysis
EVANS, PEACOCK and HASTINGS • Statistical Distributions, *Second Edition*
FISHER and VAN BELLE • Biostatistics: A Methodology for the Health Sciences
FLEISS • The Design and Analysis of Clinical Experiments
FLEISS • Statistical Methods for Rates and Proportion, *Second Edition*
FLEMING and HARRINGTON • Counting Processes and Survival Analysis
FLURY • Common Principal Components and Related Multivariate Models
GALLANT • Nonlinear Statistical Models
GHOSH • Estimation
GLASSERMAN and YAO • Monotone Structure in Discrete-Event Systems
GNANADESIKAN • Analysis, *Second Edition*
GOLDSTEIN and LEWIS • Assessment: Problems, Developments and Statistical Issues
GOLDSTEIN and WOOFF • Bayes Linear Statistics
GREENWOOD and NIKULIN • A Guide to Chi-squared Testing
GROSS and HARRIS • Fundamentals of Queuing Theory, *Second Edition*
GROVES • Survey Errors and Survey Costs
GROVES, BIEMER, LYBERG, MASSEY, NICHOLLS and WAKSBERG • Telephone Survey Methodology
HAHN • and MEEKER • Statistical Intervals: A Guide for Practitioners
HAND • Construction and Assessment of Classification Rules
HAND • Discrimination and Classification
*HANSEN, HURWITZ and MADOW • Sample Survey Methods and Theory, Volume 1: Methods and Applications
*HANSEN, HURWITZ and MADOW • Sample Survey Methods and Theory, Volume II: Theory
HEIBERGER • Computation for the Analysis of Designed Experiments
HELLER • MACSYMA for Statisticians
HINKELMAN and KEMPTHORNE • Design and Analysis of Experiments, Volume 1: Introduction to Experimental Design
HOAGLIN, MOSTELLER and TUKEY • Exploratory Approach to Analysis of Variance
HOAGLIN, MOSTELLER and TUKEY • Exploring Data Tables, Trends and Shapes
HOAGLIN, MOSTELLER and TUKEY • Understanding Robust and Exploratory Data Analysis
HOCHBERG and TAMHANE • Multiple Comparison Procedures
HOCKING • Methods and Applications of Linear Models: Regression and the Analysis of Variance
HOEL • Elementary Statistics, *Fifth Edition*
HOGG and KLUGMAN • Loss Distributions
HOLLANDER and WOLFE • Nonparametric Statistical Methods
HOSMER and LEMESHOW • Applied Logistic Regression
HØYLAND and RAUSAND • System Reliability Theory: Models and Statistical Methods
HUBERTY • Applied Discriminant Analysis
IMAN and CONOVER • Modern Business Statistics
JACKSON • A User's Guide to Principle Components
JOHN • Statistical Methods in Engineering and Quality Assurance
JOHNSON • Multivariate Statistical Simulation
JOHNSON & KOTZ • Distributions in Statistics
• Continuous Multivariate Distributions
JOHNSON, KOTZ and BALAKRISHNAN • Continuous Univariate Distributions, Volume 1, *Second Edition;* Volume 2, *Second Edition*
JOHNSON, KOTZ, and BALAKRISHNAN • Discrete Multivariate Distributions
JOHNSON, KOTZ and KEMP • Univariate Discrete Distribution, *Second Edition*
JUDGE, GRIFFITHS, HILL, LÜTKEPOHL, and LEE • The Theory and Practice of Econometrics, *Second Edition*
JUDGE, HILL, GRIFFITHS, LÜTKEPOHL, and LEE • Introduction to the Theory and Practice of Econometrics, *Second Edition*
JURECKOVÁ and SEN • Robust Statistical Procedures: Asymptotics and Interrelations
KADANE • Bayesian Methods and Ethics in a Clinical Trial Design
KADANE and SCHUM • A Probabilistic Analysis of the Sacco and Vanzetti Evidence

*Now available in a lower priced paperback edition in the Wiley Classics Library

KALBFLEISCH and PRENTICE • The Statistical Analysis of Failure Time Data
KASPRZYK, DUNCAN, KALTON and SINGH • Panel Surveys
KHURI • Advanced Calculus with Applications in Statistics
KISH • Statistical Design for Research
*KISH • Survey Sampling
KOTZ • Personalities
KOVALENKO, KUZNETZOV and PEGG • Mathematical Theory of Reliability of Time-dependent Systems with Practical Applications
LAD • Operational Subjective Statistical Methods: A Mathematical, Philosophical and Historical Introduction
LANGE, RYAN, BILLARD, BRILLINGER, CONQUEST, and GREENHOUSE • Case Studies in Biometry
LAWLESS • Statistical Models and Methods for Lifetime Data
LEE • Statistical Methods for Survival Data Analysis, *Second Edition*
LePAGE and BILLARD • Exploring the Limits of Bootstrap
LESSLER and KALSBEEK • Nonsampling Error in Surveys
LEVY and • LEMESHOW • Sampling of Populations: Methods and Applications
LINHART and ZUCCHINI • Model Selection
LITTLE and RUBIN Statistical Analysis with Missing Data
LYBERG • Survey Measurement
McLACHLAN • Discriminant Analysis and Statistical Pattern Recognition
McLACHLAN and KRISHNAN • The EM Algorithm and Extensions
McNEIL • Epidemiological Research Methods
MAGNUS and NEUDECKER • Matrix Differential Calculus with Applications in Statistics and Econometrics
MALLER and ZHOU • Survival Analysis with Long Term Survivors
MALLOWS • Design, Data, and Analysis by Some Friends of Cuthbert Daniel
MANN, SCHAFER, and SINPURWALLA • Methods for Statistical Analysis of Reliability and Life Data
MASON, GUNST, and HESS • Statistical Design and Analysis of Experiments with Applications to Engineering and Science
MILLER • Survival Analysis
MONTGOMERY and MYERS • Response Surface Methodology: Process and Product in Optimization Using Designed Experiments
MONTGOMERY and PECK • Introduction to Linear Regression Analysis, *Second Edition*
MORGENTHALER and TUKEY • Configural Polysampling
MYERS and MONTGOMERY • Response Surface Methodology
NELSON • Accelerated Testing, Statistical Models, Test Plans, and Data Analyses
NELSON • Applied Life Data Analysis
OCHI • Applied Probability and Stochastic Processed in Engineering and Physical Sciences
OKABE, BOOTS, and SUGIHARA • Spatial Tesselations: Concepts and Applications of Voronoi Diagrams
PANKRATZ • Forecasting with Dynamic Regression Models
PANKRATZ • Forecasting with Univariate Box-Jenkins Models: Concepts and Cases
PORT • Theoretical Probability for Applications
PUKELSHEIM • Optimal Design of Experiments
PUTERMAN • Markov Decision Processes: Discrete Stochastic Dynamic Programming
RACHEV • Probability Metrics and the Stability of Stochastic Models
RADHAKRISHNA RAO and SHANBHAG • Choquet-Deny Type Functional Equations with Applications to Stochastic Models
RÉNYI • A Diary on Information Theory
RIPLEY • Spatial Statistics
RIPLEY • Stochastic Simulation
ROSS • Introduction to Probability and Statistics for Engineers and Scientists
ROUSSEEUW and LEROY • Robust Regression and Outlier Detection
RUBIN • Multiple Imputation for Nonresponse in Surveys
RUBINSTEIN and SHAPIRO Discrete Event Systems: Sensitivity Analysis and Stochastic Optimization by the Score
RYAN • Modern Regression Methods

*Now available in a lower priced paperback edition in the Wiley Classics Library

RYAN • Statistical Methods for Quality Improvement
SCHOTT • Matrix
SCOTT • Multivariate Density Estimation: Theory, Practice, and Visualization
SEARLE • Linear Models
SEARLE • Linear Models for Unbalanced Data
SEARLE • Matrix Algebra Useful for Statistics
SEARLE, CASELLA and McCULLOCH • Variance Components
SKINNER, HOLT, and SMITH • Analysis of Complex Surveys
STOYAN, KENDALL, and MECKE • Stochastic Geometry and Its Applications, *Second Edition*
STOYAN and STOYAN • Fractals, Random Shapes and Point Fields: Methods of Geometrical Statistics
THOMPSON • Empirical Model Building
THOMPSON • Sampling
TIERNEY • LISP-STAT: An Object-Oriented Environment for Statistical Computing and Dynamic Graphics
TIJMS • Stochastic Models: An Algorithmic Approach
TITTERINGTON, SMITH and MARKOV • Statistical Analysis of Finite Mixture Distributions
UPTON and FINGLETON • Spatial Data Analysis by Example, Volume 1: Point Pattern and Quantitative Data
UPTON and FINGLETON • Spatial Data Analysis by Example, Volume II: Categorical and Directional Data
VAN RIJKEVORSEL and DE LEEUW • Component and Correspondence Analysis
WEISBERG • Applied Linear Regression, *Second Edition*
WESTFALL and YOUNG • Resampling-Based Multiple Testing: Examples and Methods for *p*-Value Adjustment
WHITTLE • Optimization Over Time: Dynamic Programming and Stochastic Control, Volume 1 and Volume II
WHITTLE • Systems in Stochastic Equilibrium
WONNACOTT and WONNACOTT • Econometrics, *Second Edition*
WONNACOTT and WONNACOTT • Introductory Statistics, *Fifth Edition*
WONNACOTT and WONNACOTT • Introductory Statistics for Business and Economics, *Fourth Edition*
WOODING • Planning Pharmaceutical Clinical Trials: Basic Statistical Principles
WOOLSON • Statistical Methods for the Analysis of Biomedical Data
*ZELLNER • An Introduction to Bayesian Inference in Econometrics

Tracts on Probability and Statistics
BILLINGSLEY • Convergence of Probability Measures
KELLY • Reversibility and Stochastic Networks

*Now available in a lower priced paperback edition in the Wiley Classics Library